Electrochemistry at Semiconductor and Oxidized Metal Electrodes

Electrochemistry at Semiconductor and Oxidized Metal Electrodes

S. Roy Morrison

SRI International
(formerly Stanford Research Institute)
Menlo Park, California

PLENUM PRESS · NEW YORK AND LONDON

Library of Congress Cataloging in Publication Data

Morrison, Stanley Roy.
Electrochemistry at semiconductor and oxidized metal electrodes.

Includes bibliographical references and indexes.
1. Electrochemistry. 2. Semiconductors. 3. Electrodes, Oxide. I. Title.
QD553.M6 541.3′7 80-20416
ISBN-13:978-1-4613-3146-9 e-ISBN-13:978-1-4613-3144-5
DOI: 10.1007/978-1-4613-3144-5

First Printing–November 1980
Second Printing–October 1984

A Division of Plenum Publishing Corporation
233 Spring Street, New York, N.Y. 10013

Preface

The objective of the present volume is to develop the theory and practice of nonmetal electrochemistry from first principles, emphasizing energy level models, in particular the fluctuating energy level model of Marcus and Gerischer. A single volume emphasizing these models, and the interpretation of experiments based on these models, has not been available. Yet this area of electrochemical technology, where the use of such models is required, has developed a great deal of interest. This is not only because of the interest in photoelectrochemical solar cells, but also because of the importance of the concepts in corrosion, sensors, coated metal electrodes, and, indeed, to the general theory of electrode reactions. This book is an attempt to fill the void—to develop in a single volume the basic description of electrode reactions on nonmetallic electrodes and oxide-covered metal electrodes.

The development of the fluctuating energy level model to describe electrode reactions on nonmetals (as described in Chapters 1 through 3) has permitted a significant forward step in the understanding of such reactions. The power of the model is illustrated by the simple methods available to determine the energy levels of interest—the conduction and valence bands of the nonmetals (Chapter 5), and their relation to the energy levels of oxidizing or reducing agents in solution. In Chapter 6, we illustrate the ability of the simple models, based on these parameters, to describe successfully electrode reactions at an inert electrode. The application of the model to more complex systems is described (much more qualitatively, naturally) in Chapters 7 and 8. Finally, the important applications of the technology are discussed in Chapter 9. The application of the technology to solar energy conversion has not been given the space that it may seem to warrant for two reasons. First, there are many good reviews of this small branch of the field, and references to such reviews are provided. Second, we are in the midst of rapid changes in the form of solar cells, and a book written say two or three years from now will be much more useful, because it can emphasize the theory of any emergent practical devices. Thus, here we have emphasized the background theories and experiments in the electrochemistry of nonmetals, and only have one chapter devoted to applications.

The subject of semiconductor electrochemistry is interdisciplinary even in its name. Thus, one can expect scientists with semiconductor physics background on the one hand, and with metal electrochemistry background on the other, to become interested in the area. The surface physicist will presumably have little difficulty with the concepts of the band model, but may often have difficulty with the concepts of current flow and ion interactions in solution. The difficult theoretical concepts will be: first, that in the solution the electron does not move (as it does in the solid) but the host ion moves; and second, that the electron does not jump from energy level to energy level, but the energy level itself fluctuates in energy, carrying the electron with it. The latter concept is similar to the description of polarons in semiconductor physics, but few physicists are very familiar with polaron theory. In order to help the surface physicist understand ion interactions in solution chemistry, we have included discussions that will seem very elementary to the electrochemist. The electrochemist entering this area of science may have some difficulty with the band model, but these days most scientists, of whatever bent, have some familiarity with the band model. The relevant features are, however, reviewed in Chapter 1. The electrochemist will have more difficulty in thinking of electrode reactions in terms of energy levels. Thus, in Chapters 2 and 3, we try to show clearly how the energy level model relates to the classical electrochemical models, and how the metal electrode (which is more familiar to the electrochemist) can be described by the fluctuating energy level model.

A substantial selection of references is included for supplementary reading. In particular, a list of reviews and books is provided separately from the list of specific references, and the titles of the reviews and books are included. It is hoped this will help to identify supplementary reading of interest.

I would like to express my indebtedness to my co-workers in the Semiconductor Electrochemistry Group at SRI, Karl W. Frese, Jr., Marc J. Madou, and Boon H. Loo, who have critically read the manuscript and offered many valuable suggestions and corrections. In addition, I would like to thank my wife, Phyllis, who helped substantially in editing the book, and who typed most of the manuscript.

S. Roy Morrison

Contents

Notation

Roman Letters

A . . . Area
a . . . Ionic radius
a_i . . . Activity of species i
C . . . Capacity
c . . . Concentration
$\bar{c}$ . . . Thermal velocity of electrons
c_{ox}, c_{red}, c_i Concentration of oxidizing agent, reducing agent, ion i
D . . . Diffusion coefficient
$D(E)$ . . Density of ionic energy levels at energy E
d . . . Thickness; distance
E . . . Energy; electric field
E^o . . . Standard electrode potential (redox potential, oxidation potential)
E^e . . . Energy of an electron in vacuum at infinity
E_c . . . Energy of the conduction band edge
E_{cs}, E_{vs} . Energies of the conduction band edge, valence band edge at the surface
E^o_{cs} . . . Energy of conduction band edge (hydrogen scale) when $V_H = 0$
E_{decomp} . . Thermodynamic energy level for decomposition
E_F . . . Fermi energy
${}_nE_F$, ${}_pE_F$. Quasi-Fermi-energies for electrons, holes
E_g . . . Energy gap of the semiconductor ($E_c - E_v$)
E_{ox} . . . Energy level on oxidizing agent in lowest free energy configuration
ΔE_p . . . Thermal energy to shift an ionic energy level
E_{red} . . . Energy level on reducing agent in lowest free energy configuration

E^o_{redox}	$= \frac{1}{2}(E_{red} + E_{ox})$
E_{ref}	Fermi energy of reference electrode
E_t	Energy of surface state, energy of trapping level
E_v	Energy of valence band edge
E^o_{vs}	Energy of valence band edge (hydrogen scale) when $V_H = 0$
e^-	Electron
F	Faraday constant
f	Fermi distribution function [Equation (1.1)]
G	Conductance; free energy
h^+	Hole
I, I_0	Light intensity
J, j	Current density
J_0	Exchange current density
K	Equilibrium constant
k, k_a	Rate constant
L	Length; ligand
L_p, L_n	Diffusion lengths for holes, electrons
M	Metal ion
m^*	Effective mass
m	Mass of electron
N	Density (usually density of electronic energy levels)
N_A	Acceptor density
N_c	Effective density of states in the conduction band (circa 10^{25} m^{-3})
N_v	Effective density of states in the valence band
N_D	Donor density
N_{sc}	Density of immobile charge in the semiconductor space charge region
N_t	Trap density
N_s	Density of charges at the surface
n	Density of electrons in the conduction band
n_i	Concentration of carriers in an intrinsic semiconductor
p	Density of holes in the valence band
pzc	Point of zero charge (usually where $V_H = 0$)
Q	Charge, Q/A charge per unit area
q	Charge on the electron
R	Resistance; two-equivalent species; organic molecule
S_1, S_2	Bonding sites for ion, configuration of polar medium around an ion
s	Surface recombination velocity
U	Activation energy; current of carriers to the surface (in recombination theory)
V	Voltage, potential difference; qV = band edge energy relative to the bulk semiconductor

V_{fb} . . . Flat band potential (electrode potential when $V_s = 0$)
V_H . . . Helmholtz double layer voltage
V_m . . . Measured voltage, applied voltage
qV_s . . . Value of qV at the surface
v . . . Velocity
$W(E)$ . . Probability that the energy level on an ion has fluctuated to the energy E
X . . . Anionic component of a compound semiconductor
Z . . . Charge on an ion, valence, oxidation state; thermal velocity of ions in solution; impedance
z . . . Charge on an ion

Greek Letters

α . . . Anodic transfer coefficient; absorption coefficient for light
β . . . $= 1 - \alpha$
γ_i . . . Activity coefficient of ion i
ϵ_0 . . . Permittivity of free space
η . . . Overvoltage
η_{sc} . . . Overvoltage when it appears across the semiconductor space charge region
κ . . . Dielectric constant; transmission coefficient
λ . . . Reorganization energy of an ion, rearrangement energy; mean free path of a carrier in a solid
μ . . . Mobility of carriers; energy difference between bulk conduction band edge and Fermi energy
μ_i . . . Chemical potential of ion i
$\bar{\mu}_i$ . . . Electrochemical potential of ion i
$\bar{\mu}_e$ . . . Electrochemical potential of electrons (Fermi energy)
ν . . . Integer (number of atoms involved in a reaction)
ρ . . . Charge density
σ . . . Conductivity; electron capture cross section
τ . . . Minority carrier lifetime
ϕ . . . Potential
χ . . . Electron affinity of conduction band edge

Subscripts and Superscripts

A . . . Acceptor
an, a . . Anodic
aq, aquo . No ligands on ion except solvent molecules (water)
b . . . Bulk

the greater overlap of the orbital wave functions. The bands associated with higher excited states normally overlap the conduction band, so there are allowed electronic energy levels from the bottom of the conduction band to infinite energy.

A forbidden region of energy called the "bandgap," or simply the "gap," is present in insulators and semiconductors between a filled valence band and an empty conduction band. For a perfect crystal there are no allowed energy levels for electrons in the bandgap. The picture is analogous to atomic energy levels, where there are forbidden energy regions between the discrete levels. In the discussions to follow in this book we will be interested only in the valence band and conduction bands of the solid because these bands are the primary sources and sinks of electrons when one is considering the transfer of charge to or from a solution.

In Figure 1.1 we illustrate the band model. In Figure 1.1a is shown a sketch of the density of allowed energy levels, $N(E)$, on the abscissa, as a function of the electron energy, E, on the ordinate. Only the two bands of interest here are shown: the conduction band, normally unoccupied by electrons; and the valence band, normally completely occupied by valence electrons of the solid. In Figure 1.1b is shown the more useful form of the band model (for our purposes), a plot of electron energy vs. distance through the crystal. In this plot, because we are plotting distance on the abscissa, a surface can be indicated, as shown on the left-hand edge of the band diagram. The key parameters are the energy of the bottom of the conduction band E_c, and the energy of the top of the valence band, E_v, and the bandgap energy, $E_c - E_v$, or E_g.

A hole is an empty level in the valence band, a valence bond with the electron missing. One can visualize the motion of a hole as a series of steps wherein valence electrons from neighboring atoms occupy the holes, resulting effectively in motion of the hole. However, it is accurate and simpler to visualize the hole as a positively charged entity moving freely through the valence band.

The lowest energy level for a hole is at the top of the valence band. To illustrate that this is the case, consider the situation when an electron from somewhere in the valence band is captured by an ion in solution. This creates an empty level—a hole. An electron with energy near the top of the valence band will promptly drop down to this empty lower energy level. In other words, the hole will lose energy by rising to the top of the valence band. It is convenient, at times, to use the term "hole injection into the valence band" rather than "electron capture from the valence band" to describe such valence electron capture, consistent with visualizing the hole as a separate entity. We will use the two expressions interchangeably.

Mobile electrons in the conduction band and holes in the valence

cat, c	Cathodic
comp	Complexed species
D	Donor
e	Electronic; equilibrium
fb	Flat band
H	Helmholtz
L	Lattice ion
n	Electron
op	Optical
ox	Oxidizing agent; oxide layer
p	Hole
red	Reducing agent
s	Surface; static (in dielectric constant)
sc	Space charge region (of semiconductor)
sol	Solution
ss	Surface state
solv	Solvated species
*	Nonequilibrium value (under illumination, usually)

The Solid and the Solution

1.1. The Solid

1.1.1. Donors, Acceptors, and Traps

1.1.1.1. The Band Model. In order to describe electron transfer to or from a nonmetallic solid electrode, whether the electrode is a semiconductor or is an insulating layer on the surface of a metal, it is desirable to utilize the concepts of the band model. In the case of a semiconductor electrode, the need is obvious, for the source and sink of electrons for the electrochemical process are the bands of the solid. In the case of an oxidized metal electrode, the electrons could be transferred by current carriers in the bands of the oxide, could be transferred by tunneling through the oxide (and tunneling depends upon the energy of the allowed states—the bands), or could be transferred by imperfections in the oxide. Any of these transfer mechanisms can best be understood using the band model, where one describes the electronic energy levels of the solid and its neighboring phases. Thus a qualitative, or better, a quantitative knowledge of the band description of solids is a definite requirement in this particular area of electrochemistry.

The band model of solids stems directly from the picture of atomic energy levels. As described by the Pauli exclusion principle, the energy levels of identical overlapping electronic orbitals cannot be equal, and analysis shows that the discrete energy levels of the atoms forming th solid broaden when the orbitals overlap, forming bands of energy level with the order of 2×10^{22} levels in each band when there are the ord of 10^{22} atoms in the crystal. The orbitals occupied by the valence electro is called the "valence band" of the solid, normally a few electron vo (eV) broad. The band associated with the first excited state, norma unoccupied by electrons, is called the "conduction band" of the so This band is normally much broader than the valence band because

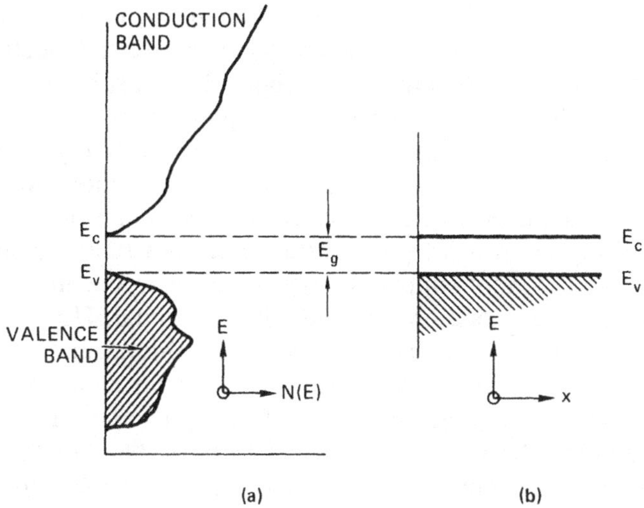

Figure 1.1. The band model of a semiconductor, where electron energy E is plotted on the ordinate. In (a) the density of allowed states $N(E)$ is plotted on the abscissa, in (b) distance through the crystal is plotted. E_c is the bottom edge of the conduction band, E_v the top edge of the valence band.

band normally carry electrical current in the solid. Mathematically the wave function of each extends through the crystal, so the electrons and holes are not localized on a single atom. If the band is very narrow, the electronic wave functions show maxima near the atoms and the electrons or holes are in a real sense somewhat localized and have a low mobility; if the bands are broad, as is generally true for the conduction band, the electrons are almost completely delocalized and have a high mobility.

1.1.1.2. Donors and Acceptors. Figure 1.1 shows with some simplifications how the band model appears for the perfect crystal, but for the real crystal with imperfections, the model is quite incomplete. Impurities and imperfections in the crystal often give rise to localized energy levels such that if an electron or hole occupies one of these levels (a) it is localized—it no longer has a wave function extending through the crystal, and (b) it has an energy level in the gap. Such impurity levels can be conveniently separated into donors and acceptors. Donors are donors of electrons, often to the conduction band. Acceptors are acceptors of electrons, either creating a hole in the valence band or extracting an electron from the conduction band.

For an example of a localized donor impurity level, consider a phosphorus atom substituting for a silicon atom in an otherwise perfect silicon crystal. Four of the five valence electrons of phosphorus are utilized in the valence band. The fifth electron has no valence bond to occupy and so must be at a higher energy level than electrons in the valence band.

The extra electron is, however, still attracted to the +5 charge on the phosphorus nucleus, so it is not a free conduction band electron. It thus has a localized orbital (energy level) about the phosphorus nucleus, the energy of which will be in the gap. The phosphorus atom is a typical donor: neutral if the energy level is occupied; positively charged if the electron is excited to a nonlocalized orbital in the conduction band.

For an example of a localized acceptor impurity level, consider boron in silicon. The boron has fewer valence electrons than the silicon atom it replaces, so it cannot supply enough electrons to occupy the available orbitals, and there is one unoccupied valence orbital. However, this unoccupied orbital is at an energy higher than the normal valence band orbitals, because the boron inner shell only carries a charge of +3, and so is less attractive to a fourth valence electron than are the silicon nuclei in the lattice. Such an impurity is an acceptor. The empty orbital can capture an electron from the valence band (can emit a hole into the valence band). Typical for an acceptor level, the region of the crystal is electrically neutral with the hole on the acceptor level (the extra orbital unoccupied) and negatively charged when the hole is excited to the valence band.

In the above discussion, examples were chosen where donor impurity levels are found near the conduction band, acceptor impurity levels are found near the valence band. It will turn out in many of our later discussions that the energy levels associated with bulk or surface impurities or imperfections can be anywhere in the bandgap. In many cases indeed the acceptor levels are near the conduction band and donor levels are near the valence band. It is well to keep in mind that the distinction between donor levels and acceptor levels is not their position relative to the bands, but it is that donor levels are electrically neutral when occupied by an electron, and acceptor levels are neutral when occupied by a hole.

Standard terminology has been developed in solid state physics describing semiconductor properties. If the solid is dominated by donors, so that conduction band electrons outnumber valence band holes, the semiconductor is *n* type; if the solid is dominated by acceptors so that holes outnumber electrons, the solid is termed *p* type. Some solids that are always found *n* type are ZnO, TiO_2, V_2O_5, and MoO_3. Some solids that are always found *p* type are Cu_2O, NiO, and Cr_2O_3. And there is a large class of semiconductors, including some group IV semiconductors such as Ge and Si, and III–V semiconductors such as GaP, GaAs, and InP, that can be made either *n* or *p* type depending on the impurity introduced. If a semiconductor is *n* type, dominated by electrons, the electrons are called the "majority carrier"; the holes are called the "minority carrier." Analogously, with a *p*-type semiconductor the holes are the majority, the electrons the minority carrier.

With a solid of reasonably wide bandgap (say $E_g \gtrsim 1.5$ eV) that has a reasonable density of majority carriers (say greater than 10^{19} m^{-3}) there are negligible minority carriers at equilibrium. However, high nonequilibrium densities can be induced by illumination, by injection at the surface, or by injection at a *p–n* junction. However, when minority carriers are induced in wide bandgap materials, they are often extremely reactive chemically. Thus minority carriers flowing to the surface are of great importance in semiconductor electrochemistry.

1.1.1.3. Fermi Statistics and the Density of Holes and Electrons. The Fermi energy and the Fermi distribution function describe the distribution of electrons and holes among the available energy levels in a solid. To a first approximation, all energy levels below the Fermi energy are occupied by electrons, all levels above the Fermi energy are unoccupied by electrons. More accurately, the Fermi function f, the probability that a level at energy E is occupied by an electron, is given by

$$f = 1/[1 + \exp(E - E_F)/kT] \tag{1.1}$$

where E_F is the Fermi energy, k the Boltzmann constant, and T the temperature (Kelvin). If the energy level E is at the Fermi energy, the probability of occupancy, by Equation (1.1), is exactly $\frac{1}{2}$, and the probability decreases rapidly for levels above E_F, increases rapidly for levels below E_F.

From a chemical point of view, the Fermi energy is the electrochemical potential of electrons in the solid. Thus from a statistical point of view, electrons moving from the solid to ions in solution originate with a free energy of E_F. From a very practical physical point of view, a voltage measurement is a measure of the difference in Fermi energies of the metal leads to the voltmeter. At equilibrium all electronically conducting materials in contact have the same electrochemical potential for electrons, so the same Fermi energy. Thus later we will use the fact that a voltage measurement gives the Fermi energy difference between two electrodes in solution, where there is a continuous electronically conducting path between each of the electrodes and the voltmeter.

The position of the Fermi energy relative to the bands of a solid is the dominant feature distinguishing a metal from a semiconductor or insulator. If the Fermi energy is located in a band, in a region of the band with a high $N(E)$ (Figure 1.1a), the material is a metal. Atomic sodium, for example, has its 3*s* levels only half occupied; solid sodium therefore has the corresponding band only half occupied. Thus the Fermi energy of sodium is in the middle of the band and the material is a metal. In Figure 1.1b only the valence band would be drawn in a typical band model for a metal, with the levels occupied up to E_F, empty above E_F [neglecting the small transition region required by Equation (1.1)]. On the other hand, if the valence band of a material is almost completely filled with electrons, and the conduction band is almost completely unoccupied, the material is a semiconductor or an insulator (see Section 1.1.3.1). To describe such a distribution of electrons by Equation (1.1), the Fermi energy must be in the bandgap region in a diagram such as Figure 1.1b. In Figure 1.6 of Section 1.2.4 are shown typical band models for a semiconductor and a metal with the Fermi energies indicated.

A "degenerate semiconductor" is a material which would normally be considered a semiconductor but is quasimetallic because large densities of electrons have been added to the conduction band or large densities of holes to the valence band. A degenerate semiconductor is defined as one where the Fermi energy has been moved into the band—either the valence or conduction band. Such addition of electrons or holes can be made, for example, by very heavy doping with donors or acceptors, respectively. Degenerate semiconductors will be discussed very often in this book as the transition between a metal and a semiconductor. A *highly* degenerate semiconductor has the two electronically important attributes of a metal: the Fermi energy is in the band region, and the Fermi energy is in a region of high density of states, such that there is a high density of unoccupied levels for $E \gtrsim E_F$, and a high density of occupied levels for $E \lesssim E_F$.

With nondegenerate semiconductors, the Fermi energy conveniently describes the density of electrons in the conduction band and the density of holes in the valence band. The description semiquantitatively follows the following arguments. As was discussed earlier, the electrons in the conduction band settle to the bottom of the band, and, because f varies exponentially with energy, we are really only interested in the density of states within one or two kT from the bottom of the band. We describe the conduction band as N_c energy levels at an energy E_c (see Figure 1.1), where N_c is the effective density of energy levels in this region of the conduction band. Its value is normally about 10^{25} m^{-3}. Similarly we introduce a parameter N_v that is the effective density of valence band energy levels at an energy E_v.

If the Fermi energy is deep in the bandgap, more than $2kT$ below E_c or more than $2kT$ above E_v, so $(E_c - E)$ or $(E - E_v)$ values of interest all exceed $2kT$, Equation (1.1) is simplified and we obtain

$$n = N_c \exp[-(E_c - E_F)/kT] \tag{1.2}$$

and

$$p = N_v \exp[-(E_F - E_v)/kT] \tag{1.3}$$

where n is the density of electrons in the conduction band, p the density of holes in the valence band.

Combining (1.2) and (1.3) we obtain

$$np = N_c N_v \exp[-(E_c - E_v)/kT] = N_c N_v \exp(-E_g/kT) \quad (1.4)$$

From Equation (1.4), if E_g is large, the product np is small, indicating why at equilibrium both n and p cannot be large. Thus if n is high, perhaps because there is a high density of donors, and if E_g is large, then p must be small.

The Fermi energy describes the occupation of energy levels at thermodynamic equilibrium, but an important case in semiconductor electrochemistry is that wherein thermodynamic equilibrium is not attained. In particular, we will often be referring to the case where the sample is illuminated by photons of energy greater than the bandgap. Then electrons are excited from the valence band of the semiconductor to the conduction band, leading to an excess of both electrons and holes. Nonequilibrium can also be caused by carrier injection from a p–n junction near the surface or carrier injection by certain oxidizing and reducing agents in solution at the surface. In any of these cases there can be far more minority carriers than equilibrium conditions [e.g., Equation (1.4)] allow. The exact density of such minority carriers depends on the kinetics of the system—how rapidly the system is deexcited, the "lifetime" of the excess minority carriers in the solid.

A concept that has been used to describe nonequilibrium conditions is the "quasi-Fermi-energy." For example, if the sample is illuminated, leading to excess photoproduced holes and electrons, a quasi-Fermi energy for electrons, ${}_nE_F$, and a quasi-Fermi-energy for holes, ${}_pE_F$, are used to describe the density of electrons in the conduction band and of holes in the valence band, respectively. In other words ${}_nE_F$ replaces E_F in Equation (1.2) to describe the nonequilibrium density of electrons in the conduction band, and ${}_pE_F$ replaces E_F in Equation (1.3) to describe the nonequilibrium density of holes in the valence band. In general, the density of majority carriers is not appreciably increased by the illumination, so the quasi-Fermi-energy for majority carriers is almost the same as the equilibrium Fermi energy, E_F. However, it is easy to change the minority carrier density substantially. For example, with an n-type semiconductor under illumination, the density of holes (because the density is so small in the dark) may be increased many orders of magnitude by the illumination, so ${}_pE_F$ will be greatly affected. In fact, ${}_pE_F$ may be very near the valence band of the semiconductor, where with an n-type semiconductor the equilibrium, E_F, is near the conduction band.

The point of the use of the quasi-Fermi-energies is that if we are interested, for example, in the reaction of holes with species in solution, we are interested in the density of holes in the valence band, and the quasi-Fermi-energy is a convenient parameter to describe this density.

It matters little to the reactants in solution whether the holes in the valence band arise because the semiconductor is *p* type or because an *n*-type semiconductor is illuminated.

However, it must always be borne in mind that the quasi-Fermi-energy describes only the density of carriers in the band. The nonequilibrium population of carriers in any other energy level, be it a bulk or a surface energy level, is not defined by this quasi-Fermi-energy. The actual electron density in such energy levels becomes dependent on kinetic parameters.

1.1.1.4. Defects, Traps, and Recombination Centers. Some of the types of defects in the solid that have importance in the study of electrode reactions are vacancies, interstitials, dislocations, and grain boundaries. For example, anodic oxidation of an electrode (Section 8.2) occurs by motion of vacant lattice sites (vacancies) or of atoms or ions in interstitial positions. Etching of an electrode by a corrosive solution or deposition of atoms from a solution often occurs preferentially at dislocations in the crystal. Grain boundaries and dislocations are chemically active at an electrode surface and these reactions can cause circulating electrochemical currents. Additionally, any of the above imperfections can act as donors or acceptors affecting the electronic properties of the electrode.

A vacancy is, of course, a missing atom in the crystal, and such sites can be electronically active. It is easily shown that cation vacancies are acceptors, anion vacancies are donors. Consider a strongly ionic oxide, where the valence electrons are in orbitals close to the oxygen ions. Thus the valence band is the energy band occupied by electrons on O_L^{2-} (lattice oxygen) orbitals. In order to remove a neutral metal atom from the solid and form a cation vacancy, leaving the region of the imperfection neutral, electron(s) must be taken from neighboring oxygen ion(s). However, the oxygen ions are still there with only slightly perturbed orbitals, and electrons can be accepted by these orbitals. By the discussions in Section 1.1.1.2 the cation vacancy can thus be identified as an acceptor. In order to remove a neutral oxygen atom from the solid and form a neutral anion vacancy, two electrons (formerly on the O_L^{2-}) must be left behind with no valence orbital to accommodate them. The anion vacancy thus acts as a donor.

An interstitial atom is one that is squeezed between the atoms of the normal crystal. Either a host atom or a foreign atom can become located interstitially in the lattice. If such an interstitial atom tends to give up electrons, it will be an active donor. If it tends to accept electrons it can, in principle, be an active acceptor. However, the latter case is rarely found in nature. Movement of vacancies and interstitials is of importance in such phenomena as anodic oxidation, the doping of solids by diffusion, and corrosion.

Dislocations and grain boundaries are also bulk defects that are of great importance in the behavior of the solution/solid interface. Both may provide energy levels in the gap.[1] Both may allow current flow along these energy levels (Sections 5.2 and 8.2). Both will have double layers associated with them because charge is trapped there (the origin of double layers is discussed in Section 2.2). Both may provide paths for easy diffusion of atoms during anodic oxidation processes. Finally, both dislocations and grain boundaries provide active sites at the point where they emerge at the surface—active sites for reactions[2] or for the removal or precipitation of atoms from, or onto the solid. Electrochemical studies of trapping and current flow due to dislocations are discussed in Section 5.2.

The dislocation, a linear defect, is most easily discussed in the form of an edge dislocation, pictured in Figure 1.2. An edge dislocation is the line of the final atoms in an extra half plane of atoms in the otherwise perfect crystal. The dislocation line, or better, the dislocation "pipe," must normally emerge in two places at the surface of the crystal. At the place where it emerges there is a poorly coordinated surface atom, an atom without the normal number of nearest neighbors and so highly re-

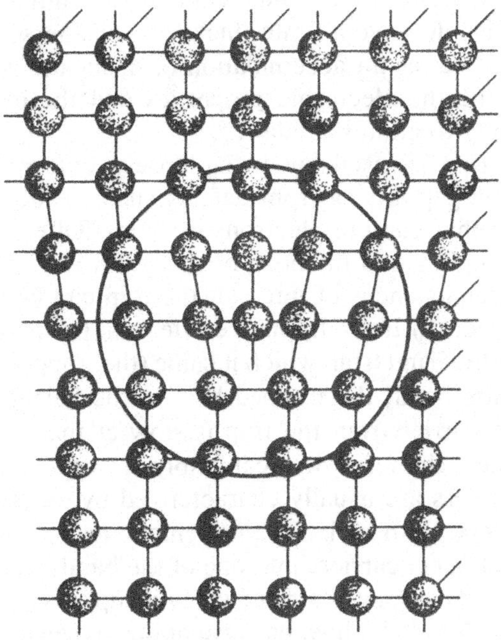

Figure 1.2. An edge dislocation intersecting a surface, where the plane of the paper is the surface. The location of the emerging dislocation pipe is encircled, with the circle approximately indicating the region of the crystal that is strained.

active. Inside the crystal there is a line of poorly coordinated atoms in the crystal, with corresponding orbitals that will often provide donor or acceptor (or both) energy levels in the band gap.[3]

Because, as indicated in Figure 1.2, the lattice is strained in the region of the dislocation, atoms or vacancies can move much more easily along a dislocation than in the perfect part of the crystal. This higher defect mobility may become important, for example in anodic oxidation.

Dislocations arise primarily in two ways. First, they arise because of mechanical damage to a crystal. If a crystal is stressed beyond its yield point, dislocations are created and move to allow the plastic flow of the crystal. Mechanical abrasion (polishing or lapping) of a crystal surface will generate dislocations. These surface dislocations will be in the form of "dislocation loops"; the extra half plane only extends a short distance into the face, and the dislocation pipe emerges a short distance from where it entered the crystal. The second origin of dislocations is their formation during crystal growth. If a dislocation is present at the surface of the seed crystal, it will normally continue on through the newly grown material. Or, a lattice mismatch between the substrate crystal and the growing material (say an anodic oxide) may cause dislocations to initiate.

Grain boundaries are essentially two-dimensional defects, the interface where two regions of a crystal meet that have different orientations. (Grain boundaries do have a finite thickness, of course.) In many ways grain boundaries act as an accumulation of dislocations, and the above remarks regarding the electronic properties and the behavior of dislocations also apply to grain boundaries.

The term "trap" is used somewhat loosely to describe energy levels deep in the band gap arising from defects such as the above or arising from impurities that capture electrons or holes. Often, but not always, the term trap implies one or more of the following characteristics: (a) Under some circumstances of interest, the trap can be, or can be made to be, unoccupied. (b) If the trapped carrier can be thermally reemitted, it is returned to the band from which it came (the trapped electron returns to the conduction band; the trapped hole to the valence band). (c) Reemission of the carrier from the trap is slow at the temperature of the experiment, where the term slow can imply seconds to days.

Good insulators are usually characterized by an effective spectrum of trapping centers. With such trapping centers, even if an electric contact is made that will inject carriers into one of the bands, the carriers will be lost to traps, and the trapped carriers will build up a space charge repelling further injection.[R1,R2,R3]* Thus, because of the presence of traps, carriers cannot be passed through the insulating material.

* The letter R preceding a reference number indicates "Review." There is a separate listing of references to books and review articles.

A recombination center is a site that tends to capture carriers readily from either band. Thus, if an electron is optically excited from the valence band to the conduction band, creating an electron–hole pair, the solid will often become deexcited through a recombination center. Most recombination centers will normally be fully occupied by majority carriers. The center first captures the minority carrier and then is reoccupied by the majority carrier, and the electron–hole pair are "recombined," the excited electron being restored to the valence band. The "lifetime" of the minority carrier, τ, is the average time excess minority carriers remain in their excited state after generation. The decay rate is usually proportional to their density. For n-type material,

$$dp/dt = -(p - p_e)/\tau_p \tag{1.5a}$$

and for p-type material,

$$dn/dt = -(n - n_e)/\tau_n \tag{1.5b}$$

where the subscript e implies "equilibrium." It will be noted that if $n < n_e$, or $p < p_e$, the formulas imply generation of carriers. In electrochemistry this condition can often occur, for the minority carriers at the surface can be consumed by the electrolyte and be of density less than the equilibrium value associated with the bulk density. The formulas are also valid for this situation and give the rate of minority carrier generation.

1.1.2. Energy Levels at the Surface

In this section surface states, electronic energy levels of the solid that are localized at the surface, are discussed. They can be donors or acceptors. Naturally, we are most interested in energy levels at the solid/liquid interface, but energy levels at the solid/gas interface will be emphasized in this section, primarily because more is known about the latter as surface states. Differences between the energy level description at the gas/solid versus liquid/solid interfaces will be discussed in more detail in Section 2.1.2, after the effect of the polar liquid on energy levels is reviewed.

There are several sources of surface state energy levels.[R4] Surface states at a clean surface are termed intrinsic states. Surface states also arise due to adsorbed species, both those adsorbed from the gas phase and those adsorbed from solution. In fact, in solution nonadsorbed ions close enough to the surface for electron exchange with the bands can be considered to provide surface states, for they obey the mathematics of electron exchange with the solid and in principle can behave in most ways as the normal surface state. Energy levels at the solid/solid interface are essentially surface states. They will be described in Section 5.2.2.1 and

Section 8.1. We will use the term "interface states" to describe levels at the solid/solid interface, and restrict the term "surface states" to refer to levels at the solid/liquid or solid/gas interfaces.

Energy levels at the surface, be they donor or acceptor, have a dominating influence on both the chemical and physical properties at the gas/solid interface and a substantial influence on these properties at the liquid/solid interface. The space charge double layer that often controls the rate of semiconductor electrode reactions, described in more detail in Section 2.2, is induced in part, and at times totally, by charge on surface states. Bonding of adsorbed species to the surface is determined to a great extent by surface energy levels because the adsorbing species interact with the surface states (orbitals) on the clean surface. Recombination of minority carriers will often occur through surface states and prevent desired chemical reactions between minority carriers and ions in solution. Surface state energy levels can act directly in electrochemical reactions by acting as intermediate states in the electron transfer between the bands and ions in solution. Finally, the capture of carriers at surface states is a dominating step in electrochemical corrosion, as will be described in Section 7.4.

Thus an understanding of surface energy level models, that is the understanding of the origin and behavior of surface energy levels, is of prime importance in the understanding of electrode processes on nonmetallic solids.

1.1.2.1. Energy Levels on a Uniform Surface. There are several different forms of energy levels at a uniform surface. They include intrinsic surface states, Lewis sites (orbitals that can share electron pairs, which can be acidic or basic), and surface states arising from foreign species becoming adsorbed.

The intrinsic surface states on the clean surface of a reasonably covalent solid will arise due to "dangling bonds" and are called "Shockley surface states." The valence bonds in the bulk phase of a covalent material such as silicon are hybridized to have orbitals concentrated between neighboring atoms. The surface atoms, therefore, having no neighbor on the one side, have one or more dangling orbitals directed outwards, and these orbitals are normally occupied by valence electrons. These dangling bonds provide Shockley surface states, surface states [(4)] that can act as either donor surface states (the dangling electron being excited to the conduction band, leaving a positively charged surface) or as acceptor surface states (electrons being captured to pair with the electrons in the dangling bonds, resulting in a negatively charged surface). A significant overlap occurs between neighboring dangling orbitals causing splitting of the resulting "molecular orbitals" into bonding and antibonding states. Thus there are normally two energy levels available with the donor bonding levels often below the acceptor antibonding levels, as indicated for

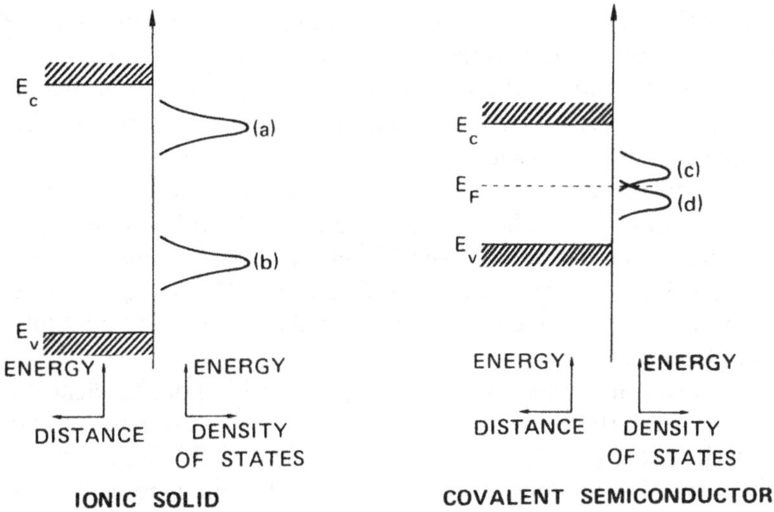

Figure 1.3. Intrinsic surface state bands. On the ionic solid, (a) represents a band of surface state levels of conduction band character, (b) a band of valence band character. On the covalent solid, (c) represents dangling antibonding orbitals, and (d) bonding orbitals.

the covalent solid in Figure 1.3. The indicated broadening of such levels into bands is expected by quantum effects either because the orbitals overlap with each other or overlap with the bands of the solid.[R4] Surface heterogeneity can also give rise to a spectrum of energy levels appearing as a band (Section 1.1.2.2).

An analogous case is that of transition metal oxides with partially occupied *d* orbitals. Here the *d* orbitals of the surface cations can be available for covalent or ligand field bonding to surface ions, and can act as donor or acceptor surface states.[5,6]

The surface of a more ionic compound semiconductor or insulator will have "ionic surface states," or "Tamm states," on its clean surface. A surface lattice oxygen ion, O_L^{2-}, is surrounded by less than its normal complement of cations. Thus the energy of the electrons on this surface O_L^{2-} is higher than that of the bulk (valence band) electrons simply because of the lower electrostatic attraction from the neighboring cations. Because they are occupied when the surface is neutral these are donor surface states. These surface orbitals on the O_L^{2-} can be located in the gap region above the valence band, as indicated in the model for ionic solids in Figure 1.3b. The orbitals are essentially valence band orbitals with higher energy than the valence band because of the electrostatic perturbation. They will probably form a band of levels for the same reasons discussed above for Shockley states. Analogously the unoccupied energy levels associated with the surface cation can be acceptor states, which can be located below the energy of the conduction band simply because the surface cation has fewer neighboring anions than a bulk cation. An exact analysis[7] of this electrostatic effect is the "Madelung potential analysis." However, because of the uncertainty about how much charge to assign to the cations and anions in a typical ionic solid, and the neglect of quantum effects by the simple model, the exact position of the donor ionic surface states on the anion and the exact position of the acceptor ionic surface states on the cation cannot be calculated by these electrostatic arguments.

Quantum calculations of both Shockley- and Tamm-type levels are providing valuable insight for the solid/gas case[R4] but quantum calculations of intrinsic surface state energy levels at the solid/liquid interface have not been made.

The Lewis site is another type of surface orbital on the clean surface of an ionic solid, that becomes of importance at the liquid/solid interface particularly because of its chemical reactivity.[8,R4,R5] A Lewis acid site is a site attractive to electron pairs from an adsorbing molecule; a Lewis base site is a site able to donate electron pairs to an adsorbing species. The Lewis site is not normally considered a surface state—it is distinguished from the ionic surface state in that for the ionic surface state we are interested in the exchange of unpaired electrons. However, the sites for ionic surface states and Lewis sites may be the same.[9] Because the measurement of the properties of Lewis sites requires the adsorption of an electron pair donor molecule or an electron pair acceptor molecule, respectively, Lewis sites are not observable on the clean surface. But at the solid/liquid interface they are of great importance. For example, a Lewis acid site M may adsorb an OH^- ion from water:

$$M + OH^- = M{:}OH^- \qquad (1.6a)$$

forming a strong bond with the electron pair from the hydroxide ion. By thus removing OH^- groups from an aqueous solution, the solid with a high density of Lewis acid centers will cause the solution to become acidic. A Lewis base site on a solid may adsorb a proton. For example, a common basic site is a surface oxide ion, O_L^{2-}, and a reaction such as

$$O_L^{2-} + H^+ \rightleftharpoons O_L{:}H^- \qquad (1.6b)$$

can occur, with the electron pair from the basic site shared on the otherwise unoccupied $1s$ orbital of the proton.

Although acid/base interactions of the kind illustrated by Equation (1.6) do not themselves donate or accept electrons from the bands of a semiconductor, their effect at the solid/liquid interface is significant in two ways. First, they contribute to an ionic double layer at the surface, the Helmholtz double layer, to be discussed in Section 2.2.3. In accordance with Equation (1.6a), for example, a highly acid surface will tend to become negatively charged. Second, Lewis activity causes adsorption and passivation of surface sites that otherwise would be active ionic surface states, electron donors or acceptors. For example, the surface O_L^{2-} ions would tend to be donor surface states, but the presence of a proton [Equation (1.6b)] will tend to stabilize the electron at the site.

Adsorbed electroactive species, species that tend to give up or accept electrons, when bonded to the surface form a different class of surface states. They can act as donors or acceptors and they will have surface state energy levels associated with their bonding and their antibonding orbitals. If the bonding interaction of the adsorbing species with the surface is negligible, the species will behave as the free molecule or ion. Thus, for example, at the solid/gas surface the adsorption of O_2 is expected to be weak, so the O_2 will form a strong acceptor state characteristic of the free molecule. If a strong bond forms between the surface and the adsorbed species, the energy of the bonding and antibonding orbitals available and the resulting surface states are determined by the complex formed and only slightly related to the properties of the isolated gas phase (or liquid phase) species.

Broadening of the adsorbate surface states into a band may not be appreciable if there is little overlap of the wave functions of the adsorbed species. However, the levels may appear to form a band because surface heterogeneity, as discussed below, makes the energy level of each adsorbate molecule somewhat different from that of its neighbors.

Ions in solution can be considered to be another class of surface states in the sense that ions within tunneling distance from the surface can exchange electrons with the bands of the solid. Such "surface states" are, of course, the main interest in electrochemistry. The energy levels of ions in solution are discussed in detail in Section 1.2.1.2.

1.1.2.2. Surface Heterogeneity and Energy Levels. The influence of surface heterogeneity on adsorption, catalysis, and electrode processes is just beginning to be understood in a reasonably satisfactory way. Special sites will have more activity, both chemically,[R6] and electronically as donors or acceptors,[5] than the uniform surface. A surface will show heterogeneity effects, for example, when different crystal faces are simultaneously exposed, when surface steps are present, or when a nonuniform deposition of foreign species is present. Such heterogeneity will make otherwise monoenergetic energy levels appear as bands because the exact chemical form of the surface group will vary from point to point on the surface so the energy of the corresponding surface orbitals will vary.

Heterogeneity associated with the exposure of many different crystal planes is obvious and needs no discussion.

The surface step has received considerable attention in recent years[R4] because it has been possible to make measurements with controlled surface step configurations. A step is the edge of an extra half plane of atoms that only partially covers an exposed plane. Low-energy electron diffraction (LEED) measurements in ultrahigh vacuum have been used to determine annealing treatments that lead to regular arrays of surface steps. Then one can measure the adsorption characteristics of a highly stepped surface, compare it to a surface comparatively free of steps, and thus identify the role of steps in adsorption. It is found that the step often provides preferred adsorption sites; in fact, at times adsorption occurs on the step and then the adsorbate diffuses over the surface.

It is understandable that steps act as particularly active adsorption sites, for the atoms of the crystal at the step are less completely coordinated than the atoms on the smooth surface (the "terraces"). Thus, strong bonding configurations at steps can be visualized. Also, the adsorbing atom finds sites in the hollows of the steps where it can bond to several atoms simultaneously, more than possible at the smooth surface. An especially attractive site is the kink site, a site where the surface step makes a turn, for this can be a site where an adsorbing atom will find even more nearest neighbors than at a simple step.

The role of steps in corrosion of electrodes will be discussed in more detail in Section 7.3.3.3.

Finally, heterogeneity can arise by the presence of patches of another phase. For example, it is found in the anodic oxidation of GaAs that the oxide nucleates at certain points at the crystal, and thereafter the edge of the resulting "island" of oxide are the active sites for further anodic oxidation. Finally, the islands meet and form a coherent, yet amorphous oxide film on the surface.

1.1.3. Conductance in Solids

1.1.3.1. Conductance by Electrons and Holes. If an energy band is either completely empty or completely filled with electrons, that band will not contribute to the electronic conductance in the solid. The reason there can be no conductance in a completely filled band is that none of the electrons can gain energy by moving to a higher energy level in the band. Thus, none of the electrons can gain velocity when an applied field is imposed, and so cannot move in response to the field.

Conductance can occur in any partially filled band. An electron at the bottom of a nearly empty conduction band can easily move to higher energy levels to increase its energy (and velocity) in response to an applied electric field. A hole near the top of the valence band increases its energy by moving to a lower level in the band (electrons from lower levels move to higher energy). The average velocity of carriers v increases with electric field, with their "mobility" μ the proportionality constant:

$$v = \pm\mu(d\phi/dx) \tag{1.7}$$

where ϕ is the electric potential.

If the bands are narrow, arising from orbitals that do not appreciably overlap, then the electrons or holes do not move as freely; they have a lower mobility. One interesting case of a narrow band with low carrier mobility is the impurity band, where enough impurities of one kind are present so that the orbitals associated with the impurities overlap. For reasons beyond the scope of the present discussion, the orbitals associated with many impurities in semiconductors are often very large, extending tens of angstroms from the impurity. Thus, orbital overlap (and hence band formation) can occur with moderate impurity concentrations, say $10^{25}\ \mathrm{m}^{-3}$. A requirement for impurity band conduction, as with all conduction in bands, is that the impurity band be partially occupied. A second interesting case of a narrow band is that of partially occupied inner shell d orbitals that occur with some transition metal oxide semiconductors.

The electronic conductivity σ in a solid is given by

$$\sigma = nq\mu_n + pq\mu_p \tag{1.8}$$

where n is the density of free electrons, p the density of free holes, μ_n the electron mobility, μ_p the hole mobility. If more than one band boasts free electrons or holes, naturally Equation (1.8) must be summed over all such bands. The conductance G and resistance R are, of course, given by

$$G = R^{-1} = \sigma A/L \tag{1.9}$$

where A is the cross-sectional area of the sample, L is the length in the direction of current flow.

1.1.3.2. Conductance by Charged Defects. Interstitial ions and charged vacancies can move through the crystal under the influence of an electric field (field-aided diffusion) or a concentration gradient (diffusion). An interstitial can jump to successive equivalent sites. Vacancies move through the crystal by a mechanism involving an atom jumping into a neighboring vacant site, then the next atom jumping into the vacant site, and so on. The net result of the process is motion of the vacancy through the crystal.

The motion of a vacancy or interstitial is an activated process, with a diffusion constant D described by an equation

$$D = D_0 \exp(-E_D/kT) \tag{1.10}$$

The rate of transport is given by a diffusion equation:

$$J = -qN\mu(d\phi/dx) - qD(dN/dx) \tag{1.11}$$

with the first term representing field-aided diffusion, the second term representing diffusion due to a concentration gradient. Here J is the current per unit area, $-d\phi/dx$ is the electric field, N is the density of vacancies or interstitials, and μ is their mobility. The relation between D and μ is given by the Einstein relation:

$$Dq/kT = \mu \tag{1.12}$$

Such defect diffusion is of direct importance in anodic oxidation.[R7] For example, in the anodic oxidation of GaAs, cation vacancies are formed[10] at the oxide/solution interface when a surface cation changes its position and forms a bond with a newly assimilated O^{2-} ion from solution. The resulting cation vacancy then migrates, under the influence of the anodic electric field and the concentration gradient, to the GaAs/oxide interface, where the vacancy is annihilated by being occupied by another cation. In general, field-aided diffusion dominates in anodic oxidation; the influence of the concentration gradient is greater in high-temperature oxidation where there is no electric field.

Such diffusion can be important also at a semiconductor electrode because ions from solution can diffuse in, or vacancies can diffuse out. For example, Haneman and Steenbeeke find[11] that with an applied cathodic voltage, ions diffuse into TiO_2 from aqueous solutions, as discussed further in Section 5.2.2.2.

1.2. The Solution

1.2.1. Introduction

1.2.1.1. Chemical Transformations. This section provides a short discussion of chemical interactions of ions in solution with the solvent and with each other, and how a change in ionic charge on the ion leads to a change in chemical structure surrounding the ion. Such chemical transformations are important in semiconductor electrochemistry in two ways. First, when we try to describe the ions in terms of their electronic energy levels, as will be done in the following subsection, these chemical transformations greatly complicate the description. Second, when describing the exchange of electrons between ions in solution and an electrode, such chemical changes that precede and/or follow the reaction greatly complicate the description of electron transfer.

An ion in solution interacts both with the solvent molecules and with other ions or molecules in solution. The overall interaction is termed the "solvation" of the ion. The interactions can be conveniently separated into "inner sphere" and "outer sphere" interactions, where the inner coordination sphere is made up of nearest neighbors, the outer solvation sphere is made up usually of solvent molecules that are farther removed. Inner sphere complex formation is the strong interaction where the ion and its neighbors form a group that can be considered almost a compound. Outer sphere complex formations are possible but are much weaker interactions, almost always between ions of opposite charge, where a shell of solvent molecules separates the ion from the counter-ion. In general we will not discuss outer sphere complex formation. The chemical "reorganization" or "rearrangement" processes discussed will be limited to two possibilities, possible inner sphere reorganization and the inevitable change in the polarization of outer sphere solvent molecules following electron transfer.

The solvation of ions in water provides illustrative examples of possible inner sphere reactions. In water the inner sphere solvation interaction may be hydration, hydrolysis, or complex formation. Normally, when water molecules are the only species in the inner sphere the process is termed "hydration" or the formation of an "aquo complex." Normally, when OH^- or HO^+ ions are in the inner sphere it is termed hydrolysis. The term hydrolysis is often used also for processes such as:

$$U(H_2O)_6^{6+} \rightarrow UO_2^{2+} + 4H^+ + 4H_2O \tag{1.13}$$

The stabilization energy of the inner sphere complex can arise from several sources. It can be associated with electrostatic forces as in hydration, which is analyzed by the Born electrostatic model discussed below. Alternately, it can arise due to other interactions, as for example, crystal field effects with transition metal ions. The interaction (stabilization) energy of strong chemical bonds can be estimated by relating the energy of complex formation or hydrolysis to the known energies of compound formation. Alternately, the energy of hydrolysis involving the attraction of OH^- ions to a multivalent metal ion can be estimated by evaluating the energy of a process where H^+ ions are released from the hydrated metal ion.

As mentioned above, the outer sphere reorganization associated with the polarization of the solvent is inevitable. In the Born electrostatic model for hydration the energy is derived by considering the ion as a rigid sphere of radius a and charge Z to be placed in a medium of dielectric constant κ. The energy of hydration is calculated as the difference in work necessary to charge the ion in vacuum and in the medium. Simple electrostatics gives as the work to add a charge dq to a sphere of charge q and radius a

$$dW = \left[\int_{\infty}^{a} \left(\frac{q}{4\pi\kappa\epsilon_0 r^2}\right) dr\right] dq \tag{1.14}$$

so the total work is obtained by integrating from $q = 0$ to $q = Ze$. The energy to charge the sphere in vacuum is given with the dielectric constant unity and the energy in the medium is given using a dielectric constant of κ so the solvation energy per ion is

$$W = -(Z^2q^2/8\pi\epsilon_0 a)(1 - 1/\kappa) \tag{1.15}$$

In most effective solvents for electrochemical use, the dielectric constant is high (for water κ is about 80). Thus, clearly, solvation on ions with a low a and a high Z can be a very energetic process. For example, Hunt[R8] lists the solvation energy for F^-, Cl^-, and Li^+ as about 2 eV, for Be^{2+} as about 8 eV, and for Al^{3+} as about 19 eV. More elaborate and accurate models for the energy of hydration have been developed,[R8] but the order of magnitude of the energy is the same.

In the case of an ion completely surrounded by inner sphere complexing species, such as $Fe(CN)_6^{4-}$, one can consider the ion plus the inner sphere as a complex with net charge $Z'e$, (in ferrocyanide above Z' being -4), and the Born equation gives the energy of hydration for the complex.

In the discussions later in the chapter we will be interested in the change in solvation energy when an electron is transferred to or from an ion. We will find the analysis in the simplest case bears a strong resemblance to the Born analysis. If there is no change in the inner sphere when electron transfer (oxidation or reduction of the ion) occurs, then the Born equation is directly related to the change in energy of the system.

As was done with solvation energies above, it is convenient to classify the chemical transformations that occur preceding or following electron transfer between an ion and an electrode into two types. First is the solvent polarization effect, the outer sphere reorganization deriving from a Born model, which must always be recognized and accounted for, and which is susceptible to mathematical analysis. Second is the stronger inner sphere chemical transformation, which does not always occur, but which can take many forms. Above we have discussed only one form of inner sphere changes, namely, the ligand shell of an ion changing—the complex changing. Other classes of chemical transformations possible following electron transfer are, for example, the dimerization when H^+ is reduced to form H_2, the precipitation or adsorption of the atom, such as when Ag^+ is reduced to form Ag metal on the electrode, and so on through all the possible chemical transformations that can be visualized. It will be shown in later discussions that (a) the form of such complexes strongly affects the energy levels calculated for the ions, and (b) changes in the inner shell during electron transfer strongly affect the kinetics, and indeed, make the use of the band models very difficult.

1.2.1.2. Energy Levels Associated with Ions in Solution. The electronic energy levels on an ion (or a molecule) in solution reflect, of course, the tendency of that species to give up or to accept an electron when the species approaches the electrode. The remainder of Section 1.2 will be devoted to a discussion of such energy levels. In the present introductory section, a qualitative picture will be presented in preparation for the more quantitative discussions to follow.

Consider the following series of chemical reactions:

$$K = K^+ + e^- \tag{1.16a}$$

$$V^{2+} = V^{3+} + e^- \tag{1.16b}$$

$$Fe^{2+} = Fe^{3+} + e^- \tag{1.16c}$$

$$O_2^- = O_2 + e^- \tag{1.16d}$$

$$Ce^{3+} = Ce^{4+} + e^- \tag{1.16e}$$

$$2F^- = F_2 + 2e^- \tag{1.16f}$$

With the exception of the F^-/F_2 case, these are all chosen to be chemically relatively simple one-electron (one-equivalent) reactions, with each species having two stable oxidation states, and with chemical transformations avoided to a large extent. They are listed in order of the tendency to proceed to the right. Potassium metal is an extremely strong reducing agent; Fe^{2+} a relatively moderate reducing agent; and Ce^{3+} a very weak reducing agent. Reading from the bottom up, the reactions are listed in order of their tendency to proceed to the left. Ce^{4+} is a very strong oxidizing agent; Fe^{3+} is a moderate oxidizing agent; and K^+ is an extremely weak oxidizing agent. Naturally, the energy levels can be expected to follow in the appropriate order: the occupied energy level of the K/K^+ couple should be very high, indicating an electron on this energy level is energetic and energy will be released if it transfers to most other species. The unoccupied energy level associated with the Ce^{3+}/Ce^{4+} couple should be very low, as would be expected for a strong oxidizing agent that will capture an electron from most species.

Figure 1.4 is an oversimplified model to show where such energy levels will fall relative to the energy levels associated with a semiconductor (ZnO is used as the example). It is oversimplified because it indicates that the energy levels for the reducing agent and the oxidizing

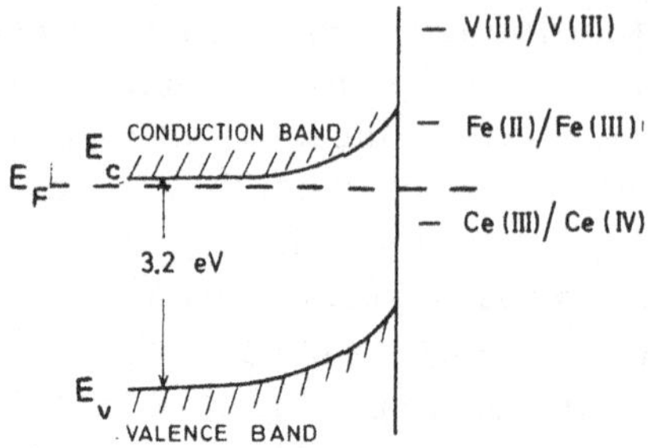

Figure 1.4. Energy levels associated with redox couples in solution. The more negative the redox potential, the more tendency the couple to give electrons to the solid, so a higher energy level is anticipated.

agent of the "redox couple" are the same. Thus, we show Fe^{2+} and Fe^{3+} as having the same energy level, with the species being Fe^{2+} if the level is occupied, and Fe^{3+} if the level is unoccupied. This assumes that if the electron were to transfer to the solid [Equation (1.16c) proceeding to the right] and then a second later the electron were to transfer back to the ion, it would come back to the same energy level. We will see below that this is not entirely accurate—because of solvation differences between the oxidized and reduced ions in solution the electron level on the oxidized ion is different from the energy level on the reduced form of the ion. However, for the present qualitative discussion, and indeed for many such discussions in the book, the assumption that the energy level when unoccupied is the same as that when occupied, as shown in Figure 1.4, is quite satisfactory.

The energy levels expected for the ions can be related quantitatively to their chemical behavior. From classical metal electrochemistry the tendency of a redox couple in aqueous solution to give up an electron to, or accept an electron from, a metal electrode is measured as the potential of the electrode and tabulated in handbooks as the "standard electrode potential" or "redox potential" of the redox couple. The word "standard" refers to standard conditions, among which is the requirement of equal activity (concentration) for the two oxidation states (see Section 1.2.2 below). The point is, that if the reducing agent of the couple is strong, so electrons tend to be given up, then, as in Equation (1.16a) or (1.16b), a metal electrode will tend to become highly negatively charged, so the "redox potential" will be large and negative. Conversely, if the couple boasts a powerful oxidizing agent that will extract electrons from the metal electrode, the metal will become highly positively charged and have a large positive redox potential. Thus, one expects a correlation between the measured redox potential at a metal electrode and the energy level of the redox couple in solution. And, since these redox potentials are tabulated in handbooks, to a first approximation one need only to glance at the handbook to find out the expected energy level of an ion in solution. The quantitative analysis showing this relation is completed in Section 1.2.4.4.

Unfortunately for the above simple model, the use of the redox potential to calculate electronic energy levels can only be approximation, except in carefully selected cases, because the redox potential reflects the energy of any chemical transformation. The redox potential reflects not only differences in the electronic energy levels, but the total energy involved in electron transfer reactions such as those in Equation (1.16), including chemical transformation such as in the fluoride case. As will be seen, the electronic energy levels often dominate the measured redox potentials for simple reactions such as those in Equation (1.16b), (1.16c), and (1.16e), but when one desires an accurate quantitative analysis, chemical transformation energies are so important that they can only be neglected in a few cases. Thus the electronic levels, as calculated from the redox potential, can only be considered a first approximation. Such energy levels must be used with caution, recognizing chemical transformation energies have been neglected.

It is convenient at times to indicate on a band model the overall energy change in a redox reaction, where by overall energy change we include both the change of electronic energy level, and also chemical changes (solvation or precipitation). Then the tabulated redox potential is exactly the property needed. The redox potential E^o, as tabulated, reflects the overall change in free energy when, under standard conditions, the electron leaves the ion or arrives at the ion, as in Equation (1.16). For nonstandard concentrations the free energy change must be modified from E^o as discussed in Section 1.2.2 below. The "energy level" of the ion thus determined directly from E^o, and used to calculate the free energy change for electron transfer, will be called the "thermodynamic energy level" for the couple. Of course, to obtain the free energy change for the electron transfer from E^o we must also locate the Fermi energy of the solid because, as discussed above, the Fermi energy is the free energy of the electron in the solid. And the Fermi energy must be known on the same energy scale as used in tabulations of oxidation potentials. Methods of locating the Fermi energy of the solid on the standard electrochemical energy scale are discussed in the last section of this chapter, Section 1.5. Knowing these energies, one can prepare a band diagram that includes both the band edges and Fermi energy for the semiconductor on the one hand, and the thermodynamic energy level for the redox couple on the other hand. Then one can observe by inspection whether it is thermodynamically favorable to transfer electrons one way or the other. Actually, such a diagram will resemble that of Figure 1.4.

To reemphasize, the thermodynamic levels of energy are not real electronic energy levels. They can be used to describe thermodynamic equilibrium conditions but not to describe kinetic parameters. Both the thermodynamic level of energy and the real electronic energy levels of ions in solution will be used in discussions in later sections of this book.

1.2.2. The Electrode Fermi Energy as a Function of the Redox Couples in Solution

It is beyond the scope of this book to delve deeply into thermodynamic derivations, but it is of value to indicate how, at equilibrium, the electrode potential (the Fermi energy of the electrode) in a solution stems from known thermodynamic functions describing the active species in solution.

1.2.2.1. Activity Coefficients. For an ideal solution, the chemical potential of an ion i is

$$\mu_i = \mu_i^o + RT \ln c_i \tag{1.17}$$

where c_i is the molarity of the solute. This relation stems from the expression of the chemical potential of a component of an ideal gas and Raoult's law stating that with an ideal gas and liquid, the vapor pressure in the gas is proportional to the concentration in the liquid. For a real solution, a nonideal solution, where ion–ion interaction occurs, the chemical potential is not described accurately by Equation (1.17). We define parameters γ_i, the activity coefficient of the ion, and a_i, the activity of the ion, such that

$$\mu_i = \mu_i^o + RT \ln \gamma_i c_i = \mu_i^o + RT \ln a_i \tag{1.18}$$

In practice, except for the present section, we shall in general use concentrations [Equation (1.17)] instead of activities [Equation (1.18)], because the substitution is reasonable within the accuracy in the analysis of electrode processes (see Reference R9, p. 53). However, it is of interest to note that the correction (the shift in γ from unity), is mainly associated with simple electrostatics. The tendency of ions to react, their "activity," is affected by the electrostatic forces due to neighboring ions.

The Debye–Hückel theory describes such electrostatic interactions and, to a good approximation, gives values for γ_i where an intermediate concentration of ions is assumed. Essentially the potential at an ion is shown to be affected by the ion cloud it attracts. Thus, if we can represent the ion as a point charge, it is found by such calculations that the value of γ_i depends on the valence Z_i of ion i, and the ionic strength of the solution, S:

$$\ln \gamma_i = 1.13 Z_i^2 S^{1/2} \tag{1.19}$$

where

$$S = \tfrac{1}{2} \sum_i c_i Z_i^2 \tag{1.20}$$

1.2.2.2. Electrochemical Potentials. The electrochemical potential of a species, $\bar{\mu}_i$ is

$$\bar{\mu}_i = \mu_i + Z_i F \phi \tag{1.21}$$

where F is the Faraday constant (96489 C/mole), Z_i the charge on the ion, and ϕ is the potential in volts. Thus $Z_i F \phi$ is the potential energy (in coulomb volts/mole) of the charged species in a region of potential ϕ. We will not define the zero of potential or the origin of the potential ϕ. The chemical potential μ_i or electrochemical potential $\bar{\mu}_i$ have properties stemming from thermodynamic theory that we will use in developing the expression for the Fermi energy of an electrode. For example, the basic definition of the electrochemical potential is

$$\left(\frac{\partial G}{\partial n_i}\right)_{T,p,n_j} \equiv \bar{\mu}_i \tag{1.22}$$

where G is the Gibbs free energy, n_i the concentration. Equation (1.22) holds if the concentrations of all species other than i are held constant, and if the temperature and pressure are held constant. A useful feature of $\bar{\mu}_i$ is that at equilibrium over all contacting phases

$$\bar{\mu}_i = \text{const} \tag{1.23}$$

Thus for example the electrochemical potential of electrons (the Fermi energy) is constant over all contacting solids in equilibrium that conduct by electronic carriers.

The chemical potential is primarily an indication of the driving force for a reaction. In any reaction, for example

$$\nu_1 A_1 + \nu_2 A_2 = \nu_3 A_3 + \nu_4 A_4 \tag{1.24}$$

where the A's are species in a reaction, the ν's are integers (with the ν's for the products carrying a negative sign), the rate of the reaction must be equal in both directions at equilibrium, and it can be shown that at equilibrium the electrochemical potentials of the reactants and products must be related by

$$\sum \nu_i \bar{\mu}_i = 0 = \Delta G_{\text{equil}} \tag{1.25}$$

Thus, for example, in the reaction describing the dissociation of acetic acid in aqueous solution

$$CH_3COOH = H^+ + CH_3COO^- \tag{1.26}$$

equation (1.25) requires at equilibrium

$$\bar{\mu}_{CH_3COOH} = \bar{\mu}_{H^+} + \bar{\mu}_{CH_3COO^-}$$

or (1.27)

$$\mu_{CH_3COOH} = \mu_{H^+} + \mu_{CH_3COO^-}$$

The latter form of Equation (1.27) is more often seen, but the two forms are equivalent in a region of constant ϕ, and are related by Equation (1.21), noting that the terms in Z_i cancel because charge is conserved in any reaction. Because we are interested in the electrochemical potential of electrons, we will be using the first form of Equation (1.27) below.

The equilibrium constant K is defined as

$$\ln K = (-\sum \nu_i\mu_i^o)/RT \tag{1.28}$$

If the potential ϕ is identical for all species, we have with Equations (1.18) and (1.25)

$$\ln K = \sum \ln a_i^{\nu_i} \tag{1.29}$$

Thus, in the example of reaction (1.26)

$$K = a_{H^+}a_{CH_3COO^-}/a_{CH_3COOH} \tag{1.30}$$

1.2.2.3. The Fermi Energy of the Electrode. The application of the above well-known equations to electrode reactions is straightforward and rewarding. Consider the reaction

$$\nu_1 A_1 = \nu_2 A_2 + ne^- \tag{1.31}$$

where the electrons e^- are transferred between the electrode and the redox couple A_1/A_2. For example, the reaction could be

$$H_2 = 2H^+ + 2e^- \tag{1.32}$$

or

$$Zn = Zn^{2+} + 2e^- \tag{1.33}$$

Application of Equation (1.25) to reaction (1.31) yields

$$\nu_1\bar{\mu}_{A_1} = \nu_2\bar{\mu}_{A_2} + n\bar{\mu}_e \tag{1.34}$$

with $\bar{\mu}_e$ the electrochemical potential of electrons in the electrode, which is equal to [12] the Fermi energy in the electrode.

We can calculate the Fermi energy as determined by the redox potential of the ions and by the potential in solution. We begin by rewriting Equation (1.34), using chemical potentials as defined in Equation (1.21):

$$\bar{\mu}_e = (1/n)(\nu_1\mu_{A_1} - \nu_2\mu_{A_2}) - F\phi_{\text{sol}} \tag{1.35}$$

and then, with the help of Equation (1.18),

$$\bar{\mu}_e = (1/n)(\nu_1\mu^o_{A_1} - \nu_2\mu^o_{A_2}) - (RT/n)\ln(a^{\nu_2}_{A_2}/a^{\nu_1}_{A_1}) - F\phi_{\text{sol}} \tag{1.36}$$

where ϕ_{sol} is the potential in the solution and where we note that

$$n = \nu_2 Z_{A_2} - \nu_1 Z_{A_1}$$

We define

$$E'_A = \frac{-1}{nF}\sum \mu^o_{A_i}\nu_i \tag{1.37}$$

whence

$$\bar{\mu}_e = +FE'_A - (RT/n)\ln(a^{\nu_2}_{A_2}/a^{\nu_1}_{A_1}) - F\phi_{\text{sol}} \tag{1.38}$$

Now the preferred units for describing electrochemical potentials are joules/mole and the preferred units for the Fermi energy are electron volts/electron, so the Fermi energy is

$$E_F = q\bar{\mu}_e/F \tag{1.39}$$

so

$$E_F = +qE'_A - (qRT/nF)\ln(a^{\nu_2}_{A_2}/a^{\nu_1}_{A_1}) - q\phi_{\text{sol}} \tag{1.40}$$

or

$$E_F = +qE'_A - (kT/n)\ln(a^{\nu_2}_{A_2}/a^{\nu_1}_{A_1}) - q\phi_{\text{sol}} \tag{1.41}$$

If we make our usual approximation that activities equal concentrations:

$$E_F = +qE'_A - (kT/n)\ln\{[A_2]^{\nu_2}/[A_1]^{\nu_1}\} - q\phi_{\text{sol}} \tag{1.42}$$

The parameter E'_A (in volts) is a property of the redox couple A_1/A_2 (H_2/H^+ or Zn/Zn^{2+}, for example) and describes its tendency to inject or extract electrons from the metal.

Thus, the Fermi energy is determined, as per Equation (1.42), first by the tendency of the couple to inject electrons, E'_A; and second, by a term reflecting the relative concentrations of the (electron-injecting) reducing agents and (electron-accepting) oxidizing agents; and finally by a third term, reflecting any potential applied to the solution and reflecting our choice of the zero of potential (and hence energy).

Until a zero of potential or some other reference potential is defined, the equations (1.41) and (1.42) are of limited value. The relative values of E'_A for various couples are determined in principle through Equation (1.37), but absolute values of μ_i^o cannot easily be measured. What can be measured are differences in E_F, voltage differences between two metal electrodes, one metal in contact with one redox couple, A_1/A_2 and the other metal in contact with another redox couple, B_1/B_2, where ϕ_{sol} is constant throughout the solution. Then the measured voltage difference V_m is, from Equation (1.41),

$$-qV_m = E_{F1} - E_{F2}$$
$$= qE'_A - (kT/n_A)\ln(a_{A_2}^{\nu_{A2}}/a_{A_1}^{\nu_{A1}}) - qE'_B + (kT/n_B)\ln(a_{B_2}^{\nu_{B2}}/a_{B_1}^{\nu_{B1}}) \quad (1.43)$$

where subscripts A and B have been used to distinguish the two contacting systems, A_1 and A_2 being the redox couple in contact with metal #1, and B_1 and B_2 being the redox couple in contact with metal #2.

In order to have a practical quantity, we choose one redox couple as the standard, defining a zero, and measure the value of V_m of all other redox couples against it. This then enables us to order the tendency of various species to inject electrons.

Specifically, the hydrogen couple [Equation (1.32)] is chosen as the reference system, with the corresponding Fermi energy chosen as our arbitrary zero. If the couple being tested tends to inject electrons into a metal more than the hydrogen couple under standard conditions, the redox potential of that couple is negative. If the hydrogen couple (under standard conditions) tends to inject electrons more, then the redox potential of the couple under test is positive. Thus, using activities of unity for H^+ and H_2 (essentially one molar acid with H_2 at a pressure of 1 atmosphere bubbled through), Equation (1.43) becomes

$$-qV_m = E_{F1} - E_{ref}(\text{hydrogen})$$
$$= q(E'_1 - E'_{ref}) - (kT/n)\ln a_{A_2}^{\nu_2}/a_{A_1}^{\nu_1} \quad (1.44)$$

where the redox couple denoted as B in Equation (1.43) is now the hydrogen couple under standard conditions—so the last term of Equation (1.43) becomes zero. This simplifies to

$$-qV_m = E_{F1} = -qE^o - (kT/n)\ln(a_{A_2}^{\nu_2}/a_{A_1}^{\nu_1}) \quad (1.45)$$

where E_{F1} is the Fermi energy of the electrode in contact with the redox couple A_1/A_2, where we have defined our zero as E_{ref}(hydrogen), and where we have introduced the parameter E^o, the "standard redox potential" of the redox couple A_1/A_2. The parameter E^o is tabulated in handbooks* as the potential measured relative to the hydrogen reference electrode when any specified redox couple of interest is present at unit activity ($a_{A_2} = a_{A_1} = 1$).

Thus, Equation (1.45) gives, relative to the hydrogen reference electrode, the Fermi energy of an electrode when it is in contact and in equilibrium with a redox couple, A_1/A_2. The reducing agent is present with activity (concentration) a_{A_1}, and the oxidizing agent is present with activity (concentration) a_{A_2}.

To illustrate further, the values of E^o are determined using the measurement shown schematically in Figure 1.5. The electrode indicated as "metal" may be the reducing agent of the redox couple, as in Equation (1.33) or may be inert, as desired in reactions (1.16b)–(1.16f). The connection indicated in Figure 1.5 as the "bridge" is designed such that, while blocking passage of active species, the bridge passes current by movement of an inert ion, such as K^+, whose movement from one cell to the other has negligible effect on the potentials. Thus the potential ϕ is constant throughout the solution. The intent of the picture is to illustrate that with such standard conditions, the measurement of the electrode potential V_m will indeed provide a measure of the tendency of the reducing agent to inject electrons into the electrode, or of the oxidizing agent to extract them. Such a simple measurement will thus provide information about the energy levels of ions in solution.

In summary, the tabulated redox potentials for species in solution describe their tendency (when both members of the redox couple are present) to give electrons to a solid or to extract electrons from a solid. A high negative redox potential such as tabulated for K/K^+ indicates that the couple, if available, will tend to give electrons to the solid (cause the electrode to assume a negative potential) while a high positive redox potential indicates that the couple will tend to extract electrons from the solid (cause the solid to assume a positive potential). The hydrogen couple is defined as the reference, zero potential.

1.2.3. The Relation between the Hydrogen and the Vacuum Scales of Energy

In the chemical physics of surfaces, there are two common zeros of energy used. At the gas/solid interface, the usual zero of energy is the energy of an electron at infinity, E^e. In electrochemistry the usual zero of energy is the Fermi energy of the hydrogen reference electrode under

* The symbol E^o is the only case in this book where the symbol E represents a voltage rather than an energy or an electric field. We use the symbol to conform to common usage, but its use results in some odd-looking equations.

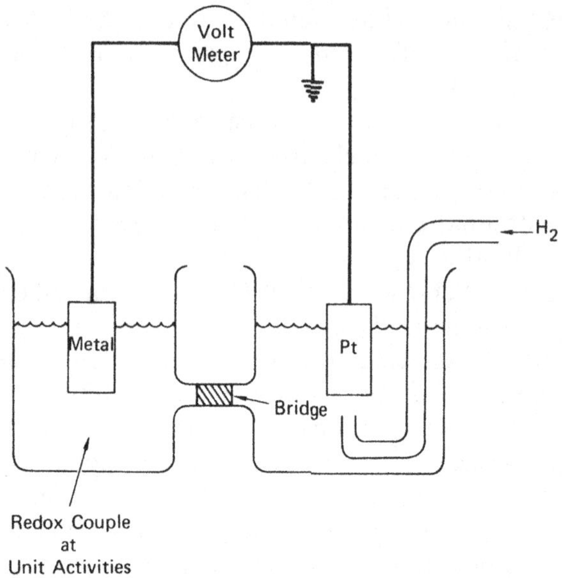

Figure 1.5. Schematic of measurement of standard electrode potential E^o. The voltmeter measures E^o with respect to ground.

standard conditions, as described in the preceding subsection. It is helpful to visualize the energy level diagrams when the two zeros of energy are compared. Several derivations of the relation between the two zeros of energy have been made, and have been reasonably consistent. The hydrogen reference electrode energy is considered to be very close to 4.5 eV below E^e.

Many authors in recent electrochemical papers have chosen to refer all energies to E^e rather than to the hydrogen zero. In this book we have chosen not to do so in general for three reasons. First the difference between the two zeros, 4.5 eV, is not a measured quantity, as will be indicated below. It is obtained by extrapolation, and to some extent empirical models. This is not in itself a great problem—the value 4.5 eV seems useful, and enables useful correlations of semiconductor bands to be made, as will be discussed in Section 5.1. However, the value may be improved in the future if somebody develops a better model. A second reason we have not used E^e as the reference zero is because we will normally be dealing with energy levels near the energy of the hydrogen reference electrode, and using the latter as the zero thus leads to conveniently small numbers, easily remembered. Thirdly, and most important, it will be shown that the energies on the hydrogen reference scale are measured quantities (at least in principle), available in handbooks, and so are convenient to use.

The best determinations of E^e on the hydrogen scale were made by

Noyes[13] and Lohmann,[14] plus a more recent determination (resulting in E^e at 4.69 eV above the hydrogen reference energy) by Trasatti.[15] Trasatti studied mercury, and used the work function of mercury to define a cycle yielding the value for E^e. Noyes[13] made an extrapolation as a function of ionic radius from the alkali ions and the halide ions down to hydrogen, using an empirical formula for the extrapolation. The formula was based on the Born approximation for hydration energy, as in Equation (1.15), but Noyes empirically introduced an additional term to permit better self-consistency. He concluded the value of E^e was 4.54 eV above the hydrogen reference energy.

We will describe the approach of Lohmann[14] which was, in principle, quite straightforward. He examined free energy changes associated with the reduction of silver:

$$
\begin{array}{rcll}
Ag^+(g) + e^- &=& Ag(g) & -7.64\text{ eV} \\
Ag(g) &=& Ag(s) & -2.60\text{ eV} \\
Ag^+(aq) &=& Ag^+(g) & +4.96\text{ eV} \\
H^+(aq) + Ag(s) &=& Ag^+(aq) + \frac{1}{2}H_2 & +0.80\text{ eV} \\
\hline
H^+ + e^- &=& \frac{1}{2}H_2 & -4.48\text{ eV}
\end{array}
$$

In this analysis, g indicates the gas phase, s, the solid, and aq the solution (in aqueous solution) phase. The first, second, and fourth free energies indicated are obtained directly from experiment. The critical energy is the third, where silver from solution becomes a gas phase silver ion. Lohmann obtained this value for the free energy change from a semiempirical analysis of Case *et al.*,[16] which we shall not reproduce, as it is very complex.

As will be found in later discussion, relating electronegativities to band edges for example, the value about 4.5 eV seems to be satisfactory using several approaches, although the most recent studies[15,17,18] suggest that a value of 4.7 eV may be more accurate.

1.2.4. Fluctuating Energy Levels in Solution

1.2.4.1. Introduction. When discussing semiconductor electrodes it is particularly important to describe the ions in solution in terms of their energy levels. It will be found, for example in this book, that we will normally use an energy level description rather than a chemical potential description. The reason is that one can conveniently connect the properties of semiconductor or insulator electrodes with the properties of the solution only when both are described with the same electronic energy level model. And the only good way to describe a semiconductor or insulator is with an energy level model.

Unfortunately, the description of the energy levels associated with ions in solution is more complex than is usually encountered in semiconductor physics because of the effect on electronic levels of the polar solvent (say water) surrounding the ions. The dipoles associated with the solvent molecules are able to move and to rotate much more easily than normally found in solids. As discussed in Section 1.2.1.1, an ion introduced into a solvent causes the solvent to polarize. To a good approximation the equilibrium polarization can be calculated by simple electrostatic theory for a spherical charge in a medium of dielectric constant κ. Now the polarization of the solvent affects the potential at the location of the ion, and the energy level on the ion will be determined in part by this calculable equilibrium polarization. However, in addition to the effect of the equilibrium polarization, we must consider thermal fluctuations in the polarization. Because the solvent dipoles undergo thermal fluctuations, the potential at the ion fluctuates, and the energy level on the ion will fluctuate around its most probable value. We will show that the probability of the energy level being at an energy E will have a Gaussian distribution.

The probability of electron transfer between an electrode and an ion in solution will depend on the energy level of the ion at the instant of transfer, and so the fluctuations are of importance. However, the fluctuations affect the electron transfer only if they are "slow," dependent on ion or molecular motion rather than electronic motion. If the fluctuations stem from the polarization of electron shells, the fluctuations will be so rapid they will average out during an electron transfer process. Only when the polar effects stem from slow motion of molecules is the electronic energy level on the ion "frozen" during electron transfer. Thus, in the mathematics below, the factor $(\kappa_{op}^{-1} - \kappa_s^{-1})$ appears, where κ_{op} is the optical dielectric constant of the solvent representing high-frequency polarization, κ_s the static dielectric constant which includes the slow dipole movement of interest.

Figure 1.6 shows the accepted model of energy levels in solution, which will be developed in subsections below. The energy is plotted on the ordinate; the probability of finding the energy level of an ion at that energy is plotted on the abscissa. Thus E_{ox} is indicated as the most probable energy level for an oxidizing agent (an electron acceptor), as determined by its chemical properties and representing the condition where the polar medium is at its equilibrium (lowest energy) polarization. A Gaussian function is sketched, centered at the level E_{ox}, that shows the probability of finding the level at another energy due to thermal fluctuations of the dipoles of the solvent. The analysis of these fluctuations is presented in Section 1.2.4.2. We will also show below that when the energy level becomes occupied (viz., when we are describing the reducing

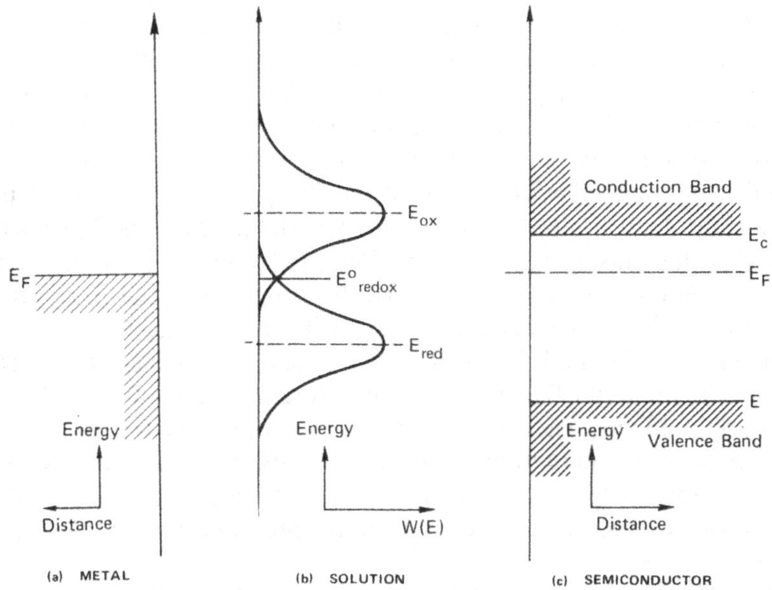

Figure 1.6. Fluctuating energy levels of states in a polar solution. Here $W(E)$ is the probability that the state E_{ox} or the state E_{red} has fluctuated to the energy E. The ordinate shows the approximate position of E_{ox} and E_{red} at equilibrium for equal concentrations of oxidizing and reducing agents.

agent of the same couple) the mean energy level is at a different energy, E_{red}, as indicated in the figure. The analysis of this shift of energy level is presented in Section 1.2.4.3. The energy E^o_{redox} defined at $E^o_{redox} = \frac{1}{2}(E_{ox} + E_{red})$ is of significance (see Section 1.2.4.4) and is indicated in the diagram.

It should be noted that although in this section we describe the effect of the polar solvent on the energy levels of ions in solution, similar effects may be expected at the solid/solution interface. These are emphasized in Section 2.1.2.

1.2.4.2. Analysis of Energy Level Fluctuations. There have been two derivations of the energy fluctuations of levels on ions in solution that arise due to polarization fluctuations, and both have yielded essentially the same result. In both calculations the authors were interested in the rate of electron transfer, based on a model (see next section) where the energy level of the ion must fluctuate to the energy of the electron in the solid, and then electron transfer occurs by tunneling between levels of equal energy. In both calculations it was assumed in the first approximation there are no changes of bonding of the species (in its inner coordination sphere) associated with the electron transfer process, and the temperature is reasonably high.[19] The first derivation of electron transfer by the fluctuating energy level model was a classical calculation by Marcus.[20] The analysis involved a complicated method of evaluating the energy of the nonequilibrium polarization that was causing the energy level fluctuation. He used a variational technique which permitted him to prove that the activation energy for electron transfer that he calculated was the minimum possible with this one "coordinate," the polarization. He found that the medium can be considered to be always polarized in a way that would be associated with a central charge $q(Z + \gamma)$, where qZ is the equilibrium central charge and γ is the variable. Marcus' later articles expanded this classical theory.[21,22] Dogonadze *et al.*[19] and Christov[23] used a quantum mechanical analysis, describing both the electrons and the ions by quantum mechanical formalism, but noting that a good approximation is obtained[24] using a semiclassical approximation to describe the ions. Efrima and Bixon[25] provide references to many modifications that have been suggested in more recent work. Gerischer[26] has applied the energy level fluctuation model to describe processes at a semiconductor surface. Christov's review provides a good history of the development of the fluctuating energy level model and a review of its limitations.

We will develop the model using a classical formalism similar to, but less rigorous than, that of Marcus.

Consider the energy ΔE_p necessary to cause fluctuations from the equilibrium polarization of the dielectric around an ion without changing the charge on the ion. We introduce a parameter γ such that the polarization of the dielectric corresponds to $qZ \pm q\gamma$, where Z is the actual charge on the ion. Thus, we are defining a polarization that corresponds to a central charge differing by $\pm q\gamma$ from the central charge actually present. In other words, in its lowest energy configuration, the dielectric medium is polarized an amount corresponding to the charge qZ on the ion. We want an expression for the energy increase in the polarized dielectric medium when the central charge remains qZ but the polarization fluctuates from its equilibrium configuration to a new configuration which would normally correspond to a charge $(Z \pm \gamma)q$ on the ion. Thus, for example, if $\gamma = 1$, the polarization would be exactly the equilibrium polarization configuration with an extra electron on the ion. The energy ΔE_p is given to a good approximation by

$$\Delta E_p = \gamma^2 \lambda \tag{1.46}$$

where the "reorganization energy" λ is given by

$$\lambda = (q^2/8\pi\epsilon_0 a)(\kappa_{op}^{-1} - \kappa_s^{-1}) \tag{1.47a}$$

Here a is the ionic radius, κ_{op} is the optical, κ_s the static dielectric constant of the medium, and ϵ_0 is the permittivity of free space. This expression was derived by Marcus[20] and by Dogonadze *et al.*[19] We note the close analogy to the Born expression, Equation (1.15). As mentioned above, Marcus showed that such a collective mode of polarization leads to the lowest activation energy for the process. We will omit the details of the derivation of Equation (1.46). It derives from simple electrostatics, a calculation of the energy in a polarized dielectric with the polarization induced by a charged sphere, but the derivation is cumbersome.

Dogonadze *et al.*,[19] using his quantum mechanical approach, and Marcus,[22] by classical arguments, developed improved expressions for λ including changes in the inner sphere. For example, Marcus' value has the form

$$\lambda = \left(\frac{q^2}{8\pi\epsilon_0 a}\right)(\kappa_{op}^{-1} - \kappa_s^{-1}) + \sum_j \frac{f_j^{ox} f_j^{red}}{f_j^{ox} + f_j^{red}} \Delta x_j^2 \qquad (1.47b)$$

where the f's are force constants for the jth bond, and Δx_j is the displacement in the bond length. Such considerations are important, as the last term may not be insignificant, but they do not change the qualitative concepts of the origin of the reorganization energy, nor do they change the quantitative equations used in this book, where we treat λ as a measurable parameter. Thus, we will continue to ignore the inner sphere changes when we describe the source of the reorganization energy qualitatively, but will refer in general to Equation (1.47) as the expression for λ without specifying (1.47a) or (1.47b).

With the help of Equation (1.46), using it to calculate the energy needed to reduce an ion in solution, we can determine an expression for the net energy required to shift the energy level of the ion by polarization. We calculate the energy necessary to reduce the ion in three steps. First, we let the medium prepolarize, expending an energy as given by Equation (1.46). This prepolarization shifts the energy level from the energy E_{ox} to an intermediate energy E as yet undefined. Then we let the electron transfer from the solid, with an energy change $E_{cs} - E$. Now the ion has a new charge, and we let the polarization of the dielectric relax to its new equilibrium value, the release of energy again being determined by Equation (1.46). We use the subscript i to indicate the configuration of the system when the surface state is at the intermediate energy E. The three steps in the reduction described above become

Reaction step	Thermal energy required	
$S_0A^Z \rightarrow S_iA^Z$	$\gamma_i^2\lambda$	(1.48)
$S_iA^Z + e^- \rightarrow S_iA^{Z-1}$	$(E - E_{cs})$	(1.49)
$S_iA^{Z-1} \rightarrow S_fA^{Z-1}$	$-(1 - \gamma_i)^2\lambda$	(1.50)

Reaction (1.48) represents the first step, the change in the polarization, or change in dipole configuration around the ion A, shifting the energy level from its equilibrium value E_t (= E_{ox} if the ion is the oxidizing agent) to a value E. This shift is accomplished by a change in polarization γ_i, requiring, by Equation (1.46), an energy $\gamma_i^2\lambda$. Reaction (1.49) is the electron transfer, during which the polarization is frozen at the configuration denoted as S_i. The initial energy of the electron is E_{cs}, the final energy is E, so the energy required is as indicated. Equation (1.50) represents the subsequent relaxation of the polarization of the medium to its new equilibrium value. The initial polarization of the surrounding medium, the polarization during the electron transfer, corresponds to an ionic charge $(Z - \gamma_i)q$. But the central charge on the ion has become $(Z - 1)q$. Thus the change in polarization required to bring the system to its new equilibrium configuration corresponds to a change in central charge of $1 - \gamma_i$. When the polarization field shifts to the new equilibrium value, [from Equation (1.46), where we now have $(Z - 1)q$ as the equilibrium configuration, so γ in Equation (1.46) equals $1 - \gamma_i$] the energy released is $(1 - \gamma_i)^2\lambda$.

The total energy change of Equations (1.48)–(1.50), which represents the energy required to transfer an electron to an ion in solution with an appropriate change of equilibrium polarization of the medium, is given therefore by

$$\Delta = \gamma_i^2\lambda - (1 - \gamma_i)^2\lambda + E - E_{cs} \tag{1.51}$$

The energy Δ must be independent of the energy E of the intermediate energy level, and we can therefore use the relation to calculate the relation between E and ΔE_p. First, we determine Δ by calculating the energy change for the case where $\gamma_i = 0$ and $E = E_t$, where E_t is the mean energy of the surface state (either E_{ox} or E_{red}) and find Δ is $-\lambda + E_t - E_{cs}$. Inserting the value for Δ into Equation (1.51) and solving for γ_i, we find

$$\gamma_i = (E_t - E)/2\lambda$$

whence the energy to complete step (1.48) is

$$\Delta E_p = (E_t - E)^2/4\lambda \tag{1.52}$$

Equation (1.52) gives the thermal energy necessary to shift the energy level from E_t to an arbitrary energy E. This is the result of interest.

The reader must make a careful distinction between the energy of the system (free or thermal energy) and the energy level of the electron. On the energy level diagram (Figure 1.6) we represent only changes in electronic energy levels and are not able to represent the accompanying changes in polarization energy. Equation (1.52) is the thermal energy needed to shift the electronic energy level, and usually the thermal energy needed is much less than the resulting energy level shift.

We can determine the expected thermal fluctuations of the electronic energy level where the thermal energy fluctuation is given by $\Delta E_p \sim kT$. Then we obtain from Equation (1.52) a mean square fluctuation

$$(E - E_t)^2 = 4\lambda kT \tag{1.53}$$

Because λ is so large [from Equation (1.47) it is of the order of 1 eV for an ion of about 3 Å radius dissolved in water], the electronic energy levels will fluctuate by energies large compared to kT. By substituting $\lambda = 1$ eV into Equation (1.53) it is seen that mean square surface state fluctuations of the order of 0.3 eV or more may be anticipated.

The probability that the energy level of the species will have a value E is less than the maximum probability by an amount given by the Boltzmann factor $\exp[-\Delta E_p/kT]$ or $\exp[-(E_t - E)^2/4\lambda kT]$, and the probability function as illustrated in Figure 1.6 becomes

$$W(E) = (4\pi\lambda kT)^{-1/2} \exp[-(E_t - E)^2/4\lambda kT] \tag{1.54}$$

where the preexponential is a normalizing constant introduced to make the integrated probability unity. An equation of the form of Equation (1.54) was used by Gerischer[27] on the basis of a harmonic oscillator approximation and provided the basis of his theory of electrode reactions.

Gerischer, in his theories, considers the time-fluctuating energy levels (Figure 1.6) as bands of energy levels. Such a model, equating the temporal fluctuation to a band of energy levels, is very convenient but must be used at times with caution. Such "bands" arising from polarization fluctuations have different properties from the fixed bands usually encountered in solid state models. There is an essential difference in concept between, on the one hand, electron/phonon interactions causing a fluctuation of electronic energy in a static distribution of levels and, on the other hand, ion/phonon interactions causing a fluctuation of the energy levels themselves. For example, one cannot have optically induced transitions between the occupied and unoccupied levels shown in Figure 1.6, because the levels do not overlap in real space.

1.2.4.3. Franck–Condon Splitting of Levels. It is common both in solid state physics and in chemistry to find that ions move after an electron transfer occurs. Such ion or atom motion could be a simple shift of position because of electrostatic forces, or could be a major chemical transformation such as dimerization or the acquisition of a new charged ligand by the central ion that lost or gained the electron. In the present section we are interested in the simple change in the equilibrium polarization of the solvent, the reorientation of the dipoles of the solvent that occurs when a new charge appears on the central ion. This continues the discussion of the effect of solvent dipoles on ionic energy levels that was started in Section 1.2.4.2. Specifically, here we will develop an expression for the difference in energy level depending on whether the level on an ion is occupied or unoccupied by an electron, where the difference stems only from outer sphere polarization effects.

The principle used to analyze the effects of ion movement that occurs after electron transfer is the Franck–Condon principle, which essentially states that during an electron transfer from one level to another the atoms and ions are effectively frozen in their position, so that atom or ion motion can only occur before or after the electron transfer. The justification is simply that the period of ionic vibrations is a factor of 100 or so longer than the characteristic periods for electron oscillation, so ions simply cannot adjust rapidly enough to follow electron jumps. With this in mind, we can analyze the effect of the polarization of solvent ions on the energy levels of ions and show that the reduced form has a substantially different energy level from the oxidized form of the ion.

For illustration, consider the reaction

$$e^- + Fe^{3+} \rightarrow Fe^{2+} \tag{1.55}$$

The solvent polarization will be different with the iron ion triply charged, Fe^{3+}, than with the iron ion doubly charged, Fe^{2+}. If, for the moment, we neglect the temporal fluctuations of the levels, the electron will be captured at an energy level E_{ox} as indicated in Figure 1.6. Following the electron capture, the dipoles of the solvent will "slowly" reorient. The extra negative charge on the central ion will induce the negative poles of the dipoles to rotate away from the central ion, and the removal of the negative poles will make the potential at the central ion higher—more attractive to electrons. The electronic energy level "slowly" moves to a lower level on the band model. Thus, in Figure 1.6 the level E_{red} is shown lower than the level E_{ox}, the unoccupied level. This shift in energy level between the unoccupied and occupied level is called a Franck–Condon shift, the terminology stemming from semiconductor physics usage.

A quantitative estimation of the Franck–Condon shift is made by considering a cycle wherein an electron is transferred to the energy level E_{ox} from an invariant initial energy level. For reasons of later simplicity we will choose an initial energy level at E_F, the Fermi energy of the electrode. Then in a later step, the electron is transferred from the energy level E_{red} back to the electrode. The total energy change around the cycle will be zero. In this subsection we are most interested in shifts due to changes in solvent polarization, but we will not restrict the system to this form of change, but simply state that the preferred configuration of the surrounding group is of the form S_1 when the site is unoccupied and the level is E_{ox}, and of the form S_2 when the site is occupied and the level is E_{red}. Thus, for the ion A^{n-} with its preferred surrounding configuration, we use the symbol $S_1 \cdot A^{n-}$, and for the ion $A^{(n+1)-}$ in its preferred surrounding, we use the symbol $S_2 \cdot A^{(n+1)-}$. Here $(-n)$ is the charge on the ion when the level is unoccupied; $(-n - 1)$ is the charge on the ion when the level is occupied.

The reaction

$$S_1 \cdot A^{n-} + e^- \rightarrow S_1 \cdot A^{(n+1)-} \tag{1.56}$$

represents the electron transfer step from the solid to the unoccupied level of the ion. The energy involved for this step is strictly electronic (by the Franck–Condon principle) and is given by $E_F - E_{\text{ox}}$. The next step, the chemical reorganization, is represented by

$$S_1 \cdot A^{(n+1)-} \rightarrow S_2 \cdot A^{(n+1)-} \tag{1.57}$$

with a free energy change we will designate as $-\Delta G_{\text{red}}$. The third step is the return of the electron to the conduction band from the characteristic energy level E_{red} of the ion when occupied:

$$S_2 \cdot A^{(n+1)-} \rightarrow S_2 \cdot A^{n-} + e^- \tag{1.58}$$

and the energy released is $-(E_F - E_{\text{red}})$. Finally, the reverse chemical transformation restores the system to its original configuration:

$$S_2 \cdot A^{n-} \rightarrow S_1 \cdot A^{n-} \tag{1.59}$$

with a free energy change of $-\Delta G_{\text{ox}}$.

Summing the energy changes around the cycle we find

$$E_{\text{ox}} - E_{\text{red}} = -(\Delta G_{\text{red}} + \Delta G_{\text{ox}}) \tag{1.60}$$

where the ΔG's are both negative, corresponding to transformations from an energetic state to a stable state, and so E_{ox} is always higher than E_{red}. The difference between E_{ox} and E_{red} depends clearly on the energy of the chemical part of the transformation.

Now, in the case of interest in this subsection, the energy of the chemical part of the transformation is known from Equation (1.46). As discussed there, the polarization energy associated with the exchange of one electron is the value of ΔE_p with $\gamma = 1$. Thus neglecting entropy effects

$$\Delta E_p(\gamma = 1) = -\Delta G_{\text{ox}} = -\Delta G_{\text{red}} = \lambda \tag{1.61}$$

if the value of λ is the same for the oxidized and reduced species (a convenient approximation with our present experimental accuracy in λ measurement). Then

$$E_{\text{ox}} - E_{\text{red}} = 2\lambda \tag{1.62}$$

Between the result described by Equation (1.54) and Equation (1.62) we have the most essential features of the energy level of a simple ion in solution as indicated in Figure 1.6. Equation (1.54) shows why we must represent the levels by a Gaussian distribution reflecting the temporal fluctuation of each level, and Equation (1.61) indicates why, and by how much, the levels are shifted by the polar character of the solvent. Both phenomena are associated with the quantity λ. It should be reemphasized that these effects are not at all minor, for λ, as can be estimated by Equation (1.47), is the order of an electron volt in general. Thus, the Franck–Condon splitting of the levels is the order of 1 to 3 eV, and the mean square fluctuations are the order of several tenths of an electron volt.

1.2.4.4. The Equilibrium Position of Ionic Energy Levels Relative to E_F on the Electrode. It is convenient to define an "effective Fermi energy" in the solution containing one dominant redox couple as E_{redox}. We define it such that if equilibrium is reached with an electrode in the solution, the Fermi energy of the electrode will move to the energy E_{redox} due to electron exchange with the redox couple. Thus we define E_{redox} to be equal to E_F at equilibrium:

$$E_{\text{redox}} = E_F \tag{1.63}$$

where E_F is the Fermi energy of the contacting electrode, usually a metal. The parameter E_{redox} is a property of the solution (if the redox couple tends to give up electrons, the Fermi energy of the electrode will be high) and E_{redox} can only be measured by its influence on the electrode.

A case of particular interest is that where there are equal concentrations (activities) of the two ions A^{n-} and $A^{(n+1)-}$, in which case we use the symbol E^o_{redox}. The two are related, from Equations (1.45) and (1.63), by

$$E_{\text{redox}} = E^o_{\text{redox}} - kT \ln[A^{n-}]/[A^{(n+1)-}] \tag{1.64}$$

The first objective of this subsection is to relate E^o_{redox} to the energy levels E_{ox} and E_{red}. Thus, below we assume equal concentrations.

At equilibrium, of course, the free energy change, as an electron moves from the Fermi energy of the solid E_F to the solution, must be zero. The free energy change for the reaction

$$S_1{\cdot}A^{n-} + e^- = S_2{\cdot}A^{(n+1)-}$$

is given by the sum of the energy change for reactions (1.56) and (1.57) or by the negative of the sum of the energy changes for Reactions (1.58) and (1.59). As these energy differences must be zero at equilibrium we have

$$0 = E_F - E_{\text{ox}} - \Delta G_{\text{red}} = E_F - E_{\text{red}} + \Delta G_{\text{ox}} \tag{1.65}$$

From Equation (1.63) in (1.65) we have

$$E^o_{\text{redox}} = \tfrac{1}{2}(E_{\text{ox}} + E_{\text{red}}) + \tfrac{1}{2}(\Delta G_{\text{red}} - \Delta G_{\text{ox}}) \tag{1.66}$$

and using the approximation of Equation (1.61), which will be used throughout this book:

$$E^o_{\text{redox}} = \tfrac{1}{2}(E_{\text{ox}} + E_{\text{red}}) \tag{1.67}$$

Equation (1.67) will be considered the fundamental definition of E^o_{redox}, always valid; Equation (1.63) is only valid at equilibrium. Thus at equilibrium for a simple redox system with equal concentrations of oxidizing and reducing agent present, the Fermi energy in the solid is at an energy half-way between the energy level of the oxidizing agent and the energy level of the reducing agent. More generally at equilibrium the Fermi energy is at E_{redox}, as given by Equation (1.64).

The energy levels for simple ions, where solvent reorganization is the only transformation, can now be related directly to each other and in particular to the hydrogen reference scale. Because the Fermi energy of an electrode at equilibrium under standard conditions equals E^o_{redox}, and because from Equation (1.45) under these same conditions

$$E_F = -qE^o \tag{1.68a}$$

where E^o is the tabulated redox potential of the couple, we have that

$$E^o_{\text{redox}} = -qE^o \tag{1.68b}$$

so the value of E^o_{redox} for a couple can be directly determined by the measurement of the Fermi energy of an electrode at equilibrium, or, more easily, by reading off the tabulated value of the standard electrode potential E^o from the literature.

Now that the various energy levels in solution have been derived, this is a good place to reemphasize the fact that for the solid/liquid interface the energy band diagram does not represent the energy of the system, but simply the available energy levels for electrons. In normal solid state band models, when an electron moves to another energy level that is the only energy change in the system. Thus one adopts the habit of equating the electronic energy with the system energy. In the solid/liquid system, however, when the energy level of the electron fluctuates (shown on the band diagram), simultaneously and necessarily the polarization energy of the medium changes (not shown on the band diagram). Thus, when one estimates the activation energy for the process, one must consider energies other than those shown on the band diagram. The results derived in the present section show this perfectly. To transfer an electron from E_{red} to E_F, a combination of Equations (1.62), (1.63), and (1.67) in Equation (1.52) show that the electronic energy level must be raised by an energy λ, but the thermal energy required is only $\lambda/4$. By Equation (1.46) if we were to add a thermal energy λ to the system, the level would be raised far past E_F, right up to E_{ox}. This illustrates that when using the energy level diagram in the polar medium, it must be kept in mind that the diagram represents only the electronic energy, not the energy of the system.

It is of value to refer again to the concept of thermodynamic energy levels, introduced in Section 1.2.1.2, used to determine whether an electron or a hole transfer is thermodynamically favored. The parameters E_{redox} or E^o_{redox} are thermodynamic energy levels.

1.2.4.5. Complex Formation and Hydrolysis. The shift in energy levels due to strong chemical interactions is difficult to allow for, compared to the polarization reorganization considered in Section 1.2.4.4, but the concepts are similar and will be described in this section. As indicated by the simplification of Equation (1.61) to yield Equation (1.62), the shift associated with polarization is always present and has more or less the same magnitude. However, the chemical shifts of interest here, such as hydrolysis or complex formation, have widely varying energies ΔG. For example in the reaction

$$m\,\mathrm{OH}^- + \mathrm{Fe}^{2+}(\mathrm{aquo}) \rightleftarrows e^- + \mathrm{Fe(OH)}_m^{(3-m)+} \tag{1.69}$$

ions with various values of m will be formed, and there should be several energy levels for the oxidizing agent corresponding to the various values of m. Granted that the energy level E_{red} for Fe^{2+} is constant, the energy E_{ox} will differ for the various complexes because of the energy of complex formation. We will show below that if each successive addition of the complexing agent OH^- to the Fe^{3+} is exothermic, and if the free energy for complexing Fe^{2+} is small, each successive energy level will be higher. The energy of the system will be lower due to complex formation, but the energy level of the unoccupied state will be higher. The percentage of the unoccupied levels expected that have a particular value of m will, of course, depend on the pH of the solution. $m = 0$ will dominate in a very acid medium, and $m = 2$ or 3 in a moderately acid medium, where the OH^- concentration is higher.

A simple analysis of the shift in energy level due to complex formation or adsorption can be made that is almost identical to the analysis of Equations (1.56)–(1.59).

Consider the reaction cycle:

Reaction	Energy	
$LA^{n-} + e^- \rightarrow LA^{(n+1)-}$	$-(E_F - E_{\text{complex}})$	(1.70)
$LA^{(n+1)-} \rightarrow A^{(n+1)-} + L$	$-\Delta G^L_{\text{red}}$	(1.71)
$A^{(n+1)-} \rightarrow A^{n-} + e^-$	$E_F - E_{\text{aquo}}$	(1.72)
$L + A^{n-} \rightarrow LA^{n-}$	ΔG^L_{ox}	(1.73)

where L is a ligand such as OH^- or is an adsorption site for the species A^{n-} or $A^{(n+1)-}$. The sum of the energies for the four steps must be zero. We let ΔG^L_{red} be the free energy when the complex is formed involving the reduced species $A^{(n+1)-}$ and the ligand L, as in the reverse direction of Equation (1.71). We let ΔG^L_{ox} be the free energy released when the complex involving the oxidized species A^{n-} is formed, as in Equation (1.73). Again, as with Equation (1.60), the ΔG's as defined are negative numbers. We assume the electron in Equation (1.70) transfers from the Fermi energy to the energy level of the complexed species, E_{complex}, thus releasing the energy $-(E_F - E_{\text{complex}})$, and we assume the electron in Equation (1.72) transfers from the energy level E_{aquo} of the uncomplexed form to the Fermi energy, releasing the energy $E_F - E_{\text{aquo}}$. (The use of the word "aquo" is generally restricted to mean uncomplexed species in aqueous solution, but for the moment we use it to refer to uncomplexed ions in any solution under study.) We are neglecting the polarization solvation discussed in the preceding sections—to be complete the levels determined here will correspond to E^o_{redox} both for the complex and for the aquo (uncomplexed) species, and each of these levels must be further split by the energy 2λ. Summing the energies through the cycle it is clear that

$$E_{\text{complex}} - E_{\text{aquo}} = -(\Delta G_{\text{ox}} - \Delta G_{\text{red}}) \tag{1.74}$$

If a species L is added that tends to form energetic complexes with the oxidized form, so that ΔG_{ox} is large and negative and ΔG_{red} is relatively small, then E_{complex} will be much higher than E_{aquo}. Thus, in the example of Equation (1.69) above, the energy level of the $Fe(OH)^{2+}$ is higher than that of the Fe^{3+}. On the other hand, if only the reduced species is complexed, by Equation (1.74) the energy level of the complex will be lower than the energy level of the uncomplexed ion. If both are strongly complexed, as with iron cyanide for example, the energy level shift (from the uncomplexed species) can be either up or down.

It should be noted that in accordance with expectation, the energy levels in solution do not, to the approximation here, depend on the concentration of the species. Although ΔG_{ox} and ΔG_{red} in Equation (1.74) each depend on the concentration $[L]$, their difference does not.

For fairly complicated systems or reactions, where the inner shell ligands are different on the stable oxidized species from those on the stable reduced species, the description in terms of an energy level model for ions becomes very difficult, and to date most of the fundamental work on semiconductor electrodes has concentrated on relatively simple systems where the ligands are not expected to change appreciably during the electron transfer process. However, it should be noted that semiconductor electrochemical methods, as described later (Section 7.1), permit the measurement of individual energy levels, and with increasing sophistication, these methods will be valuable for describing the levels present in complicated systems.

1.2.5. The Energy Levels Associated with Two-Equivalent Ions

With many ions or atoms, the removal or addition of one electron to the species results in an unstable valence state. Thus, redox reactions involving these species are usually two-electron reactions, and the species are "two-equivalent species." In many cases, for two-equivalent species, the transfer of a single electron results in an unpaired electron on the ion. An ion or atom with such an unpaired electron is called a free radical. When such a radical is formed, it is energetically favorable for the ion either to give up the unpaired electron or to capture another electron to form a pair. Thus, the unpaired electron on the radical is in a high energy level[28] (so it is lost easily), and the unoccupied state for capture of an electron on the radical is at a low energy level (so is attractive for electron capture). This is the reverse of the usual order of atomic levels, where the unoccupied levels are above the occupied levels.

During an electrode reaction on a metal electrode, such two-equivalent species when reduced or oxidized always can be expected to change their charge by two electrons. For example

$$As^{3+} = As^{5+} + 2e^- \tag{1.75}$$

is seldom, in classical electrochemistry, broken down into

$$As^{3+} = As^{4+} + e^- \tag{1.76}$$

$$As^{4+} = As^{5+} + e^- \tag{1.77}$$

because if the Fermi energy of the metal is low enough that the reaction (1.76) will proceed, then it is low enough that (1.77) will immediately follow. However, in semiconductor electrochemistry there are many cases where Equation (1.75) needs to be considered in its two steps both because experimentally the two steps are distinguishable and the second step does not automatically follow the first (the reason is discussed in Section 3.4.2 and the results are discussed in Section 6.3.1) and because on the energy level diagram used in semiconductor work, Equations (1.76) and (1.77) are described by two different energy levels.

Figure 1.7a shows the appropriate energy level diagram to describe a two-equivalent species (where for simplicity the Franck–Condon splitting of the levels and the temporal fluctuations of the levels are not shown). The conduction and valence bands of a semiconductor are also indicated to aid in the discussion. Consider the case where As^{3+} is present in the solution and will give up electrons to the electrode. The first electron [Equation (1.76)] will come from the lower level, the As^{3+}/As^{4+} level, the lower level will become unoccupied, and the ion is now As^{4+}, the radical. The radical has an unoccupied lower energy level, the As^{3+}/As^{4+} level as shown, and has an occupied energy level at a higher energy as indicated by the As^{4+}/As^{5+} level. This As^{4+} ion finds it energetically favorable either to give up an electron to the conduction band (from its occupied upper level) or to accept an electron from the conduction band (from its unoccupied lower level). Thus one way or another, a rapid exothermic reaction will occur. The radical is thus very unstable.

The model of Figure 1.7a is incomplete because we have omitted the Franck–Condon splitting and the temporal fluctuations of each level. Thus, in an actual case, as indicated partially in Figure 1.7b, four Gaussian-shaped levels must be determined when describing a two-equivalent species. In Figure 1.7b we have for simplicity only indicated the midpoints of the Gaussians. The splitting of the levels is indicated by 2λ: for simplicity no overlap of the two systems is shown, but there often will be overlapping of levels (e.g., see Figure 6.9). Figure 1.7b is only for illustration and the levels shown are not derived at all from the specific properties of arsenic.

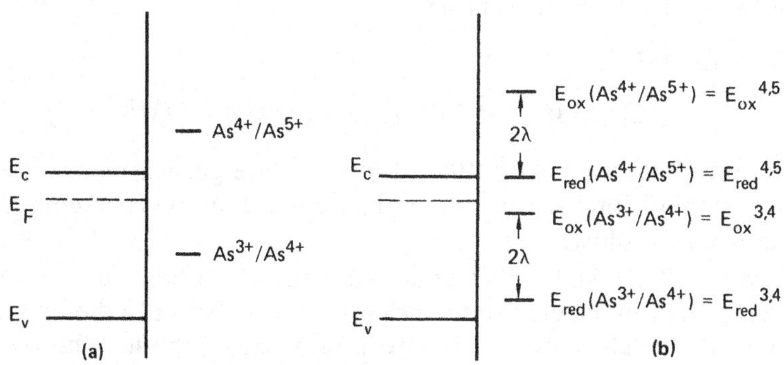

Figure 1.7. Energy levels required to describe two-electron processes (reactions with two-equivalent species). (a) The band model for electron exchange neglecting Franck–Condon splitting; (b) the band model including Franck–Condon splitting.

With the system at equilibrium, we have from Equations (1.63) and (1.64)

$$E_F = E^o_{redox}(As^{4+}/As^{5+}) - kT\log[As^{5+}]/[As^{4+}] \quad (1.78)$$

where, of course, $[As^{4+}]$ is a very low concentration. In addition we have

$$E_F = E^o_{redox}(As^{3+}/As^{4+}) - kT\log([As^{4+}]/[As^{3+}]) \quad (1.79)$$

where

$$E^o_{redox}(As^{4+}/As^{5+}) = \tfrac{1}{2}(E^{4,5}_{ox} + E^{4,5}_{red}) \quad (1.80)$$

and

$$E^o_{redox}(As^{3+}/As^{4+}) = \tfrac{1}{2}(E^{3,4}_{ox} + E^{3,4}_{red}) \quad (1.81)$$

with the superscripts 4,5 and 3,4 indicating levels as shown in Figure 1.7b. Thus the energies $E^o_{redox}(As^{4+}/As^{5+})$ and $E^o_{redox}(As^{3+}/As^{4+})$ are the midpoints of the two split levels in Figure 1.7b, and are, to an approximation sufficient for our purposes, the energies indicated in Figure 1.7a as As^{4+}/As^{5+} and As^{3+}/As^{4+}, respectively.

Now if we define

$$E^o_{redox}(As^{3+}/As^{5+}) = \tfrac{1}{2}[E^o_{redox}(As^{4+}/As^{5+}) + E^o_{redox}(As^{3+}/As^{4+})] \quad (1.82)$$

then from the sum of Equations (1.73) and (1.74) we have

$$E_F = E^o_{redox}(As^{3+}/As^{5+}) - kT\log[As^{5+}]/[As^{3+}] \quad (1.83)$$

Now the standard redox potential E^o of a two-equivalent couple such as As^{3+}/As^{5+} is defined, just as in the one-equivalent case, as the negative of the Fermi energy (in volts) of a metal electrode exposed to the couple under standard conditions, relative to the hydrogen reference electrode. That is, Equation (1.68a) applies. Thus under standard conditions, from Equations (1.68), (1.82), and (1.83)

$$\begin{aligned} -qE^o &= E^o_{redox}(M^{n+}/M^{(n+2)+}) \\ &= \tfrac{1}{2}[E^o_{redox}(M^{n+}/M^{(n+1)+}) + E^o_{redox}(M^{(n+1)+}/M^{(n+2)+})] \end{aligned} \quad (1.84)$$

In Equation (1.84) we have returned to the more general notation using M as the symbol for the ion rather than continuing with arsenic as the example used for illustration.

Equation (1.84) states that under standard conditions at equilibrium the Fermi energy in the electrode lines up half way between the two levels of the ion, the levels shown in Figure 1.7a. Clearly, the handbook value of E^o is less useful for a two-equivalent couple than for a one-equivalent couple. In Equation (1.84), as is perhaps clearer from Figure 1.7, the measured value E^o does give a value for the average of two energy levels of Figure 1.7a, but gives no information about the separation of the two energy levels. Thus in the two-equivalent case, in contrast to the one-equivalent case of Equation (1.68), the value of the standard redox potential does not give a good estimate for the energy levels associated with the ions, for the measurement of only one parameter cannot be used to determine two energy levels.[29,R10]

1.2.6. Conductance in Liquids

Conductance in liquids is normally by ions (mercury, for example, would provide an exception) and is described by Equation (1.11).

2

The Solid/Liquid Interface

2.1. Surface Ions and Their Energy Levels

2.1.1. Adsorption

2.1.1.1. Adsorption of Water on Oxides and Hydrous Oxides. When an oxide or a hydrous oxide (the distinction is sometimes nebulous) is immersed in an aqueous solution, protons and hydroxide ions, and possibly other ions, become adsorbed. Hydroxide ions are adsorbed on Lewis acid sites, and protons are adsorbed on Lewis base sites as indicated in Equation (1.6). The bonding is dominantly acid/base, the sharing of electron pairs. Actually, in general any oxide surface will have adsorbed water vapor present in the form of adsorbed protons and hydroxide ions before being immersed in the solution. However, in the case of adsorption from the vapor phase the adsorption of H^+ and OH^- are in equal quantities. When the solid is immersed in liquid water, as discussed in relation to Equations (1.6a) and (1.6b), the requirement of equal adsorption of H^+ and OH^- is no longer imposed. The surface will in general adsorb an excess of one or the other of these species and become charged positive or negative. In this section we discuss the models for adsorption of H^+ and OH^- In Section 2.1.1.2 we examine the adsorption of other foreign species.

Figure 2.1 indicates the simplest picture of adsorption of water vapor on an oxide such as TiO_2. The OH^- groups are attracted to the cations, the Lewis acid sites of the solid; the H^+ ions are attracted to the oxygen ions (Lewis base sites). The result is that the surface seems to be covered by hydroxide ions, and the surface is said to be hydroxylated. When the solid is now immersed in water, the reactions

$$H_s^+ + OH_{soln}^- = H_2O \tag{2.1}$$

or

$$OH_s^- + H_{soln}^+ = H_2O \tag{2.2}$$

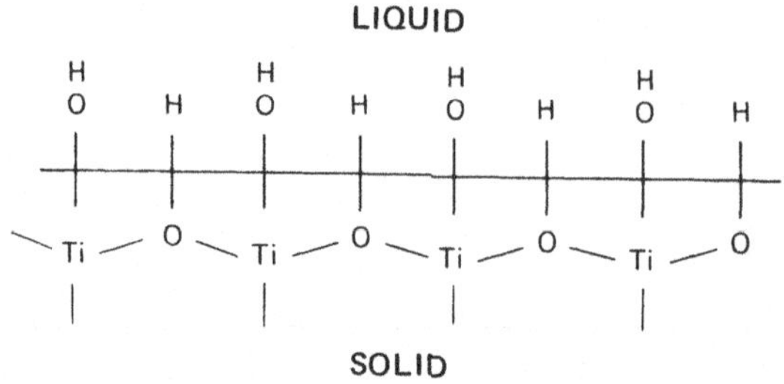

Figure 2.1. The adsorption of water on an oxide surface leads to an apparent monolayer of OH groups, a "hydroxylated" surface.

can occur, where the subscript *s* indicates an adsorbed species as indicated in Figure 2.1, and where the water resulting from the reactions is probably adsorbed (the site is hydrated).

Reaction (2.1) will, of course, tend to produce a net negative charge on the surface. Reaction (2.2) will tend to produce a net positive charge. Clearly from Equations (2.1) and (2.2) it can be expected that the resulting net charge will depend on the *p*H of the solution. This is the case; the *p*H at which the concentration of OH_s^- equals the concentration of H_s^+ is termed the point of zero charge (pzc) for the oxide. On SiO_2, for example, the pzc is at about a *p*H of 2; on TiO_2 the pzc is about 6. The pzc is discussed in more detail in Section 2.1.1.3.

Because of the tendency to adsorb more OH^- or H^+, the solids themselves are characterized as "solid acids" or as "solid bases." Consider the behavior of a fully hydroxylated surface, hydroxylated first by vapor adsorption and then immersed in solution. The proton and the hydroxide ions are not held with equal strength. If the hydroxyl ions are strongly adsorbed compared to the protons, the solid will release protons and is a solid acid. If the hydroxide ions are given up [Equation (2.2)] the solid is a base. There is a close relation between Lewis and Brønsted acid/base activity of the solid,[R4] but the name Brønsted activity implies proton exchange, while Lewis activity (Section 1.1.2.1) implies sharing of electron pairs. It is expected that the protons on the lattice oxygen (Figure 2.1) will be the ones to be given up, so these would be the Brønsted acid sites; the cationic sites are expected to be the Brønsted base sites, the sites that most easily lose hydroxide ions. It is possible to lose H^+ not only from the Brønsted acid sites in Figure 2.1, but also dissociating the OH^- ions adsorbed on the cationic sites. The latter protons are expected to be more difficult to remove. They will have a much lower "acid strength." In accordance with this picture Boehm[1] showed that half the sites on TiO_2 give up their protons much more easily than the other half (exchanging them for Na^+ in his experiments).

The above is a somewhat oversimplified model for the hydroxylated surface because it ignores heterogeneity. More complex and stronger adsorption of water and its fragments can occur associated with sites of extra chemical activity on the oxide surface, or sites of active ionic surface states. Such sites are to be expected on the heterogeneous surface. For example, in catalysis studies at the gas/solid interface it is recognized[R4] that poorly coordinated lattice oxygen ions at the surface, say on surface steps, provide strong Lewis basic sites. That is, they tend to give up electrons easily. In an energy level diagram they have a higher surface state energy level than a more completely coordinated oxygen ion has, and therefore will give up their electrons more easily, or will be more likely to capture holes.

As another example of the complexity possible on the surface of water-covered oxides, consider the model of ZnO corrosion as proposed by Pettinger *et al.*[2] They interpret corrosion on ZnO in terms of hole capture mainly by the different hydroxyl groups on the different crystalline faces. Because of the structure of the wurzite crystal only Zn ions are exposed on a (0001) surface; any residual oxygen over the surface Zn plane is weakly bonded and will be dissolved into the solution. Then a preferred adsorption of OH^- is expected for this plane because the surface sites are all cationic. Holes coming to the (0001) surface may interact with, and oxidize, these adsorbed OH^-, but lattice ions will not be oxidized. On an $(000\bar{1})$ face, on the other hand, surface oxygen ions are exposed. For this plane preferred adsorption of H^+ is expected, giving O_LH^- or O_L^{2-} groups, where O_L^{2-} is a lattice oxygen ion. Holes coming to the O_L^{2-} or O_LH^- layer will then be able to oxidize the lattice oxygen to a -1 valence or to the zero-valent (O_2) state, enabling the lattice oxygen to evolve, followed immediately by the now weakly bound underlying zinc ions. Thus the $(000\bar{1})$ plane corrodes much more readily than the (0001) plane.

2.1.1.2. Adsorption of Ions from Solution. "Specific adsorption" is adsorption with chemical bond formation. This is to be distinguished from ions "adsorbed" as a result of electrostatic attraction to a charged solid surface, that are normally farther from the surface with an intervening sheath of adsorbed water. In this subsection we review specific adsorption of foreign ions on a nonmetallic electrode. Although such specific adsorption and precipitation of species from solution are of interest in the electrochemistry at nonmetallic electrodes, they will be reviewed only qualitatively. From a theoretical viewpoint, the theories[3] of specific adsorption and precipitation are only moderately different from the theories at a metal electrode, and experimentally the number of cases in the electrochemistry of nonmetals where the phenomena have been explored is limited. The details of the interpretation for these few experimental studies will be described with the discussion of the results. Here a few generalizations and definitions of terminology will be presented to provide some background for future discussions.

In most cases of interest in the electrochemistry of oxides or oxide-covered electrodes, the adsorption of highly ionic foreign species (species that would tend to have highly ionic bonds in compound formation) is weak. Adsorption in aqueous solutions is essentially restricted to the adsorption of hydroxide ions and protons as discussed in Section 2.1.1.1. Although specific adsorption of a soluble foreign ion would involve a gain of energy due to bond formation, it also involves a loss of energy: loss of the solvation energy of the ion and the hydroxylation energy of the solid. For most ionic materials such an energy loss, associated with loss of the hydration sheath, is considerable. Thus, for example, James and Healy[4] conclude that during the adsorption of the cations they studied (particularly Co^{2+} on SiO_2 and TiO_2) the hydroxylation and hydration sheaths of the cations is not lost upon "adsorption" because its removal would be energetically unfavorable. And, by the above definition this would not be exactly "specific adsorption." James and Healy also conclude that adsorption of these ions will only occur in a limited pH region. Thus adsorption becomes essentially a precursor to precipitation as the hydroxide. In general such behavior can be expected to be typical of soluble charged ions adsorbing on oxide.

With concentrated solutions, however, adsorption of ionic species occurs and can be considered as an ion exchange phenomenon. Reactions such as

$$(OH^-)_s + Na^+ = (ONa^-)_s + H^+ \tag{2.3}$$

or

$$(OH^-)_s + F^- = (F^-)_s + OH^- \tag{2.4}$$

can be considered to occur, where the subscript s indicates the adsorbed group. It is found that the presence of adsorbed, nonreactive, anions can have a significant effect on electrode redox processes, as found for example by Micka and Gerischer,[5] possibly because such anions have a different effective charge from the OH^- or H^+ ions they are replacing and so change the surface state energy level. This work of Micka and Gerischer will be discussed in Section 6.3.2, but to date the explanation of the effect is not clear. Of course, anions, if adsorbed, can themselves also provide surface states.[6]

Adsorption with a covalent bond, where ions with covalent character adsorb on solids with covalent character, is more likely to yield a high surface coverage of adsorbed species. For example, the iodide ion is known to adsorb on the semiconductor germanium,[7] and its adsorption is suspected[8] on CdS. Hydrogen atoms and hydroxyl radicals also are known to adsorb on germanium[9] (see Section 6.4), forming a covalent bond.

Precipitated atoms or solids are also, of course, a form of "adsorption" common in any electrode reaction. One of the most interesting in semiconductor electrochemistry is the neutral sulfur, selenium, or tellurium, formed during anodic current flow through sulfide, selenide, or telluride semiconductors. Another case we will discuss (Section 5.1) is the adsorption of copper ions on germanium, where the copper is reduced to a metal.[10,R10] However, these cases are closer to formation of a second phase than to adsorption.

2.1.1.3. The IEP and the pzc. Sections 2.1.1.1 and 2.1.1.2 describe qualitatively the species adsorbed; here we will discuss the charge associated with these adsorbed species. This charge is the dominant contributor to the Helmholtz double layer V_H at an insulator or a semiconductor surface, as described in Section 2.2.3 below, and has a strong effect on charge transfer processes. Additionally, dipole effects may occur, but these will be neglected for simplicity in the discussion to come. When we begin to analyze data to determine band diagrams for the semiconductor/electrolyte system, the case where $V_H = 0$, where there is no *net* adsorbed charge, will be of particular interest. The pH that leads to no net charge on the surface is termed the IEP or the PZZP, the isoelectronic point or the point of zero zeta potential. This should not be confused with the similar concept, the pzc, the so-called "point of zero charge." As described in Section 2.1.1.1, the pH at which H^+ and OH^- are adsorbed in equal amounts is termed the pzc. If other ions than H^+ and OH^- are present and lead to additional surface charge, the pzc and IEP are not equal.[11,12] The presence of excess adsorbed cations raises the IEP; the presence of anions lowers the IEP.[R11,13]

The IEP is often measured by electrophoresis, the drift of particles of the semiconductor in an electric field. If there is a charged surface, the particles are charged relative to the surrounding solution (zeta potential) and move under the field. If the particles are charged positively, they drift in one direction; if negatively they drift in the other direction, and at the IEP there is no motion. The pzc, on the other hand, can best be determined by the pH change when the powder is immersed in the solution.[14] If the pH is that of the pzc, an equal number of H^+ and OH^- are adsorbed and there is no pH change. If the pH differs from the pzc, immersion of the sample will cause a pH shift. Cornell *et al.*[11] studied the effect of deposited Zn^{+2} and SO_4^{-2} on the pzc and IEP of rutile, and report similar studies by others. They find with no deposited ions the pzc equals the

IEP; deposited ions shift them in opposite directions. Ritter[14] also provides an interesting review of pzc measurements on rutile.

Apparently the bulk doping of a semiconductor may have an effect on the pzc. And, certainly more drastic variations in composition are expected to alter the pzc or IEP. Such effects are described by Roy and Fuerstenau,[15] who refer to several earlier studies along this line, and by Parks in his review.[R11] Parks suggests that a more *n*-type semiconductor should show a more basic pzc because adsorbed protons would be stabilized by the higher electron density. Again he suggests that dehydration of an oxide tending to replace OH^- groups by O_L^{2-} ions should shift the pzc to a more acid *p*H. If the solvent is nonaqueous, of course, other species than H^+ and OH^- may dominate the IEP.[16]

The IEP for metals and highly covalent semiconductors will be discussed in Section 2.2.3, where the origin of the double layer on these materials is described.

2.1.2. Surface States at the Solid/Liquid Interface

When, in Section 1.1.2, the electronic energy levels contributed by adsorbed species and defects at the surface were discussed, the influence of the polar solvent on the surface state levels was not considered. A quantitative analysis of the influence of the solvent on surface states is not available, but in this section a qualitative discussion will be attempted.

As discussed in Section 1.1.2, the energy levels associated with ions in solution close to the surface can be considered a form of surface state energy levels because the electron exchange laws can be expected to be analogous to those for surface states associated more directly with the solid; e.g., intrinsic surface states or adsorbed ion surface states. We will lean heavily on this analogy in this section because we wish to explore the influence of the polar solution, and for the case of ions in solution the influence of the polar solution has been quantitatively analyzed in Section 1.2. It was shown that large temporal fluctuations in energy level occur, and that a Franck–Condon shift occurs between the occupied and unoccupied energy levels.

In the case of energy levels at the solid/liquid interface, we can expect temporal fluctuations and Franck–Condon shifts due to the presence of the polar solvent just as we expect for ions in solution. However, because the surface state only sees the solvent in one hemisphere, the effective value of the reorganization energy λ will be lower, and the fluctuations will be smaller. There is a further reason why λ can be expected to be lower for intrinsic or adsorbate surface states. Near the surface, associated with the surface double layers, there is a strong electric field. This field constrains the movement of the solvent dipoles, so the effective static dielectric constant κ_s in the vicinity of the solid surface is much lower than in the bulk solution. In accordance with Equation (1.47), the lowering of κ_s leads to a much lower value for λ.

At the present time there is no experimental information available regarding the temporal fluctuations of either intrinsic or adsorbate surface state energy levels. Certainly it is to be expected, considering their location, that in general adsorbate energy levels would be affected by the solution more than intrinsic surface states are, and so would exhibit a much greater temporal fluctuation in energy. But a polar solvent may even affect intrinsic surface states enough to cause them to behave more like an ion in solution than a surface state at the gas/solid interface. The question of whether temporal fluctuations dominate the behavior of a surface state is an important one, as will be demonstrated in Chapter 3. The mathematics for charge transfer to and from the surface states differs somewhat depending on the reason for a distribution of surface state energies. If bands are formed by orbital overlap, or a distribution of surface state energies arises because the surface is nonuniform, then an electron capture model stemming from surface physics is used. If the distribution is caused by temporal fluctuations of the surface state, then the "fluctuating energy level model" of the next chapter is used.

A second reason that surface states at the solid/liquid interface have much more complicated properties than their analogs at the solid/gas interface is because of chemical transformations connected with the solvent. The complications will be discussed in more detail in Section 5.1, but they are mentioned in this section for completeness. First, upon capture or injection of an electron, complexes can form or be dissociated, so the chemical group that provides the surface state can take on a quite different chemical form. Second, the chemical groups giving rise to the surface states often will go into solution upon capturing or injecting an electron or a hole. Third, perhaps in part because of these chemical transformations, surface states may often be able to capture or be able to inject carriers, but not reversibly (Section 6.1). And, if reversible electron exchange is not possible, the "surface states" can not be in equilibrium with the semiconductor Fermi energy. This makes characterization of the surface states by normal semiconductor techniques difficult. As mentioned, such difficulties will be discussed further in later sections. The irreversible charge exchange will be discussed in Sections 3.3.3.2 and 5.1.2, the influence of complex formation in Sections 3.3.1 and 7.1, and the solubility problem in Section 7.4.

2.2. Double Layers at the Solid/Liquid Interface

2.2.1. General

Charged planes or space charge layers almost inevitably form at any interface, including the solid/liquid interface. They lead to electrical double layers, consisting of layers of positive charge, layers of negative charge, and regions of high electric field between, or within, the charged layers. Such double layers are dominant in controlling the electrical and chemical properties of the surface. They control both physical properties such as carrier recombination and photovoltaic effects, and chemical properties such as electrode reactions involving charge exchange with the solid.[17–19,R12,R4] Three distinct double layers appear at the solid/liquid interface, each involving charged planes or space charge layers and each characterized by a double layer voltage. In the present section the origin and the mathematical analysis of these double layers are described.

The three double layers at the solid/liquid interface are indicated in Figure 2.2. They include, first, a semiconductor space charge double layer; second, a Helmholtz double layer, between the solid and the "outer Helmholtz plane," the distance of closest approach of nonadsorbed ions to the surface; and third, the Gouy–Chapman double layer, in the solution

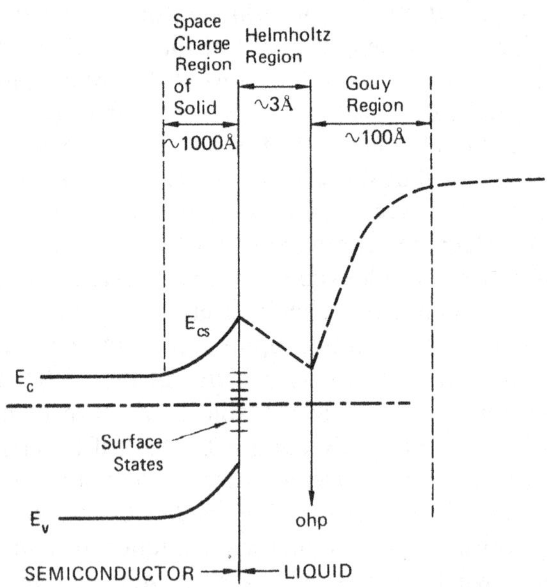

Figure 2.2. Double layers at the solid/liquid interface. The dashed line through the liquid indicates the variation in potential energy of a unit negative charge, as determined only by the double layer voltages, as it moves from the conduction band of the solid into the solution. The Gouy layer thickness indicated would represent a very dilute solution (Section 2.2.2).

near the solid, in which an excess of ions of one sign exists. Corresponding to these three double layers, there are four regions of excess charge. First, we have the space charge region in the solid. Here the charge is in the form of uncompensated impurities or trapped holes or electrons, or possibly in the form of mobile holes or electrons very near the surface. The analysis of how this layer of bulk charges, together with the countercharge located at the surface, leads to a space charge double layer within the solid, is made in Section 2.2.4. Second, we have charged planes on the two sides of the Helmholtz region that form the Helmholtz double layer. One plane is at the solid surface and the charge is in surface states or at the location of adsorbed ions. At a metal electrode the charges in the solid and on adsorbates are often distinguished, but on a nondegenerate semiconductor they may not be distinguishable, both being the same charged bonding orbital, bonding the adsorbate to the surface ions. The other plane is called the "outer Helmholtz plane" (ohp), the position of closest approach of mobile ions. At the ohp the charge arises from ions that drift up to the surface, attracted by the excess charge in or on the semiconductor. The analysis of the Helmholtz double layer will be presented in Section 2.2.3. Finally, there is an extended region of excess space charge in the Gouy–Chapman region of the solution, associated with mobile ions, a region where the field near the surface separates the positive from the negative ions. This charged region, together with the countercharge in the Helmholtz region and in the solid, form the Gouy double layer. Analysis of this Gouy region will be made in Section 2.2.2.

When a third phase is present—for example, an oxide covering on a metal or a semiconductor electrode—other double layers must be taken into consideration. Charge can be stored (a planar configuration) at "interface states" between the oxide and the substrate, of particular importance if the substrate is a semiconductor; and charge can be stored in the form of space charge in the oxide. The mathematical formulations describing these cases are, in principle, not greatly different from those to be developed in the next subsections for the three cases of Figure 2.2. Thus double layers associated with a third phase will not be discussed further until we come to the sections devoted to specific discussions of such layers.

The possibility of storing charge in these various space charge regions and Helmholtz planes is always related to electrical capacitance. Each double layer has a capacitance associated with it. The capacitance C for a "parallel plate" capacitor is given by

$$C = dQ/dV = A\kappa\epsilon_0/d \qquad (2.5)$$

Here dQ is a differential increase in stored charge in the layers, dV the differential voltage change across the double layer. A is the area, and d

is the distance separating the charge being stored. Each of the double layers above has a capacitance associated with it, generally defined by Equation (2.5).

In addition to its capacitative impedance, each double layer, particularly the Helmholtz double layer and the space charge double layer in the solid, has a resistive impedance associated with a dc current flow. Such transfer of charge through the double layers is the main topic of Chapter 3. If charge cannot appreciably flow across one or more of the double layers, one has what is called a perfectly polarizable electrode. With a perfectly polarizable electrode there is no dc current flow: a change in voltage results in a transient current that charges the double layer capacitor and the voltage across that double layer shifts, absorbing all of the shift in applied voltage. If the electrode is not at all polarizable, it means there is no resistive impedance through the double layers, and any change in applied voltage appears as an IR voltage drop elsewhere in the circuit. Normally one has the intermediate case, a polarizable electrode, when, with a change in applied voltage, the major fraction occurs across one of the double layers, but some current can flow.

The analysis of surface double layers for all three cases is based on a one-dimensional analysis, assuming no potential dependence on the y and z coordinates. Under these conditions Poisson's equation for the variation of potential with distance becomes

$$d^2\phi/dx^2 = -\rho/\kappa\epsilon_0 \tag{2.6}$$

where ϕ is the potential, x the distance from the surface, κ the dielectric constant; ϵ_0 the permittivity of free space; and ρ is the charge density.

A change in coordinates is helpful in relating the mathematics to the band diagram where, rather than potential, the energy of an electron is plotted. We will define the parameter V as

$$V(x) = \phi_b - \phi(x) \tag{2.7}$$

where ϕ_b is the potential in the bulk of the semiconductor, and so $qV(x)$ then is the extra electron energy at the position x over the bulk energy due to potential differences. The Poisson equation becomes

$$d^2V/dx^2 = \rho/\kappa\epsilon_0 \tag{2.8}$$

and the "surface barrier," the value of V at the surface, becomes

$$V_s = \phi_b - \phi_s = q^{-1}(E_{cs} - E_c) \tag{2.9}$$

with E_{cs} the conduction band edge at the surface. E_{cs} and E_c are indicated in Figure 2.2.

From Equations (1.2) and (1.3) we can relate the density of holes and electrons in the bulk to their density at the surface (where $V = V_s$). Specifically

$$n_s = N_c \exp[-(E_{cs} - E_F)/kT] = n_b \exp(-qV_s/kT) \tag{2.10}$$

and

$$p_s = N_v \exp[-(E_F - E_{vs})/kT] = p_b \exp(qV_s/kT) \tag{2.11}$$

where n_b and p_b are the density of electrons and holes, respectively, in the bulk semiconductor.

2.2.2. The Gouy Layer

The Gouy–Chapman theory describes a region in the solution near the electrode in which there is a space charge due to an excess of free ions of one sign. The ions attracted to the outer Helmholtz plane do not suffice to compensate all the charges on the electrode, and the residual electric field directed normal to the surface results in a charged Gouy layer. If the solution is concentrated, it will be shown that the Gouy layer is extremely thin, and essentially can be considered to have melded into the outer Helmholtz plane. Thus the Gouy layer is only of interest at all for dilute solutions, and can normally be neglected in semiconductor studies because a high concentration of electrolyte ions is used. We will present a simplified theory of the Gouy layer; more detailed analyses are available.[20]

In the Gouy region the excess charge is determined by the Boltzmann formula relating the density of mobile charge to the potential energy at the point x. Thus

$$\rho = qc[\exp(-q\Delta\phi/kT) - \exp(q\Delta\phi/kT_0)] \tag{2.12}$$

where we are assuming the concentrations c of cations and anions in the bulk solution are equal, and where $\Delta\phi$ is the change of potential from that in the bulk solution. The first term of Equation (2.12) uses the Boltzmann relation to describe the increase of concentration of charges of one sign. The second term describes the decrease of charges of the other sign, when the potential deviates from that of the bulk solution. Inserting Equation (2.12) into Poisson's equation, (2.6), we find upon a single integration, using the boundary condition that $d\phi/dx = 0$ at $\phi = 0$,

$$(d\phi/dx)^2 = (2ckT/\kappa\epsilon_0)\{[\exp(q\Delta\phi/kT) - 1] + [\exp(-q\Delta\phi/kT) - 1]\} \tag{2.13}$$

The integration is facilitated by multiplying both sides of Equation (2.6) by $\frac{1}{2}d\phi/dx$, for the integral of $\frac{1}{2}(d\phi/dx)(d/dx)(d\phi/dx)$ is $(d\phi/dx)^2$.

With Equation (2.13) we can show the thickness of the Gouy layer is negligible with c high. If the concentration c is 10^{26} m^{-3} (just higher than 0.1 moles per liter), and the value of κ is 80 (the value for water), then the preexponential factor is the order of 4×10^{14} V^2 m^{-2}. Now if the potential at the ohp is say, 0.25 V above or below the potential of the bulk solution, one of the exponential terms will be the order of 10^4, and from Equation (2.13) $d\phi/dx$ at the ohp will be $\pm 2 \times 10^9$ V/m. Thus, most of the excess potential $\Delta\phi$ will be lost in about an angstrom. Clearly when the concentration of ions in the electrolyte is the order of 0.1 or more moles per liter, the Gouy layer will be almost indistinguishable from the Helmholtz layer, except for a very small residual voltage, of the order of 0.1 V at most.

2.2.3. The Helmholtz Double Layer

The Helmholtz double layer, an extremely important double layer in both semiconductor and metal electrochemistry, is considered to be formed between two planar sheets of charge. One is due to ions in solution attracted by the charged surface, particularly those ions that are at the distance of closest approach (at the outer Helmholtz plane, ohp). The other is due to charges at the surface of the solid. First, we will discuss the physical characteristics: the distance between the planes, the double layer voltage due to the two charged planes, and the electric field between the planes. Then we will discuss the origin of the charges on the solid surface.

The formula for the capacity per unit area, C/A, as given by the second equality in Equation (2.5), can be used to provide an estimate of the effective distance between the two planes defining the Helmholtz double layer. The dielectric constant κ to be used is, however, not clear. If we consider water to be the dielectric, we would use $\kappa = 80$. If we considered vacuum (the space between atoms), as the dielectric, we could use $\kappa = 1$. Actually, the medium is water that is strongly polarized, due to the electric field of the double layer, so the water dipoles cannot easily rotate. Then a value of κ of the order of 5 is reasonable and is adequate for our purposes. Typically, at a metal electrode the measured Helmholtz capacity, C_H, is of the order of 10^{-1} F/m^2, leading with $\kappa = 5$ to d of the order of 5×10^{-10} m or a few angstroms. It is of interest that on metallic electrodes the value of d (as measured by capacity) varies with the double layer voltage. It decreases with a double layer voltage of either sign, with a maximum in d when there is no double layer voltage, or no charge stored at the Helmholtz planes. This suggests that the stronger the field the closer the mobile ions in solution are drawn to the surface.

Equation (2.6) is solved for the simple case of two charged Helmholtz planes to yield

$$\phi = ax + b \tag{2.14}$$

where a and b are constants ($-a$ = electric field). We define the Helmholtz double layer voltage V_H:

$$V_H = \phi_1 - \phi_2 \tag{2.15}$$

where ϕ_1 is the potential at the solid surface, ϕ_2 the potential at the ohp.

Normally the charge densities (per unit area) on the surface and on the outer Helmholtz plane are much higher than the charge density of any other charged layers, so the other charged layers can be neglected in the Helmholtz double layer region. Then, if the simplifying approximation is made that C_H, the Helmholtz capacitance, is approximately constant with voltage so that Equation (2.5) can be integrated, we have from the two equalities in Equation (2.5)

$$E = V_H/d = Q/\kappa\epsilon_0 \tag{2.16}$$

which can be also obtained directly from Gauss' law. As a numerical example, if 10^{17} to $10^{18}/\text{m}^2$ excess ions of one sign (H^+ or OH^-) are adsorbed on the solid surface (about 0.01 to 0.1 monolayers), and we use $\kappa = 5$, the electric field E would be 3×10^9 V/m. If the distance between the ohp and the solid surface is a few angstroms, the Helmholtz voltage V_H would thus be about 0.1 to 1 V.

The charge on the planes is of several forms. On the solution side of the Helmholtz double layer, the charge originates by the accumulation of ions at the ohp. On the solid side of the Helmholtz double layer, the charge can arise in three forms: an accumulation of free charge, free charge trapped from the solid onto surface states, or adsorbed ions. The form that dominates varies with the solid. With these mechanisms of charging, the concentrations described above can easily be reached. Concentrations of excess charge in this magnitude almost never appear in the space charge region of a nondegenerate semiconductor (see Section 2.2.4.1).

If the solid is an insulator, the charge on the solid surface cannot stem from free charges, and must be due solely to adsorbed ions or dipoles. Of particular importance is the adsorption of protons or hydroxyl ions. Indeed, in Section 2.1.1 the adsorption of protons and hydroxyl ions on oxides was discussed in detail showing that the relative adsorption of H^+ and OH^- depends on the pH of the solution (particularly in the case of aqueous solutions[(16)]), and by varying the pH of the solution the charge adsorbed on an oxide electrode can change its sign. Other ions or dipoles can sometimes be adsorbed (Section 2.1.1.2) if the bonding to the solid is sufficient to overcome the energy of solvation, and such adsorption may also contribute to the voltage drop across the Helmholtz double layer.

With insulators, the Helmholtz double layer potential adjusts itself so the rate of adsorption equals the rate of desorption of ions. If we consider the rate of proton adsorption to be controlling, a simple derivation will indicate the relationship between the Helmholtz double layer voltage V_H and the excess surface charge $q[H_s^+]$. Here $[H_s^+]$ represents the concentration of adsorbed protons in excess over the neutral surface (protons and hydroxyl ions adsorbed equally). Square brackets as usual indicate the concentration thereof. In the reaction

$$H_3O^+ = H_s^+ + H_2O \tag{2.17}$$

where H_3O^+ is a hydronium ion (essentially a proton) in solution, the free energy of the reaction, ΔG, varies linearly with the double layer potential V_H, for the proton must acquire the potential energy (qV_H) to become adsorbed. With this proportionality, equilibrium in Equation (2.17) leads to

$$\frac{[H_s^+]}{[H_3O^+]} = \exp\left(-\frac{\Delta G}{kT}\right) = A \exp\left(\frac{-qV_H}{kT}\right) \tag{2.18}$$

where A is a constant. The double layer potential V_H in turn is proportional to the charge adsorbed, if we make the approximation that C_H, the capacity per unit area of the Helmholtz double layer, is independent of V_H:

$$q[H_s^+] = C_H V_H \tag{2.19}$$

Inspection of Equations (2.18) and (2.19) shows $[H_s^+]$ will vary slowly with $[H_3O^+]$ and, to the approximation of this analysis, we have

$$qV_H = B + kT \ln[H_3O^+] = B - 2.3kT(pH) \tag{2.20}$$

and the Helmholtz potential decreases about 60 mV per pH unit.[21,R10]

If the solid is a metal, the voltage of the Helmholtz double layer is determined by electron exchange between ions in solution and the metal. Part of the charge on the solid may be due to dipoles or adsorbed species. But much of the charge on the solid is located normally within the metal, arising due to a change in occupancy of electronic levels near the Fermi energy at the surface. When a single metal electrode is introduced into a solution, the Helmholtz double layer potential adjusts itself to a value such that the current flow, the rate of electron transfer, is equal in the two directions, into the metal and out of the metal. This concept was implicitly used in the development of Section 1.2.2 where we discussed the equilibrium case. The Helmholtz voltage adjusts to make the electron transfer processes come to equilibrium. Consider for example, if when

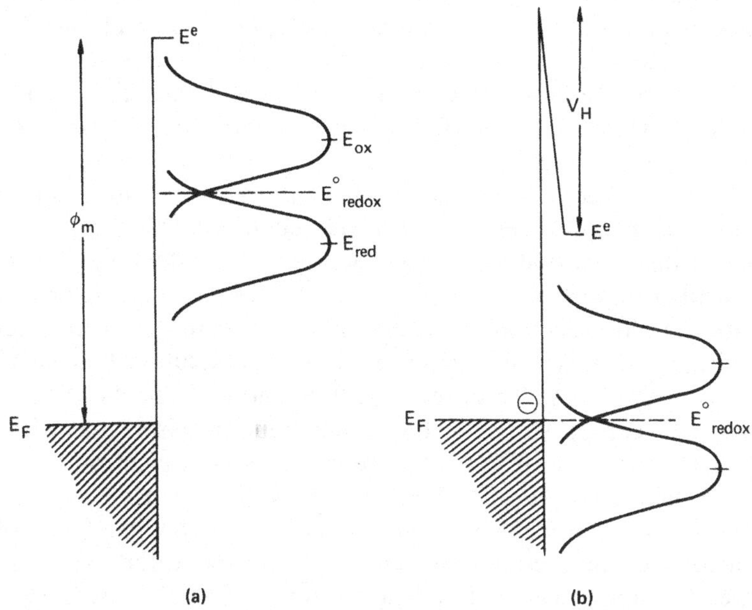

Figure 2.3. Equilibration of the metal/electrolyte interface. For simplicity, the redox couple in solution is specified to be present under standard conditions, so the symbol E^o_{redox} is used.

the electrode is first introduced, the rate of electron injection is higher than the rate of electron extraction from the electrode, as indicated in Figure 2.3a. Obviously electrons from the reducing agent will flow freely into the empty levels of the metal above E_F. A negative charge will build up in the solid, at the surface, as indicated schematically by the encircled minus sign in Figure 2.3b. A corresponding charge of positive ions, say K^+ ions in solution containing KCl, is attracted from the solution to the ohp and a double layer potential V_H is built up. The resulting V_H diminishes the rate of electron injection and increases the rate of electron extraction. The charge buildup and the corresponding V_H buildup will continue until the rates are equal; until equilibrium electron transfer is reached.

It should be pointed out as a matter of interest, that the IEP, isoelectric point, for metals—the point where $V_H = 0$—can be immediately determined by the use of the results of Section 1.2.3. If we examine Figure 2.3b, and specify that the redox couple is the hydrogen couple, so the energy of an electron at infinity, E^e, is about 4.5 to 4.7 eV above E^o_{redox} for this couple (Section 1.2.3), we see immediately that to make $V_H = 0$, we must apply an overpotential equal to the V_H shown in the figure. That is, a potential of between $(\phi_m - 4.5)$ and $(\phi_m - 4.7)$ volts must be applied, where ϕ_m is the metal work function. Experimentally, Vermilyea[R13] presents data showing that, indeed, a potential ($\phi_m - 4.7$) volts relative to the hydrogen reference electrode must be applied to reach the IEP for a series of metals. The importance of these concepts in semiconductor electrochemistry arises when we consider narrow band, highly covalent semiconductors with clean surfaces, where similar considerations may apply.

If the solid is a semiconductor, there is the possibility that either adsorption/desorption or electron transfer could control the Helmholtz potential. It turns out that for nondegenerate semiconductors V_H is dominated by adsorption/desorption processes.[19,22] As in the case of metals, there can be redox reactions at semiconductor electrodes, with electron transfer from the bulk semiconductor to ions in solution that have appropriate energy levels. However, it will be shown in Section 2.2.4 that the amount of charge stored in the semiconductor associated with this transfer is the order of 10^{16}, or in extreme cases 10^{17}, electronic charges per square meter, and usually it is much less. This density is very low compared to the significant fraction of a monolayer that can be involved in the adsorption of protons, so it is clear why the latter will normally dominate. As another calculation leading to the same conclusion, we can use Equation (2.5) to calculate the voltage drop across the Helmholtz double layer resulting from this charge of 10^{16} or 10^{17} charges m^{-2}. We assume the Helmholtz double layer thickness resembles that of metals or insulators, so the capacity is the order of 10^{-1} farads/m^2. The contribution to V_H due to bulk semiconductor charges is only about 0.016 V (using 10^{16} electronic charges/m^2), a negligible contribution.

Thus, the Helmholtz double layer on a simple semiconductor is normally, like an insulator, associated with adsorption–desorption processes. This is particularly clear where the semiconductor is an oxide or is oxidized, when normally the Helmholtz double layer potential is found experimentally (as reviewed in Section 5.1) to vary with pH in agreement with Equation (2.20).

If the solid is a degenerate semiconductor (see Section 1.1.1.3), we are in an intermediate case between a semiconductor and a metal. The value of V_H is expected to become more associated with redox processes the more degenerate the material, for the more degenerate, the closer it resembles a metal. More charge can be accommodated in the bands of the semiconductor near the surface and can contribute to the surface charge. We will observe in Section 3.3 that one of the differences between a well-behaved metal and a well-behaved semiconductor is the so-called "transfer coefficient." The transfer coefficient is usually about 1/2 for a metal, and the apparent transfer coefficient is zero for a semiconductor. As the semiconductor surface becomes degenerate it can, to some extent, be observed by the transfer coefficient approaching 1/2.

2.2.4. The Space Charge Double Layer in the Semiconductor

When the electrons are transferred in or out of the semiconductor surface, charged regions, regions of "space charge," result. The space charge can be in the form of immobile charged impurities or immobile trapped carriers near the surface of the semiconductor, or can be in the form of mobile electrons or holes in the conduction or valence bands of the semiconductor. We will analyze simple voltage/space charge layer thickness relationships where there is a uniform density of immobile charge in the semiconductor near the surface, and cases where there are mobile carriers in the conduction or valence bands. More detailed analyses are available.[R4,R14,R15,R16] Calculations are available describing cases where the density of immobile charge is nonuniform,[23,24] but as the analysis depends so much on the charge distribution, these calculations will not be reproduced.

The amount of charge in the space charge region of the semiconductor depends on several processes. A semiconductor in vacuum will almost inevitably develop a space charge near the surface simply because some of the bulk carriers near the surface will be captured by surface states. When the semiconductor is dipped into a solution, the charge in the space charge region may change further because (a) the surface states will be affected by their interaction with the solution and the charge in the surface states will therefore change; and/or (b) ions in solution with appropriate energy levels will inject or extract electrons from the semiconductor. The above will occur with an open circuit condition. Further variation in the space charge will occur if (c) a second electrode is dipped into the solution and an external voltage applied between the two electrodes different from the open circuit voltage. Then the space charge capacity will charge up in accordance with Equation (2.5), and a charge Q will be added (or subtracted) from the open-circuit case.

The ability to intentionally change the voltage in the semiconductor space charge region by controlling the applied voltage—changing it by an amount equal to the change of the imposed external voltage—is one of the key characteristics used in semiconductor electrochemistry. The voltage across the space charge will not always change an amount equal to a change in external voltage, because some of the change in voltage can appear across the other double layers in the circuit. It will be shown later (Section 3.2) that in most cases of interest the semiconductor space charge dominates—most of any change in applied voltage does appear across the semiconductor space charge region. Then the charge in the semiconductor space charge region must change by the amount calculated by integrating Equation (2.6). Because of this key characteristic, most early work in semiconductor electrochemistry were studies of the semiconductor space charge layer. However, as reviewed by Pleskov,[25] such studies are now more rare, as the behavior is well understood. To reemphasize, the most important factor to keep in mind while reviewing double layers (space charge layers) in a semiconductor electrode is that by changing the cell voltage applied, the experimenter can, in principle, produce any form of space charge layer he wishes at the semiconductor surface.

The three forms of space charge layer that can be obtained on a semiconductor are illustrated in Figure 2.4. The depletion layer, Figure 2.4a—sometimes also termed an exhaustion layer—forms if majority carriers are extracted in moderate amounts. The surface region is "depleted" of majority carriers, and minority carriers are not present, so the surface region is depleted (exhausted) of both forms of mobile carriers. Unless V_s is made very small, the surface region is essentially insulating. The space charge to insert into Poisson's equation is due only to ions or trapped charges in the depleted region. The accumulation layer, Figure 2.4b, results when majority carriers are injected into the semiconductor from the surface, and these excess majority carriers form the space charge. The inversion layer, Figure 2.4c, results when majority carriers are ex-

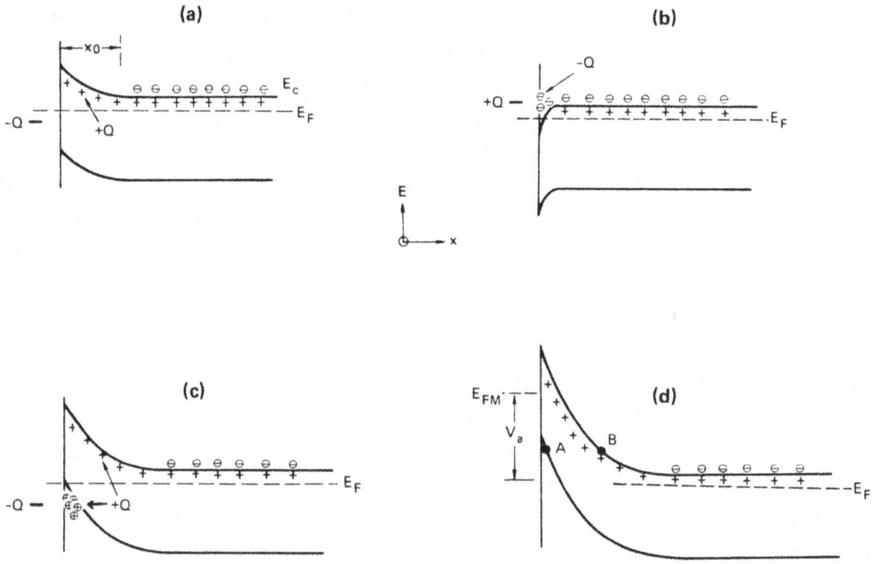

Figure 2.4. Types of space charge double layers on an n-type semiconductor. (a) Depletion layer; (b) accumulation layer; (c) an inversion layer can be induced by an adsorbate as shown or by an applied voltage; (d) a deep depletion layer, which can only be induced by an applied voltage V_a, indicated here as being applied between a metal with its Fermi energy E_{FM} and the semiconductor (the symbols A and B are for later reference).

tracted but the number extracted is excessive, so much that the majority carrier band (the impurities near the surface) cannot supply all the majority carriers required without resulting in excessive band bending. Carriers have to come from the minority carrier band. Thus, in the example of Figure 2.4c, where an *n*-type semiconductor is used for illustration, majority carriers, electrons, are not only extracted from the conduction band, but also some are extracted from the valence band (hole injection). The term "inversion" describes the fact that the semiconductor type has changed because of the process: in the case illustrated, the *n*-type semiconductor has actually been "inverted" to *p* type at the surface. The nonequilibrium deep depletion layer, shown in Figure 2.4d, will be discussed in Section 2.2.4.2.

Any of the surface layers in Figure 2.4 can be produced, in principle, at will, by varying the electrode potential. If a voltage is applied to make the semiconductor very negative relative to the solution, the bands bend down at the surface: with an *n*-type semiconductor an accumulation layer (Figure 2.4b) is induced; with a *p*-type semiconductor, an inversion layer. As the voltage applied makes the semiconductor less negative and possibly slightly positive (the transition voltage needed is discussed qualitatively in Section 2.4, especially with reference to Figure 2.6), the band picture moves to the depletion form, shown for *n*-type material in Figure 2.4a. With an extremely high positive voltage on the semiconductor, relative to the solution, the bands bend up at the surface, becoming an inversion layer (Figure 2.4c) for *n* type, or an accumulation layer for *p* type.

Although it is anticipating slightly, it is of value to note here the analogy between the models of Figure 2.4 and the Schottky barrier diode in semiconductor physics.[26] As far as current flow is concerned, the electrode of Figure 2.4 will, in general, have diode characteristics. If the band bending is such as to move majority carriers to the surface, the diode is in the forward direction and, if energy levels are available in the solution to accept them, high currents can and will flow. Thus for the *n*-type example of Figure 2.4, the accumulation layer of Figure 2.4b and probably a modest depletion layer as in Figure 2.4a, will permit copious cathodic current flow with electrons flowing out of the semiconductor. When the voltage is lowered and reversed, such that majority carriers are drawn away from the surface, we are increasing the barrier height. Not only can electrons no longer flow out of the semiconductor, but in general, states are not available to inject electrons into the majority carrier band of the semiconductor, and the minority carrier density is too low to carry appreciable current. Thus negligible current flows. This becomes the reverse direction of the diode. The terms "forward direction" and "reverse direction" of the applied voltage thus describe both important characteristics of the barrier layer at the surface (large or small barrier) and the expected flow of current. The terms will be used often in later discussions when current flow is considered.

2.2.4.1. The Depletion Layer. As described above, the space charge in the depletion layer is the immobile charge on traps or on ionized donors or acceptors. As is usually done, we will assume that the density of such charge is constant in the space charge region. Then Poisson's equation, Equation (2.8) (we prefer the V parameters to facilitate direct application to the band model), becomes

$$d^2V/dx^2 = qN_{sc}/\kappa\epsilon_0 \tag{2.21}$$

where N_{sc} is the density of positive charge in the space charge layer. (With p-type material N_{sc} is, of course, a negative number.) Upon integrating twice, this becomes

$$V = (qN_{sc}/2\kappa\epsilon_0)(x - x_0)^2 \tag{2.22}$$

after applying the boundary conditions that x_0 is the point where the field, dV/dx, becomes zero (Figure 2.4a) and that [Equation (2.7)] $V = 0$ in the bulk semiconductor at $x \geq x_0$. If we specify, as is normally done, that $x = 0$ at the surface, then the surface barrier, V_s, given by

$$V_s = qN_{sc}x_0^2/2\kappa\epsilon_0 \tag{2.23}$$

the Schottky relation.

In Figure 2.4a is shown the band model for a depletion layer using n-type material as an example. The donor ions are indicated as being of constant density throughout the sample, including the space charge region. In the bulk of the sample they are neutralized by their electrons, but near the surface the available conduction band electrons have moved to the surface to create the double layer. We indicate the net charge at the surface as Q, which, of course, equals the charge in the space charge region, $+Q_{sc}$, i.e.,

$$-Q = Q_{sc} = qN_{sc}x_0 = -qN_s \tag{2.24}$$

where N_s is the density of electronic charges captured at the surface. Equation (2.24) can be inserted into Equation (2.23) to give another useful form of the Schottky relation:

$$V_s = qN_s^2/2\kappa\epsilon_0 N_{sc} \tag{2.25}$$

It is of interest to estimate the value of x_0, the thickness of the insulating depletion layer. If we assume a reasonable impurity density, say 10^{23} m^{-3}, a dielectric constant of 8, and a value of 1 V for V_s, we find $x_0 \cong 10^{-7}$ m = 1000 Å. As x_0 is insensitive to V_s and N_{sc}, varying as the square root, it is seen that x_0 usually does not increase much above this order of magnitude. And compared to the Helmholtz double layer, estimated in Section 2.2.3 as about 3 Å thick, we see the semiconductor double layer is much thicker. We note also that these typical values yield, by Equation (2.24), an excess charge N_s of $10^{16}\ m^{-2}$. This is much less charge than typically found on the Helmholtz planes ($\sim 10^{18}\ m^{-2}$) as discussed in Section 2.2.3.

2.2.4.2. The Accumulation, Inversion, and Deep Depletion Layers. If on an *n*-type sample electrons are injected instead of extracted, or on a *p*-type sample holes are injected instead of extracted, then the space charge region is termed an accumulation layer, indicating that the majority carrier is accumulated in the space charge region in density higher than the bulk density. For *n*-type material, for example, Poisson's equation becomes

$$d^2V/dx^2 = (-n_b q/\kappa\epsilon_0)[\exp(-qV/kT) - 1] \qquad (2.26)$$

where n_b is the bulk density of electrons and V is negative. The first term on the right represents the contribution of electrons in the conduction band, equal to the bulk density multiplied by the Boltzmann factor. A second term represents the contribution of immobile positive ions. A single integration by the method used for Equation (2.13) yields

$$(dV/dx)^2 = (2n_b q/\kappa\epsilon_0)\{(kT/q)[\exp(-qV/kT) - 1] + V\} \qquad (2.27)$$

where we have used the boundary condition that $dV/dx = 0$ when $V = 0$ (in the bulk semiconductor). Equation (2.27) is similar to that for the Gouy layer, Equation (2.13). In the accumulation layer case the density of charge n_b is generally much less than the ion concentration c used in the Gouy case, and the dielectric constant of the semiconductor is usually less than the dielectric constant of the solvent. A slightly more detailed calculation[R4] suggests the order of 100 Å is a typical value for the thickness of the accumulation layer.

Figure 2.4b shows one way of acquiring an accumulation layer at the surface—by having a surface donor species sufficiently active (with a high enough surface state relative to the Fermi energy) to inject majority carriers, in this case electrons. Another way, common in electrochemistry, is to apply an appropriate voltage between the electrode and a counter-electrode to induce majority carriers at the surface. In the case of *n*-type material, a negative voltage on the semiconductor relative to the solution will, if sufficiently high, induce an accumulation layer. Obviously the electrode must be reasonably polarizable, as described earlier in Section 2.2.1, or high currents will flow and such a double layer voltage will be difficult to induce.

If the band bending suffices to force the Fermi energy into the band, we have forced the very important case of the "degenerate surface," which will be referred to many times throughout the book as the transition between a semiconductor and a metallic behavior. In particular, it is discussed with reference to Figure 3.6 in Section 3.4.3. It will be recalled from Section 1.1.1.3 that a degenerate semiconductor is one that exhibits the most important characteristic of a metal: the Fermi energy is in a continuum of levels—in a band.

Figure 2.4c shows a third type of space charge layer at the semiconductor surface, namely, an inversion layer. As indicated in the figure, two types of space charge must be accounted for with an inversion layer. First, the majority carrier band is further from the Fermi energy, the majority carriers are depleted from the surface, so there is a space charge N_{sc} present, just as in the depletion layer case. Second, near the surface minority carriers are present as a space charge, so they must be considered also in the Poisson relation.

$$d^2V/dx^2 = (\kappa\epsilon_0)^{-1}[qN_{sc} + qp_b \exp(qV/kT)] \tag{2.28}$$

is the resulting form of the Poisson equation for n-type material with a surface layer inverted to p type. The first term in (2.26) represents the immobile ions, the second term represents the density of minority carriers. The bulk density of holes is denoted by p_b, and is given from Equation (1.4) as

$$n_b p_b = N_c N_v \exp(-E_g/kT) \tag{2.29}$$

With a typical n-type semiconductor, for example one with n_b the order of 10^{23} m^{-3} and an energy gap greater than 1 or 2 eV, it is clear that p_b is very small, and relatively high values of V are required to make the last term of Equation (2.28) comparable to the first term. Thus, in Figure 2.4c a very strong band bending is indicated.

Integration of Equation (2.28) once yields an expression for the electric field dV/dx in the space charge region:

$$(dV/dx)^2 = (2/\kappa\epsilon_0)\{qN_{sc}V + (qp_bkT/q)[\exp(qV/kT) - 1]\} \tag{2.30}$$

In the inverted region, where the second term of Equation (2.28) or (2.30) dominates, the solution is similar again to that for the Gouy layer and for the accumulation layer. Again, it can be shown that the depth of the inverted region, where the density of minority carriers exceeds the bulk density of majority carriers, is the same order of magnitude as the accumulation layer thickness. A value of the order of 100 Å is typical as a rule of thumb.[R4]

An inversion layer can be obtained either with a very active species present on the surface, or by an applied voltage. Figure 2.4c indicates the case of an n-type semiconductor where a very active acceptor surface species is present, one with an energy level very close to the valence band so that a very strong band bending is necessary to bring the energy level of the acceptor surface state up to the Fermi energy. To induce an inversion layer with an applied voltage, the *n*-type semiconductor electrode must be made very positive with respect to the solution. The electrode must also be highly polarizable—in particular the rate at which minority carriers flow into solution must be less than the rate at which they arrive or are created in the inverted region, in order that their density can build up to the equilibrium values used in the above calculations.

A "deep depletion layer" is a nonequilibrium configuration that is seldom important at the gas/solid surface, but often is formed under electrochemical conditions. This surface double layer is obtained when a high voltage is applied such that an inversion layer *should* be formed, but either (1) minority carriers are just not available to collect at the surface during the time allotted; or (2) the minority carriers are consumed in an electrochemical reaction as fast as they reach the surface. Thus, comparing Figure 2.4d, consider the case where the bands are bent by an external voltage to 10 or 100 V, such that the bulk Fermi energy indicated would be well buried in the valence band. If the semiconductor has no appreciable hole density to draw on, because in Equation (2.29) either n or E_g is very large (usually a large E_g is the important factor), and if holes are generated thermally very slowly, equilibrium will not be reached during the experiment. Or, if the holes are rapidly consumed when they reach the surface, equilibrium will never be reached. The valence band will never (in the time of the experiment) collect a significant hole density. The last term of Equation (2.28) will always be negligible. Then the Schottky relation, Equation (2.23) will be valid at all positive values of V_s. An equivalent description for *p*-type materials can be given leading to the Schottky relation for all negative values of V_s. The development of a nonequilibrium deep depletion behavior with high reverse voltage is usually observed for materials with $E_g > 2$ eV, and sometimes is observed for materials with somewhat lower bandgap energies, particularly in electrochemistry where reactions can proceed that consume the minority carriers.

2.2.4.3. Breakdown in the Space Charge Layer. Electric breakdown in an insulating region such as the depletion layer occurs if extra carriers are spontaneously generated either by tunneling (Zener breakdown) or by collision (avalanche breakdown). Tunneling will be reviewed qualitatively in Section 5.1.2, and more quantitatively in Section 6.5.2. Zener breakdown occurs when tunneling is excessive; when the electric field in the semiconductor is so high that valence electrons at point A have an energy equal to conduction band states at point B, and the points A and B are close, separated only by tunneling distances, say 30 Å or less. Figure 2.4d indicates the model for the points A and B. Then electrons from the valence band can tunnel through to the conduction band and high currents can flow.

Avalanche breakdown occurs when a carrier in an electric field can gain enough energy to excite another carrier, normally exciting it from the valence band to the conduction band. Now, essentially independent of the field, a carrier travels a distance λ, a "mean free path" before colliding with a lattice vibration and losing its excess energy. Values of λ from 50 to 100 Å are common. So, if the electric field is high enough that the carrier can gain the necessary energy, E_g, within a distance λ, avalanche breakdown can occur. For example, if $\lambda = 100$ Å and $E_g = 3$ eV, a field of at least 3×10^8 V/m is needed for avalanche breakdown.

The voltage required for breakdown depends primarily on the density of charge centers in the material. At a given temperature, of the order of room temperature or above, and with a reasonably pure semiconductor, the mean free path is reasonably independent of impurity concentration. Thus the electric field required for breakdown is, in principle, independent of the impurity density. However, the applied voltage required is not independent of the impurity density. The electric field produced by a given voltage depends on N_{sc}. The electric field, given by Equation (2.30), is highest at the surface, where $V = V_s$.

When the field is high enough that avalanche breakdown occurs and new electron and hole pairs are generated, the minority carrier moves toward the surface, and the majority carrier toward the bulk of the semiconductor or insulator. Either or both carriers may still be in a region where the field is high enough for another collision and another electron–hole pair generated. Hence the name "avalanche"—one stray carrier can initiate an avalanche of current carriers and lead to high electric currents.

2.3. Theoretical Predictions of the Energy Levels of Band Edges

In this section we discuss the analysis to determine energy of the conduction and valence bands of a semiconducting or insulating solid on a useful energy scale.

As discussed in Section 1.2.3, we will almost always, in this book, use the E^o_{redox} of the H^+/H_2 couple under standard conditions (unit activity of H^+, one atmosphere H_2) as the reference zero for our energy scale. This value of E^o_{redox} is equal to the value of E_F for the metallic contact to the hydrogen reference electrode, as per Equation (1.63). This zero is used in handbook listings of standard electrode potentials (standard redox potentials) E^o, and, as discussed in Section 1.2.4, the energy levels of simple ions in solution are $E^o_{redox}(= -qE^o)$ on this scale. Then, if we place E_{cs} and E_{vs} on the same scale, we can compare the energy level of the bands of the solid with the energy levels of the ions in solution and can immediately guess, and with more effort calculate, whether hole or electron transfer between the solid and the solution is plausible. (This comparison is made in Figure 5.5.)

In our discussions of band edges of semiconductors, only the theoretical determination will be described in this section. Experimental measurement of the energy levels of the bands in the solid, which preceded theoretical predictions in most cases and are, presumably, more accurate than theoretical values, are discussed in Chapter 5.

We will describe the approaches of Butler and Ginley,[27] and of Frese,[28] who calculated the position of the conduction band edge from the formula

$$E_{cs} = E^e - qV_H - \chi = E^o_{cs} - qV_H \tag{2.31}$$

where χ is the electron affinity, the energy difference between the free electron energy and the conduction band edge; E^e is the energy of free electrons on the hydrogen scale; and V_H is the Helmholtz potential difference. Here E^o_{cs} is defined as the energy of the conduction band at the point of zero charge, or more accurately, where $V_H = 0$. E^o_{cs} is the energy of interest. The symbols are illustrated in Figure 2.5, where it is noted

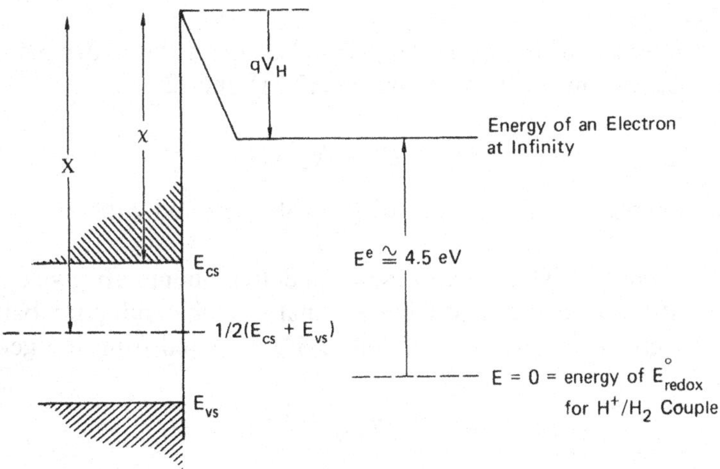

Figure 2.5. Energy diagram to illustrate the parameters used in the calculation of the conduction band edge E_{cs}. The relative energies of the hydrogen zero and the rest energy of an electron at infinity are fixed in accordance with the discussion of Section 2.2.3. The interface between vacuum and the solution is not indicated because it is not needed.

that according to the sign convention of Equation (2.15), V_H as shown is negative. Now E^e is derived in Section 1.2.3. V_H is known, or better, can be made zero by specifying the pH that corresponds to the pzc, as discussed in Section 2.1.1.3. Thus, the evaluation of the energy level of the conduction band at the pzc, E^o_{cs}, depends only on evaluation of the electron affinity χ. At any other pH, E_{cs} changes in accordance with Equation (2.31).

The electron affinity is estimated based on the "electronegativity" concept introduced by Pauling.[29] Butler and Ginley[27] used the Mulliken definition of electronegativity of an atom, the arithmetic mean of the atomic electron affinity A and the first ionization energy I:

$$X(M) = \tfrac{1}{2}(A + I) \tag{2.32}$$

where $X(M)$ is the electronegativity of the atom M. Sanderson[30] showed evidence that the electronegativity of a compound could best be described as the geometric mean of the electronegativities of its component atoms. The Sanderson electronegativity X for a semiconductor is assumed to be the midgap energy, $\frac{1}{2}(E_c + E_v)$, as is illustrated in Figure 2.5. Specifically, the electron affinity is given by

$$\chi = X - \tfrac{1}{2}E_g \tag{2.33}$$

where E_g is the energy gap of the semiconductor. Frese[28] calculated the midgap energy by relating the Pauling electronegativity to give real or virtual work functions ϕ for all the elements. He averaged the anion and cation work functions to yield the midgap energy that replaces X in Equation (2.33).

The position of the conduction band edge on the hydrogen scale at the pzc, E^o_{cs} becomes, from Equations (2.31) and (2.33),

$$E^o_{cs} = E^e - X + \tfrac{1}{2}E_g \tag{2.34}$$

and, of course, E^o_{vs}, the energy of the valence band edge at the pzc, is E_g lower.

We will not derive the expressions in detail, but merely give examples for illustration. In Butler and Ginley's analysis the conduction band edge of TiO_2 is given with X in Equation (2.34) obtained from the geometric mean:

$$E^o_{cs} = E^e - \{X(\mathrm{Ti})X^2(\mathrm{O})\}^{1/3} + \tfrac{1}{2}E_g \tag{2.35}$$

while in Frese's analysis X in Equation (2.34) is determined by $\frac{1}{2}(\phi_{Ti} + \phi_{O_2})$ where the ϕ's are obtained as above. With $\phi_{Ti} = 3.87$ eV and $\phi_{O_2} = 7.28$ eV, $\chi = 4.05$ eV. Using Butler and Ginley's value for $X(TiO_2)$ of 5.83 eV, and the energy gap for TiO_2 the order of 3 eV, we have

$\chi = 4.33$ eV. If we assume that the number 4.5 eV is correct for E^e, we have from Frese's value $E^o_{cs} = 0.4$ eV above the hydrogen reference energy, and from Butler and Ginley's value $E^o_{cs} = 0.2$ eV. Such values would suggest, for example, that electrons that have a free energy equal to that of the conduction band of TiO_2 should be expected to reduce protons in solution. However, such reduction in the dark is not observed, probably because TiO_2 does not catalyze the hydrogen atom recombination reaction, so a large overvoltage is required, as discussed in Chapter 3.

In Figure 2.7 in Section 2.4 we show the theoretical predictions for the conduction band edges E^o_{cs} of many semiconducting oxides according to the theoretical prediction. These are obtained using Equations (2.34) and, as Figure 2.7 illustrates, seem to be in reasonable agreement with the experimental values. The experimental values are obtained by the techniques described in Chapter 3 and are reviewed in Section 5.3.

2.4. The Band Model of the Solid/Solution Interface

In this short section we summarize how the models of these first two chapters are combined to provide a theoretical band model for the semiconductor/electrolyte interface at equilibrium. A slightly oversimplified band model for an electrochemical cell is compiled as in Figure 2.6.

The zero of energy is defined in Section 1.2.2.3, together with Equation (1.63), as the energy E^o_{redox} for the H^+/H_2 couple under standard conditions within the solution. The energy of the Fermi energy of the reference electrode, E_F(ref), relative to the standard hydrogen reference electrode is also obtained from the handbook value. For example, if the calomel reference electrode is used, the Fermi energy of the reference electrode will be at about -0.24 eV relative to the zero of energy (as the handbook gives $E^o = 0.24$ V for calomel). We will assume the hydrogen reference electrode itself is used, so of course its Fermi energy, E_F(ref) is defined to be at the zero of energy in accordance with Equation (1.63).

The energy levels E_{ox}, E_{red}, and E^o_{redox} of simple ions in solution, such as the ferrocyanide/ferricyanide couple, can be estimated directly from their redox potential E^o, as discussed in Section 1.2.4.4, and so can be indicated on the diagram. Specifically $E^o_{redox} = -qE^o$ on Figure 2.6 if $E_F(\text{ref}) = 0$. For example, the value of E^o_{redox} for the iron cyanide couple under standard conditions (equal activities) will be at approximately -0.35 eV relative to E_F(ref), because E^o for iron cyanide, as determined from the handbooks[R17] is 0.35 V. The value of λ is not known well for any species, but is the order of 0.5 to 1 eV.

The parameter V_H cannot be determined entirely theoretically. The

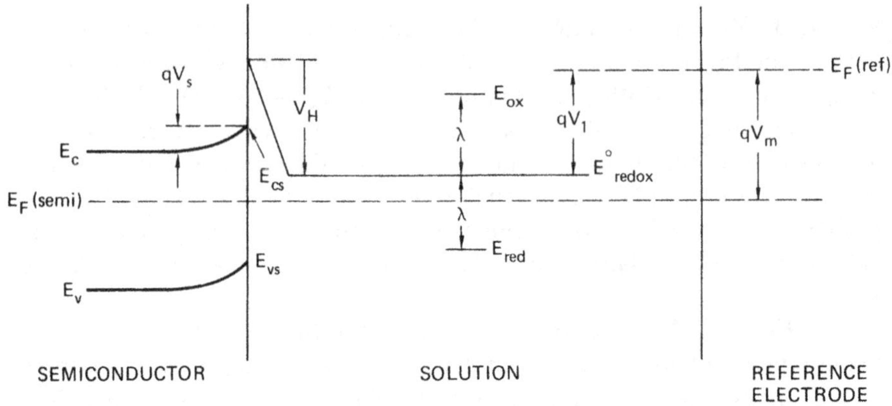

Figure 2.6. Energy level model for the semiconductor/solution interface. Here V_m is the measured voltage, the difference in Fermi energies E_F, at the semiconductor relative to the reference electrode (positive as indicated) $E^o = \frac{1}{2}(E_{ox} + E_{red})$ for the redox couple in solution. V_H is the Helmholtz potential, and V_s the surface barrier. The distances are not to scale (compare Figure 2.2).

value of V_H is zero at the pH of the point of zero charge, and this pzc is determined experimentally as discussed in Section 2.1.1.3. Then, knowing this pH, the value of V_H at any other value of the pH is determined theoretically by Equation (2.20).

The value of E_{cs} can be determined theoretically using Equation (2.35) to yield E^o_{cs} and adding the correction term for V_H as determined by Equation (2.20) (where again we assume the pH at the pzc is known).

The voltage V_m is the electrode potential, the potential difference between the working electrode and the reference electrode, as would be measured with an electrometer. At equilibrium, in principle, it is defined using Equation (1.63). If V_m is changed from its equilibrium value (e.g., by inserting a battery between the electrode and the reference electrode) a net current will flow, as discussed in Chapter 3.

This completes the band model as it is determined on a more or less purely theoretical basis. The values of E_c and V_s, in principle, follow from the equilibrium calculation of V_m, but in practice they are really the variables in semiconductor electrochemistry. The parameter V_s depends on applied voltage, on the redox species in the solution, on surface state energies, and on the illumination of the semiconductor, to name the dominant parameters.

The key to the use of such a model is the evaluation of E_{cs}. Of course, E_{cs} can be determined experimentally. To help further in appreciating the value of the band model description of the semiconductor/liquid interface, we will anticipate somewhat, and will indicate how E_{cs} is determined experimentally, starting from the measurement of the bulk Fermi energy by the measurement of V_m with a hydrogen electrode as the reference. When E_F is thus measured on the hydrogen scale, the energies of the bulk conduction and valence band edges, E_c and E_v, are immediately determined in principle because the bulk properties of the semiconductor are assumed known. The solid state methods of determining μ and E_g are beyond the scope of this book, but, for example with an n-type semiconductor of typical donor density, μ will be 0.1–0.2 eV.

Knowing E_c by the above voltage measurement, we can experimentally determine E_{cs} by measuring V_s. The parameter V_s is determined by any of several methods described in Chapter 4.

A particularly important case for the experimental determination of E_{cs} is the value of V_m when V_s is zero. This value of V_m is called the "flat band potential," V_{fb}. Now, as discussed many times, any change in the electrode potential V_m on a nondegenerate semiconductor appears across V_s. In other words, V_s varies directly with applied voltage, i.e.,

$$V_m = V_s + V_{fb} \tag{2.36}$$

Thus, an appropriate voltage can be applied to make V_s zero, or at least one can extrapolate to find the value of V_m when $V_s \to 0$. The experimental determination of V_{fb} by extrapolation is described in Section 4.1.3.1. From Figure 1.2 it can be seen that when $V_s = 0$, and the ref-

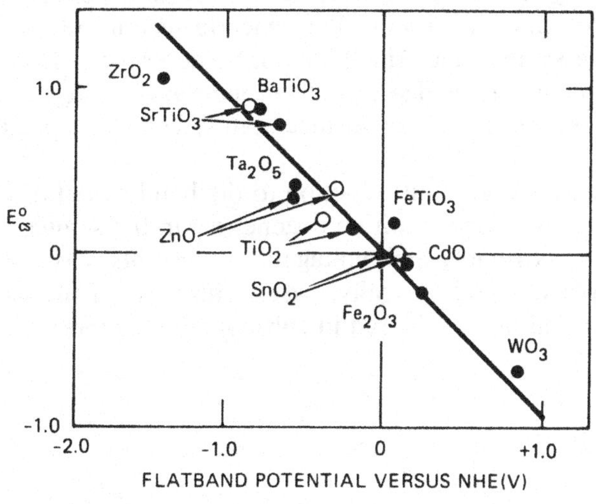

Figure 2.7. The theoretical values of E^{o}_{cs}, the conduction band edge at the pzc, versus the measured flatband potential on the hydrogen scale. The value of μ was considered zero for both the theoretical and the experimental estimate. The open circles are calculated by Butler and Ginley,[27] the solid circles by Frese.[28]

erence electrode is the hydrogen reference electrode [so $E_F(\text{ref}) \equiv 0$], then the voltage measured, V_m, is given by

$$qV_m|_{V_s=0} = qV_{fb} = E_{cs} - \mu \tag{2.37}$$

We note again that μ typically is 0.1–0.2 eV for an n-type semiconductor at room temperature. Thus, a measurement of V_{fb} immediately gives E_{cs}.

The basic parameter E_{cs}^o is obtained by extrapolating the measured value of E_{cs} by Equation (2.20) to the pzc. Then a comparison can be made between experimental and theoretical values of E_{cs}^o. Such a comparison is shown in Figure 2.7. Here, for a series of semiconductors at the point of zero charge, the measured value of V_{fb} is compared to the theoretically predicted value of E_{cs}^o, showing reasonably good agreement. More extensive tabulations of values of E_{cs} for various semiconductors are shown in Section 5.3.

Figure 2.7 is included in the present section not primarily to indicate the agreement of theory and experiment in the determination of E_{cs}^o for a series of semiconductors, but rather to give the reader at this stage an overview of some of the values of E_{cs}^o that are typical for semiconductors. Thus the reader can observe that electrons going into solution from the conduction band of ZrO_2, if such electrons were available, would be very strong reducing agents, reacting with many of the common species in solution, whereas electrons from the conduction band of WO_3 would be relatively inactive in reducing species in solution.

In summary, all the parameters are fairly well defined for a semiconductor, with the exception of V_s, which turns out to be the dominant variable of the system and which cannot be determined from handbook values, theory, or knowledge of bulk properties, but depends on many details of the system, such as surface states, voltage applied, and light intensity.

Of course, the above introduction to the band model describes only the simplest cases—cases with no degeneracy in the semiconductor due to doping or to extremely high voltages and currents, cases where there are no third phases, and basically, cases with none of the various complications that will be considered in subsequent chapters.

3

Theory of Electron and Hole Transfer

3.1. Introduction

3.1.1. General

All electrode processes include, in principle, a reducing agent in solution injecting electrons into the electrode and an oxidizing agent in solution extracting electrons from the electrode. When the currents associated with these two processes are equal, so there is no net current flow, the electrode is considered to be at equilibrium. The electrode has reached the "open circuit potential" or "open circuit voltage" (relative to the desired reference electrode) as described, for example, by Equation (1.45). The current in each direction at equilibrium is called the "exchange current." If now an "overvoltage" is applied, forcibly shifting the potential from the open circuit potential, the overvoltage tends to suppress current flow in one direction, to encourage it in the other, and a net current flow results.

Current flow between a solid that conducts by electrons and a solution that conducts by ions always implies a chemical reaction at the interface. Electrons must be captured, or holes injected, by oxidizing agents in solution, in order that cathodic current can flow. The oxidizing agent is reduced in these cathodic processes. Electrons must be injected, or holes captured, by a reducing agent in solution, in order that anodic current can flow. The reducing agent is oxidized in these anodic processes. Thus, for example, if the redox couple Fe^{3+}/Fe^{2+} dominates the electrode reaction, the cathodic reaction is

$$e^- + Fe^{3+} \rightarrow Fe^{2+} \tag{3.1a}$$

or

$$Fe^{3+} \rightarrow Fe^{2+} + h^+ \tag{3.1b}$$

with the oxidizing agent, Fe^{3+}, reduced to the reducing agent, Fe^{2+}. The reverse of these reactions would be the anodic reaction, where electrons are injected into the solid or holes extracted from the solid.

In principle, an open circuit potential is defined even for an "indifferent" electrolyte, where no species are added to the solution that actively inject or capture carriers from the solid. All solutions will, in principle, exchange carriers to the point of defining an open circuit potential, although the current at the open circuit potential may be very small. Consider an indifferent solution consisting of only water (perhaps with KNO_3 added to provide inactive current-carrying ions if desired). Then the reactions that dominate the carrier exchange could be

$$e^- + H^+ \rightarrow H \rightarrow \tfrac{1}{2}H_2 \tag{3.2a}$$

for the cathodic reaction, and

$$OH^- \rightarrow OH + e^- \rightarrow \tfrac{1}{2}H_2O_2 + e^- \tag{3.2b}$$

for the anodic reaction. The point is, that there are always species in solution that, at least in principle, can be reduced, albeit with a very high activation energy; and species that, at least in principle, can be oxidized. If all else is removed, then the electrode or the solvent can be the reactant if enough voltage is applied, or if senstive enough current measurements are made. Thus, in all cases the solution will, in principle, define an open circuit voltage, that is an equilibrium E_F in an electrode, in accordance with a formula similar to Equation (1.45), even if there are no highly active ions present.

3.1.2. The Activation Energy in Electrode Reactions

We will distinguish two models that are used to indicate the activation energy for carrier transfer, and thus to derive rate expressions. One is completely general, and the other, based on an energy level model, emphasizes a specific activation process for the reaction.

The current flow across an electrode/electrolyte boundary has historically been described in terms of electrochemical potentials of the species, using the absolute rate theory, where an activation energy for the reaction is assumed. By this general electrochemical model, the origin of the activation energy is not necessarily defined. The system must, however, reach that activated state before electron transfer can occur. At a semiconductor/electrolyte or an oxide/electrolyte boundary, however, it is usually more convenient to use the specific model, incomplete though it is, of carrier transfer between energy levels. This energy level description, which will be that most utilized in the present book, gives one very specific origin for the activaton energy. The rate of charge exchange between an electrode and a solution is proportional both to the probability that the appropriate ion is within tunneling distance of the surface, and to the probability that its energy level has fluctuated in energy to an activated state where electrons can tunnel to or from the solid with no change in energy. The general model is of much wider applicability because the electronic or chemical steps necessary to reach the activated state do not need to be explicitly defined. However, for systems simple enough that some of the specific steps can be assumed and fed into the model, the energy level model permits much more in the way of specific predictions. A more detailed discussion comparing the two approaches is given below in Section 3.2. In the present short section we offer a few remarks about activation energies in general, and the problem of defining activation energies.

The activation energy of an electrode reaction will often depend on the reaction route and on the catalytic action of the electrode or some ion in solution. For example, it would require very energetic electrons to reduce protons to hydrogen atoms in solution, as indicated in the first step of Equation (3.2a). The energy is, of course, recovered when the hydrogen atoms combine in step two. Similarly, it would require a great deal of energy to oxidize OH^- ions to OH radicals in solution, as indicated by the first step of Equation (3.2b). If these reactions proceeded as indicated, the energy to form such free radicals in solution would be the activation energy for the process, or at least, a great portion of the activation energy for the process. In practice, nature usually finds a route for the reactions that requires less activation energy than the brute force routes indicated. The electrode, or some other ion in solution, acts as a "catalyst" that lowers the activation energy. An important example of catalytic action lowering the activation energy and thereby increasing the rate of an electrode reaction, is the action of platinum. Platinum is used as an electrode to permit reaction (3.2a) to proceed with very low activation energy. Platinum forms a strong covalent bond to hydrogen atoms, and the protons can be reduced at a platinum electrode with much less energy when the resulting atoms form bonds to the platinum, rather than becoming free radicals in solution. The adsorbed atoms can then recombine on the platinum surface, or a proton neighbor of the adsorbed hydrogen atom can be reduced, and the hydrogen molecule is formed locally with very low activation energy. The lowering of activation energies by catalytic behavior is an important part of electrochemistry—in some cases intentional and the action known, in some cases unintentional, and the route of the reaction not known.

There are several steps in an electrode reaction that could require an activation energy and thus limit the rate of the reaction, the current. These rate-limiting steps can be considered types of polarization, in the sense used in Section 2.2.1. There a polarizable electrode was defined as one that blocks dc current flow to the extent that it becomes the dominant resistance in the circuit, limiting the rate of the electrode reaction. First, there is concentration polarization. The reaction rate—the current—is limited by the availability of the reactant. The limited supply can arise because the concentration of reactant provided in the solution is limited, and the current can only flow as fast as the ions can diffuse to the surface. The limited supply can alternately arise because some reaction in solution, or on the surface, must occur to provide the actual electron accepting or donating species, and the rate of the required chemical reaction limits the rate of the current flow. For example, deRiepen *et al.*[1] find the rate of electron injection into the ZnO conduction band by luminol is limited by the rate of its chemical oxidation, and Kokado *et al.*[2] find electron injection from a solution of NaOCl and H_2O_2 is limited by the rate of their reaction in solution.

A second type of polarization, and the one that we will normally be concerned with in this book, is polarization due to a limitation on the rate of charge transfer. With semiconductors, this rate-limiting step is particularly common, because the concentration of carriers in the solid is limited. As will be described, in some cases, appreciable current can only flow if the activation energy is lowered by an applied voltage, or by some other stimulation such as illumination, that increases the carrier density at the surface.

A third type of limitation on the current is termed[R9] "nucleation polarization," where a separate phase must be nucleated on the sample to support the fast reaction. If the nucleated phase remains on the surface, the limitation of current is a transient limitation; if the nucleated phase does not remain, an extra voltage (overvoltage) may be required to maintain a steady state concentration of the nucleated phase. For example, if the production of adsorbed hydrogen atoms on a metal is slow, but the reduction of H_2^+ is fast, then one would consider the adsorbed atoms as the nucleated phase, and the reduction of hydrogen, in a sense, is a process dominated by nucleation polarization. Such nucleation voltage has not been studied on semiconductors as such, but one simple case certainly occurs and can confuse semiconductor measurements—the deposition of a metal by reduction of the ion. For example, in the reduction of Ag^+ to Ag on ZnO, the resulting metal layer increases the current orders of magnitude and makes study of the semiconductor electrode parameters very difficult.[3]

Surface heterogeneity leads to a variable activation energy.[R4] Heterogeneity effects are probably much more serious in general for semiconductors than for metals even though for semiconductors electron transfer polarization usually dominates. The problem arises because the activation energy involves not only fluctuating energy levels in solution, but the activation of carriers in the solid. Flaws at the surface of the semiconductor can have quite different activation energy for carrier exchange with ions in solution than the uniform surface, and much of the current can arise at these flaws. The problem is much more serious with nonmetals than with clean metals because the currents are lower with the nonmetals, and spurious currents can provide a relatively greater fraction of the total current. Also, with nonmetals the properties of the electrode, in particular the details of the band picture, can change considerably in the neighborhood of a flaw. With both metal and nonmetal electrodes, flaws will change the surface bonding of adsorbates, or of the surface atoms of reactive electrodes, and this chemical change also will possibly lead to higher or lower rates at such flaws. In section 5.2.2 for an example, the behavior of dislocations at the semiconductor electrode surface will be described and shown to lead to quite different characteristics of the electrode.

Heterogeneity will not only lead to difficulties in interpreting electrode behavior with a net current flowing (with an applied overvoltage) but may also lead to difficulties in analyzing open circuit conditions. Circulating currents can be expected because the characteristics of the electrode differ from point to point. With the same open circuit voltage between the electrolyte and the electrode, at one point on the sample surface current will flow into the solution; at another point current will flow into the electrode; and one has a "circulating current." In fact, of course, the voltage assumed by the electrode will be such that the net current is zero, but these circulating currents can corrode or erode the electrode in various places while, in principle, the system should be unchanging.

Disregarding all these qualifications concerning heterogeneity, we will proceed in most of this chapter to discuss electrode reaction rates at a uniform surface, emphasizing cases of electron transfer polarization. In Section 3.2 we will briefly review the general electrochemical model that satisfactorily describes not only electron transfer polarization, but other sources of activation energy and polarization overvoltage. Then in the last section, Section 3.3, the energy level model is presented. By utilizing both models we can rationalize, and in some cases obtain quantitative agreement with, experimental observations of semiconductors and oxidized metal electrodes.

3.2. Classical Model

The rate at which electrons transfer between ions in solution and a solid electrode is considered proportional to the density of the appropriate ions in solution; proportional to the density of carriers in the solid (this factor is usually omitted for metal electrodes because the density of carriers does not vary); and proportional to a Boltzman factor describing an "activated state" that the system must reach before the transfer can occur. Thus, the cathodic current due to electron capture by ions in solution (reduction of an oxidizing agent) becomes

$$J_c = qk_c n_s c_{\text{ox}} \exp(-U_{\text{cat}}/kT) \tag{3.3}$$

where k_c is a rate constant, n_s the electron density at the surface, c_{ox} the concentration of oxidizing agent in solution, and U_{cat} the activation energy associated with the cathodic reaction, electron capture by the oxidizing agent. The electron density at the surface is considered a constant, and can be incorporated into k_c for p-type semiconductors or metals. Similarly the anodic current due to electron injection (or hole capture) is given by

$$J_a = qk_a p_s c_{\text{red}} \exp(-U_{an}/kT) \tag{3.4}$$

with k_a the rate constant, p_s the hole density or the density of unoccupied states (considered a constant and can be incorporated into k_a for electron injection into n-type semiconductors or metals), c_{red} the density of reducing agents, and U_{an} the corresponding activation energy for the anodic process, electron injection from the reducing agent.

In Equations (3.3) or (3.4), the activation energies U may represent the work necessary to bring an ion to the surface, or may represent the energy of a chemical transformation that leads to an active species. We will normally be interested in the case where U represents the energy required for the energy level of the ion to fluctuate [Equation (1.52)] to an energy (E_F or E_{cs} or E_{vs}) where electron transfer can occur. In the case of semiconductors, the expression for the carrier density may also involve an activation energy [Equations (2.10) and (2.11)].

Equations (3.3) and (3.4) are helpful in pointing out the major distinction between the ideal behavior of nondegenerate semiconductor and of metal electrodes. In the case of simple semiconductors, the voltage-dependent parameters are p_s and n_s; in the case of simple metals, the voltage-dependent parameter in Equations (3.3) and (3.4) is U.

This important difference between semiconductor and metal electrodes arises as follows: In the case of a semiconductor electrode with a depletion layer, there is conductivity from the battery through the semiconductor and up to the edge of the depletion layer, and there is conductivity through the solution up to the outer Helmholtz plane. In between, the material is essentially an insulator. Thus, any change in electrode potential, say by an externally applied voltage, appears across both the Helmholtz layer and the depletion layer. Because the Helmholtz double layer is only a few angstroms thick and the depletion layer about

1000 Å (see Chapter 2), the change in voltage drop across the Helmholtz double layer is a negligible fraction of the change in electrode potential. Thus, the relative position of the energy levels right at the surface of the solid and the energy levels in solution are not affected by any voltage change. Now U depends on these relative positions. So in the fluctuating energy level model as applied to semiconductors, U is independent of applied voltage. The important effect of a voltage change is to influence the density of charge carriers p_s or n_s at the surface of the semiconductor electrode, and it is through these factors in Equations (3.3) and (3.4) that J_a and/or J_c are changed.

A metal, on the other hand, is an excellent conductor right up to its surface, and the solution is a conductor up to the outer Helmholtz plane. Thus, any change in voltage appears across the very thin Helmholtz double layer, and so the energy levels in solution move relative to the energy levels in the solid. In the fluctuating energy level model, then, U depends on voltage for metals.

We will here emphasize metals and the classical approach normally used to describe them, and will develop an expression for current flow based on this approach. We will assume in the analysis that the electrode is characterized by charge transfer polarization rather than concentration polarization or nucleation polarization, as defined in the last section. We will also assume any change in the electrode potential appears across the Helmholtz double layer.

The equations for electrode current, Equations (3.3) and (3.4) (with p_s and n_s constant) are interpreted using a very general configurational coordinate model to illustrate how U varies with potential. In Figure 3.1 some unspecified coordinate is plotted on the abscissa with the energy of the system on the ordinate. In Section 3.3 for example, the coordinate will be chosen to be the polarization of the solvent, treated as a uniform dielectric. However, the coordinate could be the separation of the ion from the surface, the position of a complexing ligand, or any other variable that affects the energy of the system. Curve (a) shows the energy of the system as a function of this unspecified coordinate when the electron is in the solid. Some value of the coordinate P_1 is indicated that leads to the lowest energy of the system. If the value of the coordinate deviates from this preferred value in either direction, the energy of the system rises. Curve (b) shows the energy of the system when the electron is on the ion. The value of the coordinate that leads to the lowest energy is now P_2, a different value from that in case (a). Again, a deviation from the preferred value of the coordinate in any direction leads to a higher energy for the system. Curve (c) shows how curve (b) changes when a potential difference $\Delta\phi_s$ (which has a positive sign when the electrode is positive with respect to the solution) is applied between the solid and the liquid.

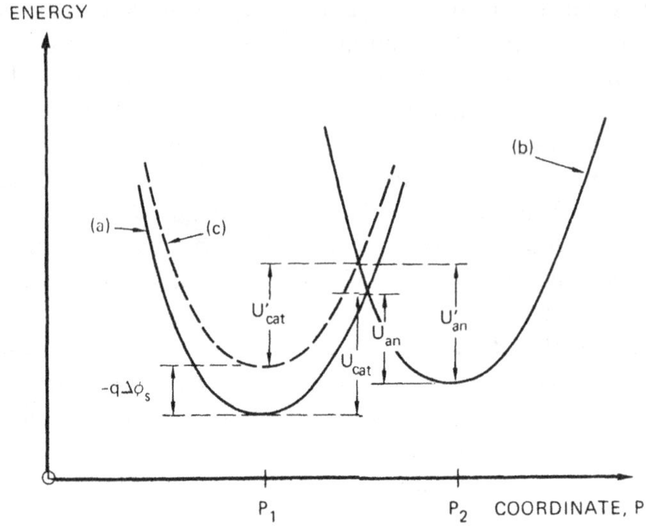

Figure 3.1. The energy of the electrode/electrolyte system as a function of some unspecified coordinate: the location of the electron, and a potential shift. On curve (a) the electron is in the solid, on curve (b) the electron is on the ion. On curve (c) the electron is in the solid and the potential of the electrode relative to the solution is changed by $-\Delta\phi_s$. Curve (b) is independent of potential because the electron is in the solution phase.

Curve (a) may also shift with an applied voltage, depending on the system, but we are only interested in the relative shift between curve (a) and curve (b), so we are at liberty to stipulate for illustrative purposes that curve (a) is independent of voltage with no loss of generality.

It could be noted here again, that with a semiconductor electrode, where any changes in applied voltage appear within the semiconductor, so the U's in Equations (3.3) and (3.4) do not change with applied voltage, a different model is needed. From this point on, the discussion is concerned with metallike electrodes.

Now, if we have chosen the correct coordinate, then when the coordinate moves from its lowest value to the value at the intersection of curves (a) and (b) [or, with $\Delta\phi_s$ applied, to the intersection of curves (a) and (c)] then electron transfer can occur with no energy change in the system. The activation energy is thus the energy necessary to reach this intersection, the energy necessary to so shift the "coordinate."

When the potential of the metal relative to the solution is changed by $\Delta\phi_s$, the energy of the system increases by $\Delta U = -qz\Delta\phi_s$ more when the z electrons are on the metal than when they are in the solution. However, ΔU does not appear entirely either in the activation energy U_{ox} for the oxidation step, or in the activation energy U_{red} for the reduction step. In both directions the change in activation energy is a fraction $zq\Delta\phi_{sol}$.

If, in the region of the intersection of the two curves, the curves are approximated by straight lines, it is a simple trigonometric exercise to show that if we define α such that

$$\Delta U_{\text{an}} = U'_{\text{an}} - U^o_{\text{an}} = zq\Delta\phi_{\text{sol}}\alpha \tag{3.5}$$

then β, defined such that

$$\Delta U_{\text{cat}} = U'_{\text{cat}} - U^o_{\text{cat}} = zq\Delta\phi_{\text{sol}}\beta \tag{3.6}$$

is related to α by

$$\beta = 1 - \alpha \tag{3.7}$$

The relative shift in energy of the two activation energies with an applied voltage is thus simply related by a quantity α termed the "anodic transfer coefficient".[R9.4] More rigorous analysis[5] shows that Equation (3.7) is actually valid in general.

As described in the preceding chapter, equilibrium at a uniform electrode occurs when the rate of electron capture equals the rate of electron injection. This will occur at some potential difference $\Delta\phi_e$, such that

$$\begin{aligned} J_0 &= qk_c c_{\text{ox}} \exp[(-U^o_{\text{cat}} - \beta zq\Delta\phi_e)/kT] \\ &= qk_a c_{\text{red}} \exp[(-U^o_{\text{an}} - \alpha zq\Delta\phi_e)/kT] \end{aligned} \tag{3.8}$$

where we have designated the current when $J_c = J_a$ as J_0, the "exchange current" that flows in both directions at equilibrium, as discussed in Section 3.1.1. From Equations (3.5)–(3.8), in (3.3) and (3.4) (with n_s and p_s taken as constant) it is immediately seen that

$$J_c = J_0 \exp(-\beta zq\eta/kT) \tag{3.9}$$

and

$$J_a = J_0 \exp(\alpha zq\eta/kT) \tag{3.10}$$

where we have defined the "overvoltage" applied to the electrode as

$$\eta = \Delta\phi_{\text{sol}} - \Delta\phi_e \tag{3.11}$$

The overvoltage η is thus the voltage difference between the actual electrode potential and the equilibrium electrode potential. If η is positive, a net anodic current flows (electrons flow into the metal). The net anodic current through the electrode is obtained from

$$J = J_a - J_c \tag{3.12}$$

$$= J_0\{\exp(\alpha zq\eta/kT) - \exp[-(1-\alpha)zq\eta/kT]\} \tag{3.13}$$

At high overvoltages, one or the other of the terms in Equation (3.13) dominates, and the expression reduces to the "Tafel equation," which for cathodic reduction becomes

$$J_{\text{red}} = J_0 \exp[-(1-\alpha)zq\eta/kT] \tag{3.14}$$

and for oxidation (anodic direction) becomes

$$J_{\text{ox}} = J_0 \exp(\alpha zq\eta/kT) \tag{3.15}$$

3.3. The Energy Level Model of Charge Transfer

3.3.1. General

Although the general classical model, as described in Section 3.2, is correct and accurate (with a slight modification in the case of semiconductors because voltage changes lead to changes in V_s rather than V_H), it is desirable to provide a more detailed model than this. Further, we wish to make this more detailed model an energy level model because, although we are interested in metals for comparison, we are interested primarily in semiconductors and insulators, and these are best described in terms of energy bands and energy levels for electrons. The development below, of the energy level model for metals and semiconductors, is done in such a way that the common factors, present in essentially all electrode reactions, can be separated from the special factors—factors that do not always appear. Under common factors we include the fluctuation of the energy levels of ions in solution, and the Fermi distribution of electrons in the energy levels of the solid. Under special factors we include such processes as changes in inner sphere configuration of the ion following electron transfer, specific adsorption of the ion, or electrical breakdown in the solid. In the present section we discuss the simple case, with only the fluctuations of ionic energy levels and the electron distribution in the solid considered. The first two subsections of the present section are somewhat mathematical. Then, to emphasize the use of the resulting models, a qualitative discussion is added in Section 3.4, where the band model description is discussed in completely phenomenological terms.

Carrier transfer by the simplest energy level model occurs in solution by a mechanism wherein the energy level of the ion in solution fluctuates (Section 1.2.4) to an energy approximately equal to that of the electron or the unoccupied state in the solid, and then the electron transfer occurs with negligible change in the energy of the electron. The simple model assumes[6] negligible phonon emission during electron transfer, and negligible coupling[7] between the initial and final electronic states. Thus the "activated state" described in the last section, now is specified to be a state wherein the two energy levels are equal: the energy level in the solution has fluctuated to the same energy as the energy level in the solid.

The electron transfer rate will be proportional to the effective number of ions available for electron transfer at the surface, proportional to the number of occupied (or unoccupied, depending on the direction of electron transfer) states in the solid, and proportional to the probability that an energy level in solution has fluctuated to the energy level of the appropriate state in the solid.

The number of ions at the surface will be proportional to their density (concentration) in the solution. It is clear, however, that the transfer probability must decrease if the ion is farther from the solid surface. In the simplest case, we can consider transfer occurs by a tunneling mechanism between the nonlocalized levels in the band and the localized level on the ion. In addition to tunneling, we could consider other more complex cases where, for example, a ligand or other ion in solution acts as a bridge for electron transfer, or equivalently as a resonance energy level to aid tunneling. The latter possibilities will not be discussed. For the case of tunneling, the ion must be within 30 Å or so of the surface for the transfer to occur at all. To determine the "effective number of ions at the surface," we must somehow average over tunneling distances. A complete analysis leading to this average distance is beyond the scope of this book, although estimates will be discussed later in Section 3.3.2.

The probability that an energy level in solution has fluctuated to an energy level E is given by

$$W(E) = (4\pi\lambda kT)^{-1/2} \exp[-(E_t - E)^2/4\lambda kT] \tag{3.16}$$

where Equation (3.16) is simply a repeat of Equation (1.54) of Chapter 1. Here E_t can be the energy level of the reducing agent, E_{red}, or the energy level of the oxidizing agent, E_{ox}, depending on which direction of electron transfer we are considering.

Thus, by the simplest model

$$J(E) \propto c_{\text{red}} d W(E) \qquad \text{or} \qquad J(E) \propto c_{\text{ox}} d W(E) \tag{3.17}$$

where d is an integrated average of distance $\int F(x)dx$ such that we can approximate that if an ion is within d of the surface, tunneling to it will occur, but any ion beyond d is nonreactive. $J(E)$ is the contribution to the current by the ions at energy E.

When applying this simple energy level model, complications can arise due to complex formation, or due to adsorption. The concentration that would normally be used in Equation (3.17) is, of course, the concentration of the redox species added to the solution. However, the energy level E_t that is fluctuating to the level where exchange can occur may not be the dominant form of the species that has been added to the solution.

As one example, we can consider the reduction of Co^{3+} in aqueous solution, where the total concentration of cobalt is divided among the species Co^{3+}, $Co(OH)^{2+}$, and $Co(OH)_2^+$. If $Co(OH)^{2+}$ happens to be the species with an energy level near the desired energy E to make $W(E)$ in Equation (3.20) high, then this form may dominate the electrode reaction even though it is not the form with the highest concentration. For a quantitative expression describing the rate of the reaction, c_{ox} should represent the concentration of $Co(OH)^{2+}$, rather than the total concentration of cobalt. An example that is harder to correct quantitatively is that when the active species is an adsorbed species. Not only will energy level of the adsorbed species differ significantly from the energy level of the species in solution, but in addition the value of λ will be low compared to its value for the ion in solution (Section 2.1.2).

The straightforward energy level model must be modified to account for such chemical transformations. We note that the model of the Section 3.2 can include such transformations as part of the activation process, which, of course, is why it is more generally applicable.

In the arguments in the present section we will assume the simplest model for the ions in solution, leaving to Chapter 7 discussion of adsorption and complex formation. A one-equivalent reaction will be assumed here. The reducing agent will be assumed to be in one form only, the oxidizing agent will be assumed to have the same inner sphere configuration as the reducing agent. A good example of such a system that is often used is the ferrocyanide/ferricyanide, $Fe(CN)_6^{-4}/Fe(CN)_6^{-3}$, redox couple. Because their ligand shells are complete and stable, they do not usually adsorb on an electrode; they do not appreciably hydrolyze; and they have only the two valence states available, so the reaction is a one-equivalent (one-electron) process. It will be found in the experimental sections of this book that much definitive work has been done using this redox couple. For our discussions of the energy level model in this section, we will visualize such a simple redox couple in solution.

3.3.2. The Metal Electrode

Marcus[8–10] has developed expressions for the current through a solid electrode using non-quantum-mechanical models. Dogonadze and his coworkers,[6,11,R22] Levich[12] and Christov[7] have studied the system more recently with a quantum mechanical analysis. Hale[13] applied and justified the model for metal electrode processes. We will present a simplified version for metals of Marcus' analysis of the fluctuating energy level model, in order to compare its use to describe a metal electrode with its use (Section 3.3.3) to describe an insulator or a nondegenerate semiconductor electrode. We will call this model the Marcus–Hale model.

Marcus[10] calculated by methods closely related to absolute rate theory that the injection current from a one-equivalent reducing agent into an inert metal electrode is given by

$$J_a = qZc_{\text{red}} \exp(-\Delta G^*/kT) \tag{3.18}$$

where Z is a constant, determined by analysis to be the thermal velocity[14] of the ions ($=[kT/2\pi m^*]^{1/2}$). The use of this constant replaces the use of a "tunneling distance" described in Equation (3.17). ΔG^* is the change in free energy to reach the activated state. Memming and Möllers[15] modified this expression slightly to include a normalizing factor with the exponential so the total number of ions of all energy would equal unity. The use of Equation (3.18) is discussed in detail by Hale.[13] It assumes a transmission coefficient κ of unity; in other words, if the activated state is reached, the energy level of an ion at the surface has fluctuated to an energy high enough for electron transfer, then electron transfer will occur with a probability of unity. Hale concludes that if an ion is within about 15 Å of the surface, $\kappa = 1$, and κ depends very sharply on distance so that beyond that critical distance κ is negligible. The more accurate form of Equation (3.18) also contains a factor depending on the mean square distance from the surface at which electron injection occurs divided by the mean square deviation from a reaction hypersurface. Since Marcus[10] and Hale[13] suggest this factor should also be the order of unity, it has been omitted from Equation (3.18).

Independently of whether electron transfer is to or from the metal, the energy level in solution must fluctuate to the Fermi energy of the metal, as can be concluded by a study of Figure 1.6. If the electron transfer is from a reducing agent into the metal, then the energy level must fluctuate to an energy above the Fermi energy in order to have an energy equal to the unoccupied levels in the metal. If the electron transfer is from the metal to the oxidizing agent, then the unoccupied level associated with the oxidizing agent must fluctuate down to an energy equal to or below the Fermi energy, in order to have an energy equal to that of one of the occupied states in the metal.

We will continue with the example of electron injection from a reducing agent. If we assume that there is no adsorption of reactants; no formation or breaking of chemical bonds; no chemical changes in the first coordination sphere of the ion; and zero work is required for the ion of interest to reach the electrode, then the energy to reach the activated state for electron injection into a metal is simply the energy necessary to shift the energy level of the reducing agent from E_{red} to E_F. When the system is at its lowest energy state (with the electron on the ion), the energy level of the electron is E_{red}. Thus, in Equation (1.52) E_t becomes E_{red}. For electron transfer, as stated above, the energy level must fluctuate to E_F. Thus, in Equation (1.52) the value of E becomes E_F. Then from Equation (1.52) we obtain the thermal energy required to reach the activated state, where the energy level of the reducing agent has fluctuated to E_F:

$$\Delta G^* = (E_F - E_{\text{red}})^2/4\lambda \tag{3.19}$$

Now with a metal electrode, the applied voltage appears across the Helmholtz double layer. This means that the difference $E_F - E_{\text{red}}$ varies directly with the applied voltage. As our reference potential is in the solution, E_{red} does not shift with voltage; only E_F changes relative to the hydrogen reference energy as the applied voltage changes. Now $E_F = E_{\text{redox}}$ when the net current through the electrode is zero (Section 1.2.2). Then

$$E_F - E_{\text{red}} = E_{\text{redox}} - E_{\text{red}} - \eta q \tag{3.20}$$

where η is the overvoltage, the voltage in excess of that required to maintain zero current. Equations (3.20) and (3.19) in (3.18) give an expression for the anodic current.

An expression for the cathodic current can be derived in an entirely analogous manner, where we consider that the probability of electron transfer to an unoccupied level in solution is unity, when the ion is at the metal surface and the unoccupied level fluctuates to an energy below E_F.

Then the net current, the anodic minus the cathodic, is given by

$$J = qZ\{c_{\text{red}} \exp[-(E_{\text{redox}} - E_{\text{red}} - \eta q)^2/4\lambda kT] - c_{\text{ox}} \exp[-(E_{\text{ox}} - E_{\text{redox}} + \eta q)^2/4\lambda kT]\} \tag{3.21}$$

where we have assumed the thermal velocities equal: $Z_{\text{ox}} = Z_{\text{red}} = Z$. At equilibrium $J = 0$, whence using the relation $E_{\text{ox}} - E_{\text{red}} = 2\lambda$ [from Equation (1.62)] we find

$$E_{\text{redox}} = \tfrac{1}{2}(E_{\text{ox}} + E_{\text{red}}) + kT \ln(c_{\text{red}}/c_{\text{ox}}) \tag{3.22}$$

which can be reformulated in a form equivalent to Equation (1.64):

$$E_{\text{redox}} = E^o_{\text{redox}} + kT \ln(c_{\text{red}}/c_{\text{ox}})$$

or (3.23)

$$E_{\text{redox}} = E_{\text{red}} - \lambda + kT \ln(c_{\text{red}}/c_{\text{ox}})$$

The introduction of Equation (3.23) into (3.21) gives

$$J = eZ(c_{\text{red}} \exp\{-[\lambda + kT\ln(c_{\text{red}}/c_{\text{ox}}) - \eta q]^2/4kT\} - c_{\text{ox}} \exp\{-[\lambda - kT\ln(c_{\text{red}}/c_{\text{ox}}) + \eta q]^2/4\lambda kT\}) \tag{3.24}$$

If $\lambda \gg \eta q$ and $\lambda \gg kT\ln(c_{\text{red}}/c_{\text{ox}})$ as is normally the case [Equation (1.47)], this simplifies to

$$J = qZ(c_{\text{red}}c_{\text{ox}})^{1/2} \exp(-\lambda/4kT) \times [\exp(\tfrac{1}{2}\eta q/kT) - \exp(-\tfrac{1}{2}\eta q/kT)] \tag{3.25}$$

$$= J_0 [\exp(\tfrac{1}{2}\eta q/kT) - \exp(-\tfrac{1}{2}\eta q/kT)] \tag{3.26}$$

which is equivalent to Equations (3.14) and (3.15) and leads to the Tafel equation. The factor J_0 as defined by Equations (3.25) and (3.26) is (at low η) the exchange current, introduced in Section 3.1.1 as the current passing both anodically and cathodically across the electrode surface at equilibrium. The first term of Equation (3.25) representing the anodic current is, as discussed in Section 3.2, written with the exponent $\alpha\eta q/kT$, with α defined as the transfer coefficient. Thus, it is observed that for the simple case evaluated here(9) the transfer coefficient is $\frac{1}{2}$. In other words, when the only contribution to the energy is the rearrangement energy λ associated with the dielectric constant of the medium, and if the approximation $\lambda \gg \eta q$ is valid, the transfer coefficient becomes $\frac{1}{2}$. Experimental values of the transfer coefficient α are commonly(R9) found to be close to $\frac{1}{2}$ for metal electrodes.

3.3.3. The Semiconductor Electrode

3.3.3.1. The Surface Physics Terminology for the Electrode Currents. In solid state physics analyses of carrier capture by energy levels in the semiconductor bandgap, it has been found rewarding to assume the capture rate is first order in the density of the carrier, and first order in the density of unoccupied states. Early in studies of the semiconductor/electrolyte interface it was shown quantitatively for one case(16) that here also such a relation is valid. Thus, it is normally assumed in semiconductor electrochemistry, as in semiconductor physics, that the rate of capture of conduction band electrons by an energy level can be expressed as

$$dn_t/dt = K_n[n_s(N_t - n_t) - n_1 n_t] \tag{3.27}$$

where n_s is the density of electrons (per unit volume) in the conduction band at the surface, n_t is the density of electrons (per unit area) in the "surface state," N_t is the total density of "surface states", n_1 is a temperature-dependent constant, determined by applying the condition that at equilibrium $dn_s/dt = 0$. Here K_n is the rate constant for electron transfer, given by(R4,R15)

$$K_n = \bar{c}\sigma \tag{3.28}$$

where $\bar{c}$ is the average thermal velocity of the conduction band electrons (10^9 m/s is an adequate approximation) and σ is the capture cross section of the surface states. The first term in Equation (3.27) is the rate of the cathodic process, electron capture on the surface state; the second the rate of the return (anodic) process.

Similarly, for the net rate of hole exchange between a surface state and the valence band we have

$$dp_t/dt = K_p(p_s n_t - p_1 p_t) \tag{3.29}$$

where analogously p_t is the density of holes in the surface state, $K_p = \bar{c}\sigma_p$ is the rate constant, p_1 is a constant determinable by applying the condition that at equilibrium the net rate is zero. The first term in Equation (3.29) is the rate at which holes are captured, the anodic process, the second term the rate at which they are injected back into the valence band, the cathodic process.

The correspondence between the variables and parameters of Equation (3.27) or (3.29), and a combination of Equations (3.3) and (3.4) is obvious by inspection. For example, with an n-type semiconductor (electron exchange with the conduction band), $k_c \exp(-U_{\text{cat}}/kT)c_{\text{ox}}$ is equivalent to $K_n(N_t - n_t)$, and $K_{a}p_s c_{\text{red}} \exp(-U_{\text{red}}/kT)$ is equivalent to $K_p p_1 p_t$. The two formalisms collect the constants in a different form, but are actually describing the same process. The reason for introducing this notation is simply usage. When dealing with semiconductors, one must be aware of the notation common in semiconductor physics.

3.3.3.2. The Tafel Equation for Semiconductors. The electrode behavior of the semiconductor with no degeneracy at the surface is relatively simple, if one requires only a partial description. As was discussed in Section 3.2, a negligible fraction of any applied voltage appears across the Helmholtz layer, the voltage all appearing across the double layer within the semiconductor. Thus, there is no appreciable voltage-dependent shifting of the energy levels E_{red} or E_{ox} in solution with respect to the band edges E_{cs} or E_{vs} of the semiconductor at its surface. This characteristic leads to a particularly simple current/voltage characteristic.

First, we will consider the possible rate-limiting steps and specify our assumptions. As in Section 3.3.3.1, we will assume that the rate-limiting step is not associated either with a chemical process in solution, or with diffusion of ions in the solution. With the semiconductor electrode, with its low carrier density, we must, however, consider a third possibility for a rate-limiting step: the rate at which carriers approach the surface. Such a possibility can particularly be expected when the reacting carriers are minority carriers, produced by illumination or by thermal generation. If the rate at which minority carriers reach the surface may be rate limiting, we will develop a "deep depletion" condition discussed at the end of

Section 2.2.4.2, where the minority carriers are consumed by the reaction and cannot build up to an equilibrium concentration. Even if the reacting carriers are majority carriers, the rate at which they can be supplied to the surface may conceivably be lower than the rate they could react, so the rate of carrier approach to the surface could be rate limiting. In the present section, we will assume charge transfer at the interface is rate limiting. We will leave such models wherein the availability of carriers is rate limiting to Section 6.3, where observations of minority carrier capture are discussed. Here we will assume that the carrier density n_s and p_s at the surface is simply given by Equation (2.10) or (2.11), whichever is applicable.

With electron transfer as the rate-limiting step, the voltage dependence of the electron transfer equations (3.27) and (3.29) arises, as discussed in Section 3.2, through n_s and p_s, respectively. The parameters n_t and $N_t - n_t$ in Equation (3.27) relate now to the density of ions in solution near the surface. Because the energy levels in solution do not shift with respect to the energy levels of the band edges at the semiconductor surface, in Equation (3.27) both the density of occupied and the density of unoccupied states, n_t and $(N_t - n_t)$, are constant, unaffected by the applied voltage. The parameters n_s and p_s vary with applied voltage as given by Equations (2.10) and (2.11), respectively.

Now the apparent Tafel equation becomes deceptively simple. Setting $J = 0$ at equilibrium in Equation (3.27) and defining $n_s = n_{s0}$ at equilibrium, we obtain

$$J = q\,dn_t/dt = qK_n n_1 n_t[(n_s/n_{s0}) - 1] \tag{3.30}$$

and, from Equation (2.10), we obtain the expression for n_s/n_{s0}:

$$n_s/n_{s0} = [n_b \exp(-qV_s/kT)]/n_b \exp(-qV_{s0}/kT) \tag{3.31}$$

$$= \exp(-\eta_{sc}q/kT) \tag{3.32}$$

where V_{s0} is the surface barrier at equilibrium, and η_{sc} is $(V_s - V_{s0})$ the overvoltage. In this case it is assumed for convenience that any change in external voltage appears entirely across V_s. Myamlin and Pleskov[R17] formulate the equation allowing both η and η_{sc} to change, but we will only discuss cases of this complexity in a qualitative way. Inserting (3.32) into (3.30), it is clear that for electron exchange with the conduction band of a simple semiconductor, we have an equilibrium equation:

$$J = J_0[\exp(-\eta_{sc}q/kT) - 1] \tag{3.33}$$

The apparent transfer coefficient is zero, as is observed by comparing Equation (3.33) to (3.13). An equivalent expression is obtained for charge exchange with the valence band, with, however, the apparent transfer coefficient now unity. The form derives immediately from Equation (3.29).

We note that Equation (3.33) is equivalent to the equation of a solid state Schottky diode, as was first pointed out in Section 2.2.4. Actually, Schottky diodes are analogous; they are constructed with a metal replacing the solution and J_0 is different, but otherwise they are mathematically the same. In the forward direction of a Schottky diode of n-type material, the bands are bent downward (from their equilibrium band bending) and a large current flows. Thus, when η_{sc} is negative in Equation (3.33), a large current flows. In the reverse direction of the diode the bands are bent up [η_{sc} positive in Equation (3.33)] and J_0 is the "reverse current" of the diode, hopefully very low for an efficiently rectifying diode.

In principle, then, if we calculate J_0, we have determined the electrode behavior of the semiconductor electrode.

Below we present analyses of J_0 based on the fluctuating energy level model, where J_0 is determined theoretically by evaluating the probability that the energy levels fluctuate to the appropriate band edge. In practice, however, the behavior of a semiconductor electrode is not so simple.

The problem is that when E^o_{redox} is in the bandgap region, far from the band edge, J_0 is extremely small. It can be negligible [see Equation (3.67) below and the accompanying discussion]. In many cases for semiconductor electrodes, particularly those where only one band is active (wide bandgap semiconductors) stray currents due to flaws in the surface (Section 5.2.3), tunneling, corrosion, or even background illumination[16] are much higher than J_0 as derived from the fluctuating energy level model. With a narrow bandgap semiconductor, thermal generation of minority carriers will exceed the current due to flaws and equilibrium will be reached. But, with a wide bandgap semiconductor with E^o_{redox} in the bandgap region, the concept of the overvoltage η_{sc} is of little value because the correct voltage at equilibrium for a particular redox couple is not measurable. The open circuit voltage is not an equilibrium voltage, but the voltage where reverse current, due to flaws, etc., is equal to the forward current due to majority carriers. For example, with an n-type semiconductor at open circuit, the anodic current will be due to flaws, and V_s at open circuit will adjust itself so the cathodic current of electrons to the surface balances this stray anodic current. Thus, η_{sc} is not in general a measurable parameter.

Normally therefore, in measurements on semiconductor electrodes, the current is presented not as a function of overvoltage η_{sc}, but as a function of the surface barrier V_s, for V_s is the only appropriate measurable parameter. A more useful formulation of the current/voltage characteristics for electron exchange with an n-type semiconductor is

$$J = J_1 \exp(-qV_s/kT) - J_a \qquad (3.34)$$

with J_1 the coefficient for the cathodic current, J_a the anodic current. It will often be found, when examining experimental results for exchange of electrons between a nondegenerate semiconductor and a redox couple, that if J_a is measurable, and arises due to simple electron injection from the reducing agent, then cathodic current is negligible, only observed with a strong accumulation layer (the material is degenerate and quasimetallic). Again, if $J_1 \exp(-qV_s/kT)$ is measurable with the surface nondegenerate, then J_a does not depend on the presence of the reducing agent. Thus, considering the currents as two independent terms, as is done in Equation (3.34), is actually desirable.

In the following discussion, various expressions for J_a and J_1 and for J_0, based on the fluctuating energy level model, are derived. All the variations will be found to predict an exponential factor involving an activation energy $(E_1 - E_2)^2/4\lambda$, where $E_1 - E_2$ is the energy difference between the conduction or valence band and E_{ox} or E_{red}, depending on the particular charge transfer under consideration. However, the various expression for J_0, J_a, or J_1 have slight differences in the preexponential factor depending on the mathematical approximation used.

Expressions for hole exchange are not directly discussed. First, the theoretical expressions are in the referenced literature; second, they are analogous to the expressions for electron exchange; and third, as mentioned later, experimental results do not seem to follow the model in any known case.

3.3.3.3. The Gerischer Model. The Gerischer model[17,18,R76] is a convenient generalization. Thus, it is of value to review it briefly here, using Gerischer's own notation. This introduces a third way of collecting the factors that is equivalent again to Equation (3.27), and to a combination of Equations (3.3) and (3.4). However, with Gerischer's approach, we integrate the contribution to the current flow over energy, and only later in the calculations do we stipulate that the only energy of interest is at the conduction band (or valence band) edge. In the formulations above this stipulation was made to begin with, and any need for integration over energy was not recognized. We will discuss only conduction band processes, exchanges of electrons between ions in solution and the conduction band, because the valence band processes are in principle analogous. Figure 3.2 is included to help identify the symbols and the model under study.

We begin with the general formulation for the current J:

$$J = q \int_{-\infty}^{\infty} \nu^-(E) n(E) D_{ox}(E) dE - q \int_{-\infty}^{\infty} \nu^+(E) N_u(E) D_{red}(E) dE \qquad (3.35)$$

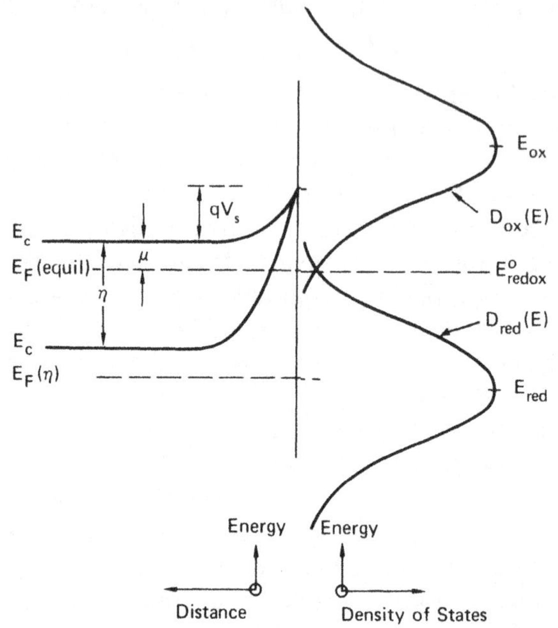

Figure 3.2. Energy level model for a simple semiconductor/redox system at equilibrium and with an overvoltage η applied. At equilibrium the Fermi energy in the solid is at an energy equal to E^o_{redox}. When the overvoltage η is applied, the bands in the bulk semiconductor shift, and the semiconductor Fermi energy shifts relative to the surface levels.

where

$$D_{ox}(E) = c_{ox}W_{ox}(E), \qquad D_{red}(E) = c_{red}W_{red}(E) \tag{3.36}$$

The left term of Equation (3.35) represents transfer of electrons from the occupied levels to the oxidizing agent in solution, the right term represents transfer of electrons from the reducing agents to the unoccupied levels in the solid. Thus, $n(E)$ is the density of occupied levels per unit energy, $W_{ox}(E)$ the probability function [Equation (1.54)], with $E_t = E_{ox}$, and $W_{red}(E)$ is the same with $E_t = E_{red}$. N_u is the density of unoccupied levels in the conduction band. The ν's, in units of m^4 eV sec^{-1}, represent frequency factors for the electron transfer plus other factors and include the transmission coefficient. The use of Equation (3.35) does not depend on the attainment of thermodynamic equilibrium.

Now, if one defines D_{redox}:

$$D_{redox}(E) = D_{ox}(E) + D_{red}(E) \tag{3.37}$$

then from Equation (1.54), (3.36), and (1.64) (where $c_{red} = [A^{(n+1)-}]$ and $c_{ox} = [A^{n-}]$) one obtains, with some manipulation, the following very convenient forms:

$$D_{ox} = D_{redox}\{1 - f(E_{redox}, E)\} \tag{3.38}$$

$$D_{red} = D_{redox} f(E_{redox}, E) \tag{3.39}$$

where $f(E_{redox}, E)$ is given by

$$f = 1/[1 + \exp(E - E_{redox})/kT] \tag{3.40}$$

This is essentially the Fermi function, Equation (1.1), providing that we have true equilibrium and Equation (1.63) holds, i.e., $E_F = E_{redox}$.

Let us assume that equilibrium obtains. Then, with the use of the Fermi distribution function to describe the electron density $n(E)$ and the density of unoccupied states $N_u(E)$, we have

$$n(E) = N(E)f(E) \tag{3.41}$$

$$N_u(E) = N(E)\,[1 - f(E)] \tag{3.42}$$

where $N(E)$ is the total density of levels per unit energy. Now when $E_{redox} = E_F$, we can insert (3.38), (3.39), (3.41), and (3.42) into Equation (3.35), and obtain

$$J = q \int_{-\infty}^{\infty} N(E) D_{redox}(E) f(E)\,[1 - f(E)]\,[\nu^{-}(E) - \nu^{+}(E)]dE \tag{3.43}$$

for the equilibrium case. By the principle of detailed balance at equilibrium J must be zero for every value of E, so

$$\nu^{-}(E) = \nu^{+}(E) = \nu(E) \tag{3.44}$$

Each of the terms in Equation (3.43) thus represents the exchange current flowing in and out of the surface at equilibrium. In an appendix, Gerischer[(18)] shows that $\nu(E)$ is proportional to the transmission coefficient $\kappa(E)$, but he does not develop a simplified approximation, so that $\nu(E)$ cannot easily be evaluated. Thus, when a quantitative estimate of the preexponential term is desired, most experimenters have used variations of the Marcus/Hale expression, Equation (3.18), described in more detail below. However, workers interested in the variation of J with E_{red} or E_{ox} have found the Gerischer formulation useful.

As we are restricting our attention to the conduction band processes in a semiconductor for illustration, it is clear that the integration of Equation (3.36) is only of interest from E_{cs} to infinity, where E_{cs} is the energy of the conduction band edge at the surface. At $E < E_{cs}$, both $n(E)$ and $N_u(E)$ go to zero. Then the exchange current, the current in each direction at equilibrium, is given, using Equation (3.44) in either the first or second term of Equation (3.43):

$$J_0 = q \int_{E_{cs}}^{\infty} \nu(E) N(E) D_{\text{redox}}(E) f(E)\,[1 - f(E)]\,dE \tag{3.45}$$

The Fermi function f in Equation (3.45) applies both to the solid and the solution because $E_F = E_{\text{redox}}$ by this analysis at equilibrium. As discussed in Section 3.2 for a nondegenerate semiconductor, all the energy levels at the surface remain in the same position relative to each other, independent of the overvoltage η_{sc}. The overvoltage η_{sc} only shifts the bands in the bulk of the semiconductor relative to the bands at the surface (see Figure 3.2), and does not affect the integration of Equation (3.45). In other words, the exponential factor of Equation (3.33) is independent of E and is not under the integral sign of Equation (3.45). Equation (3.45) provides an attractively compact form to predict J_0 when equilibrium actually obtains at a semiconductor electrode.

Alternately, more descriptive forms of Equation (3.45) can be obtained by replacing $N(E)$ and $D_{\text{redox}}(E)$ by the appropriate combinations: for the reduction process Equations (3.38) and (3.41) are used, and for the oxidation process Equations (3.39) and (3.42) are used to provide more descriptive expressions:

$$J_0 = q \int_{E_{cs}}^{\infty} \nu(E) N(E)/(4\pi\lambda kT)^{-1/2} \times \{\exp[-(E - E_F)/kT]\} c_{\text{ox}} \exp[-(E - E_{\text{ox}})^2/4\lambda kT]\,dE \tag{3.46}$$

or

$$J_0 = q \int_{E_{cs}}^{\infty} \nu(E) N(E) (4\pi\lambda kT)^{-1/2} c_{\text{red}} \exp[-(E - E_{\text{red}})^2/4\lambda kT]\,dE \tag{3.47}$$

Equation (3.46) is of particular interest when describing cathodic reduction processes, and Equation (3.47) when describing electron injection (anodic oxidation processes). To obtain Equation (3.46) and (3.47) in the simplified form shown, it was assumed that $E_c > E_{\text{redox}}$, which is equivalent to assuming the surface is nondegenerate at equilibrium. With this assumption, we can replace the Fermi functions f in Equation (3.39) by the Boltzmann function, and consider that $1-f$ is unity.

Earlier in this section the experimental observation[16,R10] was discussed that shows that for most wide bandgap semiconductors, equilibrium is not reached at open circuit between electron transfer to and from a redox couple, so usually $E_F \neq E_{\text{redox}}$. It is of interest to develop expressions for J_a and J_1 of Equation (3.34) using Gerisher's formulation, but without the requirement that equilibrium be reached at open circuit. In the development of Equation (3.47) from the second term of Equation (3.35), none of the steps actually depended on equilibrium being defined except the determination of $\nu^+(E)$. Thus, with $\nu(E) = \nu^+(E)$, Equation (3.47) defines J_a, the current associated with electron injection into the conduction band. Thus, if we ignore current due to flaws, the anodic current J_a is the calculated value of the exchange current:

$$J_a = J_0 \tag{3.48}$$

However, it will be observed experimentally (Section 5.1) that the theoretical value of J_0 so estimated is extremely small for wide bandgap semiconductors when $E^o_{\text{redox}} < E_{\text{cs}}$, so that the experimental anodic current J_a is always much higher than J_0, because of current through flaws or photocurrents due to stray illumination.

The cathodic current (for n-type material) can also be calculated independently of whether equilibrium is reached, if we make the same approximations as is done for Equation (3.46). We can return to the original expression (3.35) and calculate for the cathodic current density

$$J_c = q \int_{E_{\text{cs}}}^{\infty} \nu(E)N(E)c_{\text{ox}}(4\pi\lambda kT)^{-1/2} \times \{\exp[-(E - E_F)/kT]\}\{\exp[-(E - E_{\text{ox}})^2/4\lambda kT]\}dE \tag{3.49}$$

where, to avoid the equilibrium requirement, we express the electron density in the solid in terms of the Fermi energy in the solid rather than the Fermi energy in solution. Noting from Figure 3.2 that

$$E - E_F = E - E_{\text{cs}} + \mu + qV_s \tag{3.50}$$

where μ is the energy separation of the Fermi energy in the bulk from the conduction band edge E_c, we have

$$J_c = q \int_{E_{\text{cs}}}^{\infty} \nu(E)N(E)c_{\text{ox}}(4\pi\lambda kT)^{1/2}\exp(E - E_{\text{cs}})/kT \times \exp[-(\mu + qV_s)/kT]\exp[-(E - E_{\text{ox}})^2/4\lambda kT] \tag{3.51}$$

From Equation (3.51) in Equation (3.34) we thus obtain an expression for J_1, and comparing it to the expression (3.46) it is clear that

$$J_1 = J_0 \exp[-(E_{\text{cs}} - \mu) - E_{\text{redox}}]/kT \tag{3.52}$$

This result is to be expected; if $E_F = E_{\text{redox}}$, Equation (3.52) simply states that $J_1 = J_0$. It was indicated in Chapter 2 that E_{redox} and μ are known for any well-studied system, and it will be seen that E_{cs} can be determined either theoretically (Section 2.3) or experimentally (Chapter 5). Thus, the relation between J_1 and the theoretical parameter J_0 can be determined experimentally for any well-studied system.

The theoretical estimation of J_0 (and J_1) is more difficult.

3.3.3.4. Evaluation of the Preexponential Factors J_0, J_1, and J_a. Gerischer[18] evaluates J_0 directly from Equation (3.45), with the approximation that D_{redox} is a constant. Comparing Figure 3.2 and Equation (3.37), it is clear that this assumption cannot be justified in general.

However, it can be justified if E_{redox} is such that D_{redox} is slowly varying in the energy region near E_{cs} (see Figure 3.2). For example, it is justified when

$$2\lambda + E_{F,redox} \gg E_{cs} \gg E_{F,redox} \tag{3.53}$$

Then in Equation (3.45), the exponent associated with $f(E)$, $(E - E_{F,redox})$, will vary much more rapidly than the exponent associated with D_{redox}, $[(E - E_{ox})^2/4\lambda kT]$, in the region of integration (viz., for $E > E_{cs}$). Essentially, we are saying that in this range the dominating consideration is the rapid decrease in electron density for $E > E_{cs}$. In fact, it is reasonable to approximate that the integration only need be taken from E_c to about $E_c + 2kT$, for $\exp[(E - E_{cs})/kT]$ varies so strongly with E that beyond $E_c + 2kT$ the contribution to the integrand will be negligible. And, in this limited energy range, the value of $E - E_{ox}$ will vary negligibly, so the factor $\exp[-(E - E_{ox})^2/4\lambda kT]$ can be replaced by $\exp[-(E_{cs} - E_{ox})^2/4\lambda kT]$. Performing the integration over the narrow energy range of interest, and using this substitution, we obtain the approximation due to Gerischer:

$$J_0 \approx \tfrac{1}{2} q\nu(E_{cs}) n_{so} c_{ox} (kT/\pi\lambda)^{1/2} \exp[-(E_{cs} - E_{ox})^2/4\lambda kT] \tag{3.54}$$

where $\nu(E_{cs})$ is the value of ν at the conduction band energy, and where n_{so} is the value of n_s at E_{cs}, is obtained using Equation (1.2). As mentioned in Section 3.3.3.2, a numerical value for $\nu(E)$ is not easily obtained from Gerischer's analysis, so the preexponential in Equation (3.54) is not well defined.

Another analysis of electron exchange derives from the Marcus–Hale model (Section 3.3.2) as developed by Marcus.[10] The model was developed to describe metal electrodes; but semiconductor application seems appropriate. The method describes charge injection from the solution only, so for our n-type example, J_0 is determined by analyzing electron injection into the available empty levels (the conduction band). The Marcus–Hale formulation for electron injection into a band derives from Equation (3.18). In Equation (3.18) it was assumed that the transmission coefficient is equal to unity for ions moving to the surface. The activation energy is given simply by the energy necessary to shift the energy level of the electron on the reducing agent, E_{red}, to the energy level of the conduction band edge, E_c. Then

$$J_a = qZc_{\text{red}} \exp[-(E_c - E_{\text{red}})^2/4\lambda kT] \tag{3.55}$$

This expression can be compared to Gerischer's expression, Equation (3.54), and is clearly much simpler to apply.

Memming and Möllers[15] used an expression for the current due to electron injection into a semiconductor that is based in part on Equation (3.47) but derives more closely from the Marcus–Hale expression. Essentially the formula was derived from Equation (3.55), but is based on integrating over energies where transitions are allowed, as is done in Equation (3.47). The injection rate is given by Equation (3.55), multiplied by the probability the energy level is high enough to inject and by a transmission coefficient. The expression used was

$$J_a = qZc_{\text{red}}(\pi kT\lambda)^{-1/2} \times \int_{-\infty}^{\infty} \kappa'(E)N(E)(1 - f) \exp[-(E - E_{\text{red}})^2/4\lambda kT]dE \tag{3.56}$$

where, of course, $N(E)$ $(1 - f)$ is the density of unoccupied states to which transfer can occur and the product $\kappa'(E)N(E)$ $(1 - f)$ is the transmission probability for electrons at the energy E. It was suggested by Memming and Möllers that a quantitative treatment of Gerischer's theory [Equation (3.45)] would lead to exactly the same expression. The integration of Equation (3.56) was not performed in the cited reference because Memming and Möllers were interested in tunneling processes which added another complication, a dependence of $N(E)$ on V_s, and this complication led to the necessity of integrating with the help of a computer (see Section 6.5.2).

Vanden Berghe *et al.*[19] integrated Equation (3.46) directly. They assumed $\nu(E)$ and $N(E)$ are slowly varying with E compared to the exponential terms, and so can be replaced by average values $\overline{\nu(E)}$ and $\overline{D(E)}$ over the regions of E where the integrand is large. Then, as will be described, the integrand can be put into the form of a Gaussian in $(E_{cs} - E_{\text{red}})/2(\lambda kT)^{1/2}$ and the integral becomes an inverse error function. To evaluate the inverse error function they use the expression

$$\text{erfc}(y) = 0.56y^{-1} \exp(-y^2) \tag{3.57}$$

valid for small y. It turns out this limits the application of the theory to $E_{cs} > E_{\text{redox}}$. They begin the analysis by defining an "equilibrium" density of electrons at the surface:

$$n_{s0} = N_c \exp[-(E_{cs} - E_F)/kT] = n_b \exp(-qV_s/kT) \tag{3.58}$$

where n_b is the density of electrons in the bulk. The first equality in Equation (3.58) arises from Equation (1.2) and the second from Equation (2.10). Substitution of Equation (3.58) into Equation (3.46) yields a form resembling Equation (3.34), and we obtain for J_1

$$J_1 = q\frac{\nu(E)N(E)n_{s0}c_{ox}}{(4\pi\lambda kT)^{1/2}N_c}\int_{E_{cs}}^{\infty}\{\exp[-(E - E_{cs})/kT]\} \times \{\exp[-(E - E_{ox})^2/4\lambda kT]\}dE \quad (3.59)$$

Then the integrand of Equation (3.59) can be manipulated into the form

$$\{\exp[-(E_{cs} - E_{redox})/kT]\}\cdot\{\exp[-(E - E_{red})^2/4\lambda kT]\}$$

As the first factor does not contain E, the integral has the form of the inverse error function and if $E_{cs} > E_{redox}$ the integral can be evaluated by the help of Equation (3.57) to yield

$$J_1 = qc_{ox}n_{s0}\frac{0.56\,\overline{\nu(E)}\,\overline{N(E)}\,(4\lambda kT)^{1/2}}{4N_c(E_{cs} - E_{red})}\exp[-(E_{cs} - E_{ox})^2/4\lambda kT] \quad (3.60)$$

When the approximation enabling the use of Equation (3.57) is valid, the Vanden Berghe approach provides a reasonably accurate way of estimating the error function integral. However, the parameters $\overline{\nu(E)}$ and $\overline{N(E)}$ must still be evaluated to obtain a numerical value for the preexponential, so the preexponential factor is not known from Equation (3.60) much better than from a combination of (3.52) and (3.54).

Another approximation can be made,[R4] based on integration between E_{cs} and $E_{cs} + 2kT$, that may be of particular aid to solid state physicists who are more familiar with considering electrons captured from a band by states of a characteristic cross section. With Equation (3.28) inserted in (3.27), and approximating that all ions within a distance d of the surface are active in capturing electrons, and have the same capture cross section [Equation (3.17)] we immediately obtain

$$J_1 = \frac{q\bar{c}\sigma n_{s0}}{(4\pi\lambda kT)^{1/2}}c_{ox}d\int_{E_c}^{E_c+2kT}\exp[-(E - E_{ox})^2/4kT]dE \quad (3.61)$$

$$= q\bar{c}\sigma n_{s0}dc_{ox}(kT/\pi\lambda)^{1/2}\exp[-(E_{cs} - E_{ox})^2/4\lambda kT] \quad (3.62)$$

where n_{s0} is the density of electrons at the surface with $V_s = 0$. From Hale's analysis (Section 3.3.2) a preferred value for d of about 15 Å may be indicated. Again, of course, there is an unknown parameter, here the cross section σ, in the preexponential. But typically, capture cross sections are of atomic dimensions ($\sim 10^{-19}$ m^2), although they vary at times many orders of magnitude from this value.

In summary, various authors have suggested possible ways to obtain a quantitative expression for the current density across the semiconductor/electrolyte interface, applying the fluctuating energy level model in its simplest form. The Marcus–Hale model, Equation (3.55), is the only model that has no unknown parameters in the preexponential, and the direct application of this to semiconductors is questionable because of the assumption that the transmission coefficient is unity. With the limited density of states in a semiconductor near the band edge, the transmission coefficient may be much less than unity for $E \sim E_{cs}$.

To end this section, it is instructive to estimate the cathodic and anodic current as predicted by the fluctuating energy level model assuming various conditions. Most of the formulas given above have the unknown parameter ν present, and so cannot be evaluated. Two forms are absolute although approximate, namely, Equation (3.55) for the anodic current, and Equation (3.62) for the cathodic. Consider the special case where $E_{cs} = E^o_{redox}$.

In Equation (3.55) we estimate $Z = 10^2$ m/sec, and consider $c_{red} = 10^{27}$ ions/m^3 (approximately 0.02 M), and assume λ is 0.7 eV. If $E^o_{redox} = E_{cs}$, so the exponent becomes $\lambda/4kT$, the current is given approximately by

$$J_a = 10^{10} \exp(-\lambda/4kT) = 10^7 \text{A/m}^2 \tag{3.63}$$

In Equation (3.62), if we assume $\sigma = 10^{-19}$ m^2, $n_{s0} = n_b = 10^{23}$/m^3, $d = 15$ Å, $c_{ox} = 10^{27}$/m^3 and again assume $E^o_{redox} = E_{cs}$, we obtain

$$J_c = 10^4 \text{ A/m}^2 \tag{3.64}$$

These represent very high currents flowing to and from a redox couple with $E_{cs} = E^o_{redox}$. In the experimental discussion of Chapter 6, it will be noted that for n-type ZnO the experimental value of J_c extrapolated to the flat band case for an oxidizing agent with $E_{ox} = E_{cs}$ (where the current is maximum), is only about 10^4 A/m^2. If this value is lowered by $\exp[-(E_{ox} - E_{redox})^2/4\lambda kT] = 10^{-3}$, to compare more directly with J_c in Equation (3.64), we note the theoretical value exceeds that observed by several orders of magnitude. In the case of reducing agents, such a high anodic current J_a as predicted by Equation (3.63) has not been observed. Diffusion limitations limit the current long before the indicated current density is obtained. But, as will be discussed in the experimental chapter, for reducing agents with E^o_{redox} approaching E_{cs} or above it, current densities of 10 A/m^2 have been observed—low compared to the above estimate, but nontheless high current densities.

Throughout the book, in particular in Section 6.2.2, we will be discussing whether or not cathodic current ever flows where surface states are involved in the process. If surface states are present, electrons could first be captured by the surface states and then move from the surface state to the ion in solution. And, it is possible that the capture by the surface states could be rate limiting in this two-step process, if the energy level of the surface states, E_{ss}, is within λ of E_{ox}. Although such a model has not been discussed in detail, all the required formulas have been presented above, and it is, therefore, instructive to compare reasonable values for the current flow via surface states and the current flow directly to ions in solution. Assume again an 0.02 *M* solution, so $c_{ox} = 10^{27}$ ions/m^3, with d (the tunneling distance) 15 Å, and a capture cross section 10^{-19} m^2 for the ions. Then, Equation (3.62) yields for current flowing by direct ion capture

$$J_c(\text{ions}) = 3 \times 10^{-16} n_{s0} \exp[-(E_{cs} - E_{ox})^2/4\lambda kT] \qquad (3.65)$$

Assume there are 10^{16} surface states/m^2, and a capture cross section of 10^{-19} m^2 for the surface states. Then, from Equations (3.27) and (3.28) we have for the current to surface states

$$J_c(\text{ss}) = 2 \times 10^{-17} n'_{s0} \qquad (3.66)$$

where $n'_{so} = n_{so}$ if the surface states are below E_{cs} in energy, and where $n'_{s0} = n_{s0} \exp[-(E_{ss} - E_{cs})/kT]$ if the surface states are above E_{cs} in energy. As mentioned above, Equation (3.66) will be considered rate limiting for current flow via surface states. If the surface state is above E_{cs} in energy, the current flow via surface states is easily shown to be much less than the current flow to ions. If, however, the surface states (and E_{ox}) are both well below E_{cs}, where $n'_{s0} = n_{s0}$, surface states can dominate. To a first approximation J_c(ss) becomes dominant over J_c(ions) where E_{ox} drops below E_{cs} enough that the exponential factor, Equation (3.65), becomes much less than 10^{-2}, say. Of course, E_{ox} is above E^o_{redox} by λ (0.7 eV or so), so E^o_{redox} should be below E_{cs} by about 1 eV before one even considers surface state effects as an important possibility. But, for example, if $E_{cs} - E_{ox} \sim 1.0$ eV, say, then $J_c(\text{ions}) \sim 2 \times 10^{-22}$ n_{s0}/m^2, compared to the 2×10^{-17} n_{s0}/m^2 for transfer via surface states. Thus in this case the surface state contribution will easily dominate. In summary, it will be noted that (a) E^o_{redox} is assumed below E_{cs} by 1.7 eV or so for this case, (b) we have assumed a high-density (10^{16}/m^2) surface state (it is difficult to imagine a higher density when there is no surface state effect on V_H, as will be discussed in Section 5.1.1.3), and (c) we assume electron exchange between the surface state and the oxidizing agent is rapid. Thus, for surface states to dominate electron flow to the surface one must have rather special conditions.

Another case that is of interest to numerically evaluate, because it will be found there are difficulties with the theory, is a rather simple case: the calculation of J_0, the theoretical exchange current, when E^o_{redox} is well below E_{cs}. We will again use n-type material as the example. Consider $E_{cs} - E_{\text{red}} \sim 1.5$ eV, $\lambda = 0.7$ eV, and again consider a 0.02 M solution of the reducing agent. Then from Equation (3.55) with $Z = 10^2\ \text{m}^{-1}\ \text{sec}^{-1}$, we have

$$J_a(= J_0) = 2 \times 10^{-4}\ \text{A/m}^2 \tag{3.67}$$

Comparing this value to that of Equation (3.63), we note the dominating effect of $(E_c - E_{\text{red}})^2$ in the exponent that rapidly lowers J_a as $E_c - E_{\text{red}}$ increases. Such a current, 2×10^{-4} A/m^2, is a current well below the level where most semiconductors will show spurious current due to flaws. Thus, by the theory, if E_{red} is 1.5 eV below E_{cs}, or equivalently, if E^o_{redox} is 0.8 eV below E_{cs}, J_0 is probably already too low to be measured. In practice (see Section 6.1) J_0 often seems to become too low to be measured when E^o_{redox} is below E_{cs} at all.

3.4. Qualitative Description of Electrode Processes Using the Band Model

In most of the discussions to follow, we will be using the energy level model of electrochemistry not in a quantitative way, as was developed in the last Section 3.3, but in a qualitative way, to help explain the various phenomena observed with such electrodes. Thus, it is of value to sketch briefly how the band model varies qualitatively, and how the current varies qualitatively, as one uses different solids, applies various voltages, and uses redox couples of various E^o_{redox}.

3.4.1. The Behavior of the Metal Electrode

We begin by describing the band diagrams that could be used to describe a metal electrode, shown in Figure 3.3. As illustrated in Figure 3.3a, in the open circuit condition (in theory no current flows through a reference electrode or a voltmeter) the exchange of electrons between the redox couple and the metal is at equilibrium. Thus, the metal assumes a Fermi energy level equal to the E_{redox} [cf. Equations (1.45) and (1.68)] of the redox couple. As described in Section 2.2.3, the Helmholtz potential adjusts to make $E_F = E_{\text{redox}}$ at equilibrium. In the diagram we indicate the redox couple is present under standard conditions and then, of course $E_F = E_{\text{redox}} = E^o_{\text{redox}} = -qE^o$, where E^o is the standard redox potential of the couple. We indicate for simplicity that the reference electrode is hydrogen, so qE^o appears in the energy level diagram as indicated.

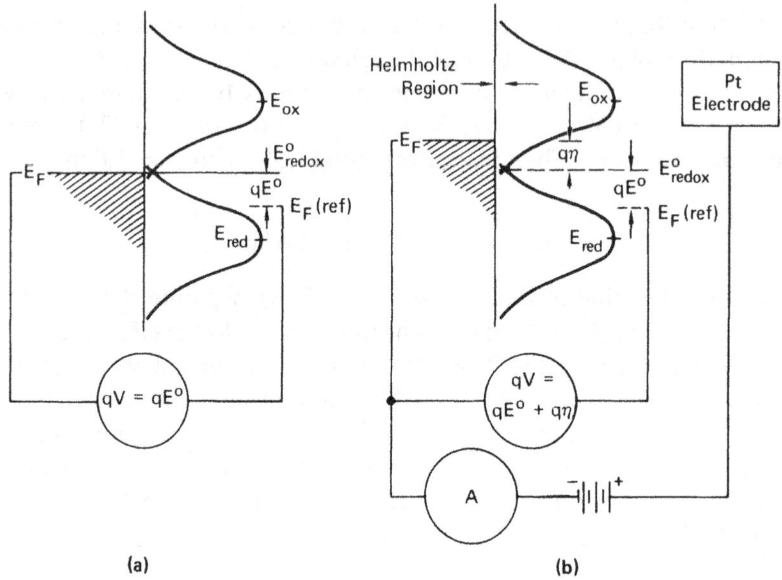

Figure 3.3. Schematic diagram combining an energy band diagram for a metal electrode and its contacting electrolyte with an external electrical circuit. In (a) the external circuit is open (the voltmeter has infinite resistance so no current flows through the reference electrode); and in (b) a voltage is applied and current is passed between the sample and a platinum counterelectrode. As always with the band model, the solid is represented by a plot of energy versus distance, the solution by a plot of energy versus density of levels at that energy.

In order that no net current flow through the surface (equilibrium), the current (the exchange current) is the same in each direction. Electrons move to the metal at a rate proportional to the overlap between the density of states of the reducing agent, and the density of empty states in the metal (those above E_F). Electrons move to the solution at a rate proportional to the overlap between the density of states of the oxidizing agent and the density of occupied states in the metal (those below E_F). We note the Helmholtz double layer region offers no obstruction to current flow—it is so thin (~3 Å) that electrons tunnel through freely.

The applicable band diagram becomes Figure 3.3b when an applied voltage different from E^o is introduced, leading to an overvoltage of η. The voltage η, negative in the figure, appears across the Helmholtz double layer. With the negative overvoltage as shown in Figure 3.3b, there is much more overlap between the filled levels in the metal and the levels of the oxidizing agent than in Figure 3.3a, so the cathodic current from the metal to the solution increases. There is much less overlap between the unoccupied levels in the metal and the levels of the reducing agent, so the anodic current decreases. Thus, there is a high net cathodic current.

Here, as will also be the case with a semiconductor electrode below, a counterelectrode, indicated as a platinum foil, is used to pass current through the solution. But, when a current passes, the surface of the counterelectrode may change (become oxidized, etc.) so that the voltage between the counterelectrode and the solution is not necessarily constant. Thus the potential of the working electrode, relative to the solution, cannot be monitored using the counterelectrode. The solution potential must be monitored, therefore, by a reference electrode. This is why in Figure 3.3 the battery is connected to the counterelectrode, but the voltmeter to a reference electrode.

If the redox couple in solution is two-equivalent, so an energy level diagram for the redox couple resembles that of Figure 1.8, then again at equilibrium the Fermi energy of the metal equals the energy E_{redox}, as indicated, for example, in Equation (1.78). Consider Figure 1.8b, viewing the solid as a metal instead of a semiconductor, so that E_F is simply the dividing energy between occupied and unoccupied states. We observe that the carrier transfer rates (the currents) will be much different from those expected for a one-equivalent redox couple. The cathodic current is determined by the rate at which electrons can move from the Fermi energy to $E_{\text{ox}}^{4,5}$, the anodic current is determined by the rate at which electrons can move from the level $E_{\text{red}}^{3,4}$ to the Fermi energy. Now the separation of these two levels can be much greater than the separation (2λ) of the corresponding levels of the one-equivalent species. To get a current in the same current range as that expected for one-equivalent species (say 10 A/m^2), the Helmholtz voltage must be shifted by a much greater amount for the two-equivalent process. The level $E_{\text{ox}}^{4,5}$ (for reduction) or $E_{\text{red}}^{3,4}$ (for oxidation) must be shifted to an energy within λ of E_F. Only then will we have an energy level configuration similar to Figure 3.3b, where the configuration necessary for a cathodic current is shown. With two-equivalent species then (barring catalytic effects that independently shift the levels) a high "polarization" can be expected before substantial currents can be drawn.

It is important to note, however, that once the "polarization overvoltage" is applied to the metal electrode, current will flow. We will observe in Section 3.4.2 that this is not true for a nondegenerate semiconductor. In principle, no amount of overvoltage will induce a large current to flow for the semiconductor shown in Figure 1.8b, unless the voltage causes the surface to become degenerate. This makes for a significant difference in the behavior of semiconductors and oxide layers from the behavior of ideal metal electrodes.

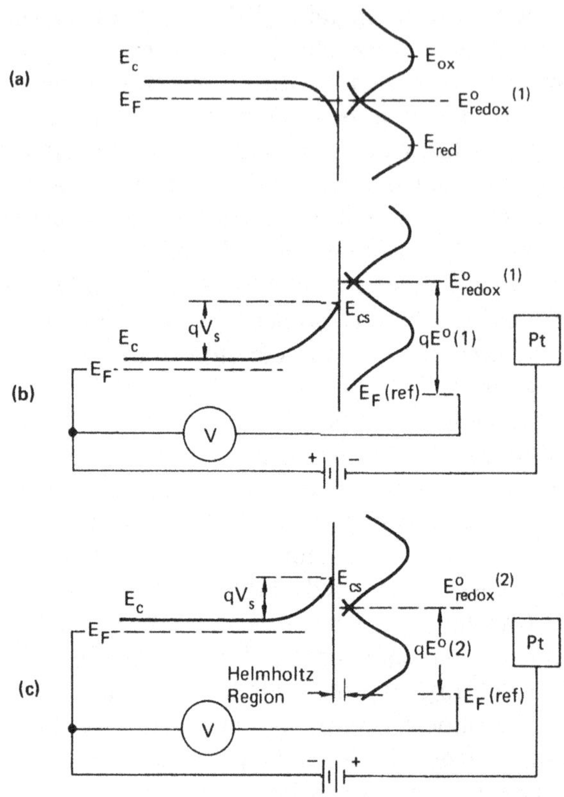

Figure 3.4. Schematic models for an *n*-type semiconductor electrode and redox couples analogous to Figure 3.3. In (a) and (b) redox couple No. 1, with $E^o_{redox} > E_{cs}$, is shown. An equilibrium diagram is shown in (a). In (b) a potential is applied to a Pt counterelectrode so electrons will be injected into the semiconductor. The electrode potential is measured relative to the Fermi energy of the reference electrode as shown. In (c) redox couple No. 2 with $E^o_{redox} < E_{cs}$ is shown, with a voltage applied so electrons can be extracted from the semiconductor. We note $E_{cs} - E_F$ (ref) is constant independant of E^o_{redox} and V_s.

3.4.2. The Behavior of the Semiconductor Electrode

The band model for a semiconductor differs from the model for a metal, partly because of the energy gap, but primarily because of the space charge double layer in the semiconductor.

Band models for an *n*-type semiconductor for different redox couples and electrode potential are sketched in Figure 3.4. As we have determined, with a nondegenerate semiconductor the Helmholtz potential does not change with applied voltage. Thus the difference between E^o_{redox} and E_{cs} is not a function of voltage. In Figure 3.4a and 3.4b are shown a case where the energy E^o_{redox} is well above E_{cs}, so the energy levels of the reducing agent, centered at E_{red}, overlap the empty conduction band levels. When the solid is immersed in the solution with no counterelectrode, electrons will be injected into the conduction band, and the surface will acquire a degenerate accumulation layer (Figure 3.4a) when at equilibrium. In Figure 3.4b we show the same case where a voltage has been applied to remove the accumulation layer. Because the Gaussian centered at E_{red} overlaps the conduction band, anodic current will flow. The anodic current should be independent of the applied voltage, because any applied voltage will appear across V_s and will not change the relative energies of E_{red} and E_{cs}. And, as the figure is drawn, negligible cathodic current will flow unless a strongly degenerate surface is induced, making the material quasimetallic so the applied voltage appears across V_H.

In Figure 3.4c is shown the more common case where the energy E^o_{redox} is in the bandgap region. Now a cathodic current may flow, electrons moving from the semiconductor conduction band to the overlapping unoccupied oxidizing agent levels. The voltage across the Helmholtz region is unchanging. Thus with $E^o_{redox} < E_{cs}$, electron injection will always be small and independent of voltage (see numerical example at the end of Section 3.3). As developed earlier, the cathodic current will be proportional to the density of levels (normally considered constant) of the oxidizing agent above the energy E_{cs}, and to the density n_s of electrons at the surface (at the energy E_{cs}). In this case the current depends exponentially on applied voltage, because as V_s decreases, n_s increases exponentially.

In principle, models analogous to those of Figure 3.4 apply to the exchange of holes with the valence band. If the levels of the reducing agent overlap the levels of the valence band, electron injection (or more accurately hole capture) will occur that is an exponential function of the applied voltage. If the levels of the oxidizing agent, the unoccupied levels, overlap the levels of the valence band, hole injection (valence electron capture) will occur independent of voltage. However, as will be discussed in Section 6.3, the simple models do not apply well for hole capture.

With a two-equivalent redox couple, it is possible that no appreciable current will flow in either direction, as long as the semiconductor remains nondegenerate, and as long as there is no overlap between the energy level E_{cs} and either E_{ox} of the oxidizing agent or E_{red} of the reducing agent of the couple. Figure 1.7b is an example of the normal situation where current cannot flow. The level $E^{4,5}_{ox}$ is far above the conduction band, so electrons will not be excited to this level. The level $E^{3,4}_{red}$ is far below the conduction band. As long as the semiconductor is nondegenerate, viz., as long as the applied voltage does not appear across V_H, the above condition will be maintained and no appreciable current will flow in either direction, independent of applied voltage. As the voltage is increased indefinitely, of course other factors become dominant, such as degeneracy or breakdown, and current will begin to flow. On the other hand, if the energy levels do overlap, viz., if $E_{ox}^{4,5}$ is within λ of the conduction band, or if $E_{red}^{3,4}$ is within λ of the conduction band, current may pass. With the overlap, the first step of the two-equivalent process is allowed by the fluctuating energy level model. The second step is not allowed in principle, but is highly exothermic, and, as discussed later in this section, exothermic processes will usually proceed even if forbidden by the fluctuating energy level model.

We can make a useful comparison of the behavior of two-equivalent species on metals, discussed in Section 3.4.1 above, and the behavior of the species on semiconductors. Specifically, we can relate the polarization overvoltage, that required to make current flow to or from a metal, to the band energy required to produce overlap of the levels and allow current flow to or from a semiconductor. As described in Section 3.4.1, when the reaction is two equivalent, a modest polarization overpotential is required with a metal to shift the energy levels in solution to a value within λ of E_F. With a semiconductor, to shift correspondingly the energy levels in solution relative to the band edges, we can only shift the bands—we must pick a more favorable semiconductor. For example, consider a reaction, say the oxidation of H_2O to oxygen, that requires a polarization overpotential η^o on a metal electrode. This means that for appreciable current flow the Fermi energy of the metal must be shifted η^o below the thermodynamic energy level of the reaction. Then, to make that same reaction kinetically active on a semiconductor electrode, where the valence band, for example, is the source of carriers, the valence band must be below the thermodynamic energy level of the reactant by an amount approximately equal to $q\eta^o$. The semiconductor must be chosen to give this low E_{vs}. This description assumes that any "catalytic" effects that affect the activation energy for the reaction are similar on the two electrodes, the semiconductor and the metal. Such a "rule" to describe the band edge necessary to induce a reaction at a semiconductor is of interest in electrocatalysis and solar energy conversion schemes, as discussed in Chapter 9. However, it should be noted that although the concept (that the overvoltage required on a metal is closely related to the band edge requirement on a semiconductor) is useful to help visualize the models, in practice catalytic effects, adsorption effects, and surface state effects will make the behavior of metals and semiconductors so different that any quantitative calculations will probably be fruitless.

We have been presenting in this section a qualitative description of the quantitative models developed earlier in this chapter. And it will be found when discussing the experimental observations, particularly in Chapter 6, that this fluctuating energy level model describes the observed qualitative behavior fairly well. However, it will be found that the simple model appears to break down in many cases and must be supplemented by refinements in theory.

One case where the model must be supplemented is with highly exothermic reactions. With hole capture, for example, the simple model apparently breaks down in every case studied to date when E_{red} is at a higher energy level than E_{vs}. This will be discussed in Sections 6.3.2 and 6.4.1. In the analogous case for electron capture, the model may break down when E_{ox} is substantially below E_{cs}, but there is insufficient evidence to establish a rule. Majority carrier capture is discussed in detail in Section 6.2, and the evidence is summarized at the end of that section. In both these cases, where the carrier capture is highly exothermic, the theoretical model of this chapter suggests that the rate of capture should be low. For example, in Figure 3.5 neither electron capture from the conduction band, nor hole capture from the valence band, should be possible with the redox couple indicated, for in neither case is there overlap between the levels of interest. However, contrary to this theoretical prediction, it will be seen that both processes will usually occur at a high rate. The higher the value of E_{red} in all cases studied, the higher the cross section for hole capture. Probably in many cases the lower the value of E_{ox}, the higher the cross section for electron capture.

Such behavior in the cases studied will be explained tentatively in four possible ways. One is to invoke intrinsic surface states; a second is to invoke excited states of the ion in solution. These two possibilities are illustrated in Figure 3.5. A third, discussed in Section 7.2, is to invoke various complexes of the ion in solution. The fourth, and perhaps most common, cause of carrier capture, discussed in Section 5.2.3.1 and illustrated in Figure 7.1, is to invoke deep bulk levels due to dislocations.

Specifically, consider hole capture involving intrinsic surface states. For compound semiconductors, the anion can take on a series of valence states. In particular with oxides, the surface oxygen ions by hole capture can take on the formal −1 valence state, corresponding to a peroxide species. On the band diagram the energy level on which the hole is captured is indicated as surface state. In Figure 3.5 we have thus indicated the possibility in the band model of surface states appearing at the surface and complicating the simple models.

3.4.3. The Transition between Semiconductor and Metallic Behavior

In the preceding Sections 3.4.1 and 3.4.2 we have described the case of a metal and the case of a nondegenerate semiconductor, and it is of interest to discuss the transition between the two cases, involving either

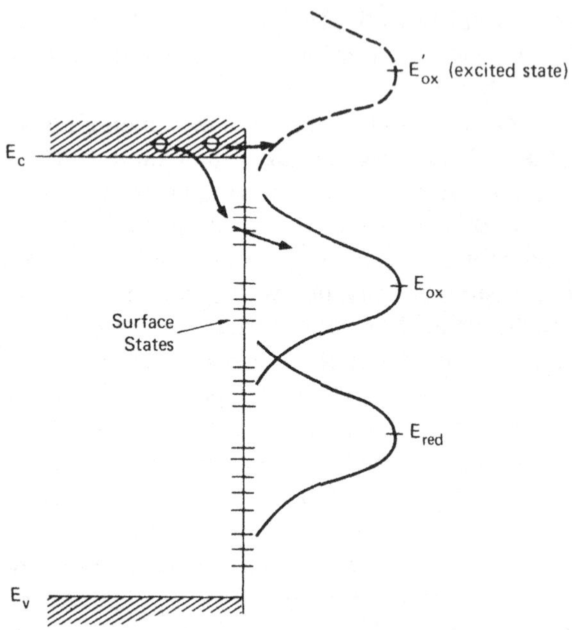

Figure 3.5. Mechanisms for electron and hole capture other than that described by the simple fluctuating energy level model. In this band picture electrons can be captured at E_{ox} by trickling down through surface states or by forming the excited species through capture at E'_{ox}. Holes can be captured by trickling up through surface states.

a thin oxide layer on a metal or involving a degenerate semiconductor. We can qualitatively summarize the characteristics that make a material act as a metal (with α close to $\frac{1}{2}$) or as a semiconductor (with α close to 0). The parameter α is the transfer coefficient, described earlier. A material acts as a metal if the applied voltage appears across a very thin double layer at the surface, such as the Helmholtz double layer in the case of a real metal, a layer thin enough that electrons can tunnel through it. On the solid side of this thin double layer, there must be a high density of states at the Fermi energy, enough states to accommodate changes in charge as given by Equation (2.16). (If such a density of states is not present, the applied voltage cannot be accommodated over the thin double layer.) Usually with a metallic electrode the thin double layer is the Helmholtz double layer. However, somewhat thicker layers than the Helmholtz double layer will also be penetrable by tunneling, and degenerate semiconductors or electrodes with thin oxides can also approach metallic behavior. Indeed, in Section 8.3 we will be discussing the behavior of such thin oxide films (less than 30 Å thick) through which carriers can tunnel to levels at the Fermi energy of the substrate metal. It will be shown that these show the same type of Tafel plot as clean metals, albeit with a low exchange current, because, in accordance with Equation (3.17), the tunneling probability is low with such a thick surface layer. In summary, a material will act as a metal if there is an adequate density of levels at the Fermi energy, and if the distance between these levels and energy levels in solution is small enough for tunneling.

A material will act as a semiconductor (or insulator) on the other hand, if the applied voltage appears across a broad ($\gg$ 50 Å) space charge region inside the solid. This characteristic will also require that (except at one unique applied voltage) the carriers that tunnel between surface ions and the solid will have an energy level far from the Fermi energy (e.g., at the energy of the conduction band edge).

There are two ways that a semiconductor can be altered so that it behaves somewhat like a metal, in the sense that applied voltages appear across the Helmholtz layer, and therefore the transfer coefficient approaches 1/2. One way is if the density of states near or at the surface is made very large and the Fermi energy becomes pinned by this high density of states. The other is if the Fermi energy gets into the band region

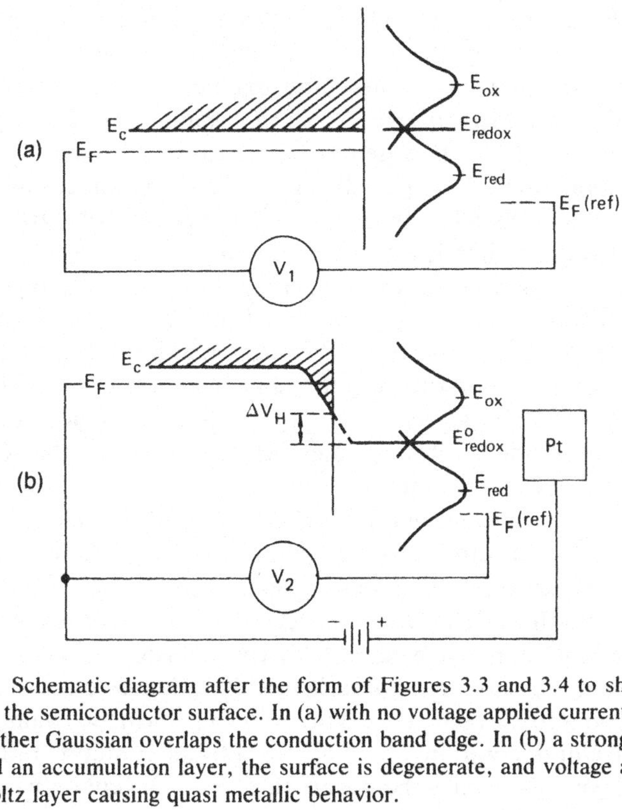

Figure 3.6. Schematic diagram after the form of Figures 3.3 and 3.4 to show forced degeneracy at the semiconductor surface. In (a) with no voltage applied current does not flow because neither Gaussian overlaps the conduction band edge. In (b) a strong cathodic bias has induced an accumulation layer, the surface is degenerate, and voltage appears across the Helmholtz layer causing quasi metallic behavior.

of the semiconductor where the density of states is high.[11] When the latter happens, the semiconductor is degenerate by definition, as described in Section 1.1.1.3.

In the preceding section, 3.4.2, we have mentioned the possibility of electron transfer to an ion in solution proceeding via a surface state. Let us assume that this is the dominant mechanism—that no direct transitions between the bands and ions in solution occur. We will also assume the density of surface states is very high (circa 10^{18} m^{-2}). It is easily shown that in this case whether the transfer coefficient is metal-like or semiconductor-like depends on whether the surface states are in equilibrium with the solution or the bulk semiconductor. We will assume thermodynamic equilibrium under open circuit conditions (no net current flow). That is, we assume sufficient electron exchange occurs between the bands of the semiconductor and the surface states, and between the surface states and the ions in solution, so that equilibrium exists across both interfaces. Then, with no net current flow, the E_F of the bulk semiconductor defines the occupancy of the surface state levels, and thus $E_{redox} = E_F$. Then two "exchange currents" can be defined: the exchange current of electrons going to and from the surface states from the band will be denoted by J_{ss}, and the exchange current of electrons going to and from the surface states from the ions in solution will be denoted by J_{soln}.

Consider first the case where the exchange current J_{soln} is much higher than the exchange current J_{ss}, so the "high-resistance" part of the circuit is in the solid. Then, when the net current J is such that $J < J_{soln}$, most of any applied overvoltage will appear across the space charge region and the material will behave as a semiconductor (with zero transfer coefficient). On the other hand, if J_{ss} is much larger than J_{soln}, then most of the overvoltage will have to appear across the Helmholtz layer and the electrode will show metallic behavior until the net current reaches J_{ss}.

Now if the electrode current is greater than both J_{soln} and J_{ss}, then some of the applied overvoltage must appear across each of the critical regions—the space charge region and the Helmholtz region. In this case the electrode will show intermediate behavior between that of a semiconductor and that of a metal.

An extreme case (and actually a more common case, as discussed in Section 5.1.1.3) where "surface states" lead to metallic behavior, is where the "surface states" are really a second conducting phase, perhaps thin but possibly thick. The same concepts apply—there will be two possible rate-limiting steps, one being electron transfer between the semiconductors and the conducting overlayer, the other electron transfer between the conducting overlayer and the solution. As in the example above, the behavior will depend on which electron transfer step is rate limiting.

Finally, we will discuss the degenerate case. Degeneracy is when the

Fermi energy in the semiconductor is in, or very near, the energy band. For example, the entire semiconductor can be degenerate, in which case the semiconductor simply acts much like a metal electrode. Or degeneracy can be induced by the applied voltage. Figure 3.6 illustrates the band picture when an applied voltage V_B bends the bands to make the surface degenerate. Two conclusions can be gleaned from comparing Figures 3.6a and 3.6b. First, because in Figure 3.6b a large charge is in the accumulated region of the surface, the conduction band is filled to a level well above E_{cs}, and now occupied levels in the conduction band overlap empty levels in solution. Second, because the charge in the semiconductor is so near the surface, a reasonable fraction of the extra voltage appears across the Helmholtz region, so the unoccupied bands are even lower; the overlap is greater. Cathodic current can now flow freely. In summary, if enough voltage is applied to the semiconductor, the surface will be driven into degeneracy (either a degenerate accumulation layer as shown, or possibly—with a narrow bandgap material—a degenerate inversion layer) and high currents will flow just as in the metal case (Section 6.2.4). Because a reasonable fraction of the voltage appears across the Helmholtz layer, the "transfer coefficient," either 0 or 1 for a nondegenerate semiconductor, starts to move toward the value 1/2 characteristic of a metal.

4

Measurement Techniques

Measurements of the electrode properties of semiconductors, or measurements on nonmetallic layers on the surface of an electrode, are usually made with one of the following three objectives in mind: One objective is to determine the properties of the solid; a second is to determine the reactions that occur upon electron transfer; and the third is to determine the properties of the solution. In order to analyze any of these, the determination of the properties of the solid is usually the first step. Only when one knows as much as possible about E_{cs}, E_{vs}, the Fermi energy, the surface barrier, and the density of majority carriers available can one hope to analyze accurately the electrochemical and chemical reactions occurring. Thus we begin this chapter with a discussion of the measurement (usually in an indifferent electrolyte) of the capacity, the measurement that leads to a reasonable knowledge about the semiconductor under study. Only then do we discuss measurements with current flow, where we hope to learn something about the reactions and characteristics of the solution.

We will emphasize techniques in semiconductor electrochemistry that are different from those used in metal electrochemistry. The latter are so well documented[(R18)] that a discussion here would represent a serious and unnecessary digression.

Semiconductors, especially well-behaved semiconductors, are unique among solids in that one has many parameters to vary when searching for a model. Significant and understandable effects are observed when changing doping, illumination, temperature; and when introducing *p–n* junctions, tunneling barriers; and when changing from one semiconductor to another. This is why remarkable progress is possible in developing theories of electrode reactions using the versatile semiconductors, just as remarkable progress was made in developing theories of the solid state in the 1950's based on studies of these same semiconductors.

4.1. Capacity Measurements

4.1.1. Introduction

As indicated above, on a semiconductor electrode the capacity as a function of electrode potential is, in general, the simplest measurement to interpret. The capacity measurement indicates the value of V_s, the surface barrier, that controls the density n_s [Equation (3.27)] or p_s [Equation (3.29)], the key parameters in determining the electrode current. In addition, the energy level E_{cs} (or E_{vs}) is determined by capacity measurements. Thus, before we go on to discuss the measurement and interpretation of current flow through the electrode, we will review the measurement of the electrode capacity.

A simple circuit for impedance measurement is indicated in Figure 4.1 for illustration. The object, of course, is to calculate the capacity from the measured impedance. A more detailed discussion of the circuit problems will be discussed in Section 4.1.2; here we present an overview of the requirements for a useful measurement. If we consider that the battery and the ammeter have zero ac impedance, and the voltmeter V has an infinite resistance, then in the circuit of Figure 4.1 the bridge measures the ac impedance between the sample and the platinum electrode. At the interface between the platinum electrode and the solution there is a Helmholtz double layer, with an associated capacity and a resistance. If the area of the platinum is made large compared to the area of the semiconducting sample (e.g., by "platinizing" the electrode to make a huge area), the Helmholtz capacity at this interface can be considered large and the impedance across this interface can be considered small. Usually with no effort the impedance across the metal/solution interface is much

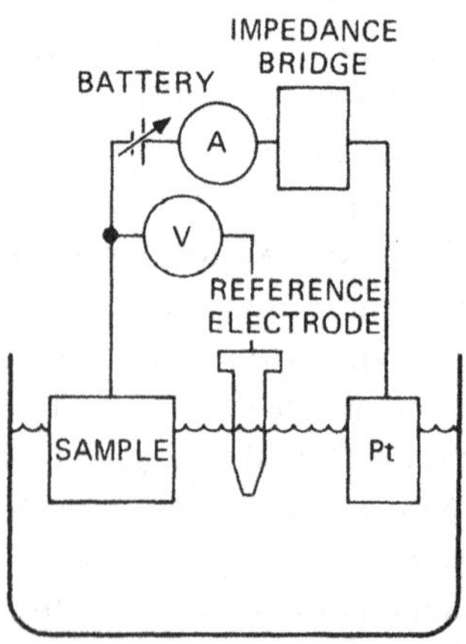

Figure 4.1. Schematic of cell featuring impedance (capacitative) measurement.

smaller than the impedance across the semiconductor/solution interface. The impedance of the solution is not necessarily so small. A resistive impedance of the order of 10^3 Ω is not uncommon. If a poorly conducting solvent/electrolyte system is used, the impedance can be much higher. Thus, in some of the discussions below, we will indicate the possibility of a finite solution resistance. The impedance between the bridge terminal and the semiconductor surface is usually considered negligible. That is, it is usually assumed that the semiconductor itself is of low impedance compared to the semiconductor/electrolyte interface, and that the contact to the semiconductor is of low impedance. A poor contact can, however, be a serious problem and it is not uncommon to find oneself measuring the capacity of a high-resistance metal/semiconductor contact. If all these impedances are indeed negligible, as desired, compared to the impedance across the double layers at the semiconductor surface, then any changes in voltage V will appear across the double layer and change the double layer capacity. By measuring the double layer capacity as a function of the dc applied voltage, we can determine many characteristics of the surface behavior.

4.1.2. Measurement Theory

As introduced above, there are many elements of capacitative impedance and of conductance in a semiconductor electrode that must be thoroughly understood before one is sure that one has a useful impedance measurement. In this section we will emphasize the development of an equivalent circuit to describe these elements, and discuss the interpretation of impedance measurements in terms of such an equivalent circuit.

As described in the discussion of Equation (2.5), the capacity measurement is a measure of the storage of charge during half an ac cycle, with the charge returned during the other half of the cycle. The differential capacity, the parameter of interest, is defined by

$$C = dQ/dV \tag{4.1}$$

where dQ charge is stored when the voltage is altered by dV. The most common case at the interface of interest here is the parallel plate configuration, where two conductors in the form of parallel plates are facing each other, and if the voltage changes by dV, the charge dQ is given by the relation

$$dQ = CdV = (\kappa\epsilon_0 A/d)dV \tag{4.2}$$

with A the area, and with d the thickness of the "insulating" layer between the two conducting plates, κ the dielectric constant, and ϵ_0 the permittivity of free space. Actually the region between the two conducting plates need not be insulating; if it has a reasonably high resistance the region can simply be described as a resistor in parallel with the capacitor. For example, on a metal the Helmholtz capacity behaves in just this way.

There are several contributions to the capacity at the semiconductor surface corresponding to various places where charge is stored: across the Helmholtz double layer, across the semiconductor space charge region, in surface states, and across an oxide film at the surface.

Upon application of a voltage change dV across the Helmholtz double layer, ions accumulate on the outer Helmholtz plane with a charge qdN_s/m^2, and a counter charge appears in or on the solid of $-qdN_s$, leading to a Helmholtz capacity

$$C_H = dQ/dV_H \tag{4.3}$$

with $Q/A = qN_s$.

With a depletion layer at the semiconductor surface, the space charge capacity, C_{sc}, also obeys Equation (4.1):

$$C_{sc} = dQ_{sc}/dV_s \tag{4.4}$$

where V_s is, as usual, the potential difference across the space charge region (positive if the bands bend up) and Q_{sc} is the positive charge in the space charge region. In this case, however, as was seen in Section 2.2.4, the charge Q_{sc} is not at the surface of the conducting layer, but is spread out, normally through the insulating (depletion) layer. The mathematics therefore is slightly different, but the principle is the same—charge is stored at the surface of the two facing conducting regions.

A third type of charge storage is quite different, the charge storage at surface states. The surface state capacity is defined as

$$C_{ss} = dQ_{ss}/dV_s \tag{4.5}$$

where Q_{ss} is the positive charge on the surface states. As V_s changes, the occupancy of the surface states changes, because, as seen in Figure 4.2, as V_s changes, E_F moves relative to the surface states. Now this charge must be stored, in the sense that the stored charge returns to the conduction band when the voltage is restored to its initial value. The charge cannot move to the solution. Then the surface states will offer a real capacity in the sense of the definition of capacity, viz., in the sense of charge storage.

Finally, if there is an oxide layer at the surface of the electrode, and the oxide layer is reasonably insulating, there can be a capacity C_{ox} associated with the thickness of the oxide layer as determined by Equation (4.2).

The measured impedance associated with the combination of the three capacities C_H, C_{ss}, and C_{sc}, can be estimated (neglecting for the

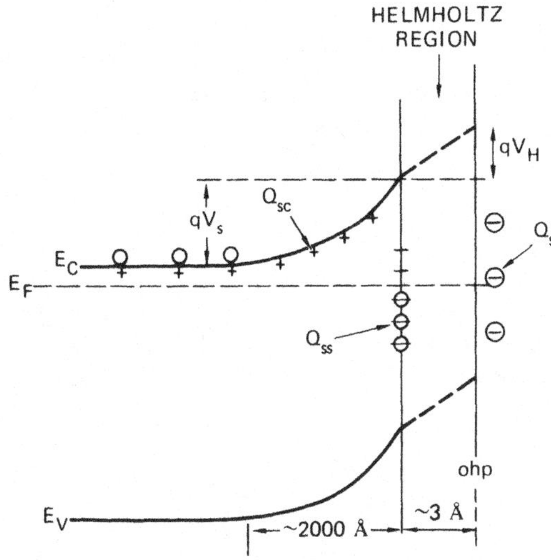

Figure 4.2. Band model showing the location of stored charge and showing voltage differences. Of particular interest is the storage of charge in surface states, Q_{ss}. If the bands are bent further down, the surface state levels will move down across E_F, and more electrons will be "stored." If, when the bands are bent up again, the charge is restored to the conduction band, the surface states contribute to the capacity.

moment any associated resistances) with the help of Figure 4.2 as follows: A change in the applied voltage ΔV is divided between ΔV_s and ΔV_H:

$$\Delta V = \Delta V_s + \Delta V_H \tag{4.6}$$

By charge neutrality:

$$Q_{\text{soln}} = Q_{\text{sc}} + Q_{\text{ss}} \tag{4.7}$$

But ΔQ_{soln} is the charge stored from the external circuit upon application of a voltage change ΔV. Thus from Equation (4.1):

$$\frac{1}{C} = \frac{\Delta V}{\Delta Q} = \frac{\Delta V}{\Delta Q_{\text{soln}}} = \frac{\Delta V_s}{\Delta Q_{\text{sc}} + \Delta Q_{\text{ss}}} + \frac{\Delta V_H}{\Delta Q_{\text{soln}}} \tag{4.8}$$

and from Equations (4.3), (4.4), and (4.5)

$$C^{-1} = (C_{\text{sc}} + C_{\text{ss}})^{-1} + C_H^{-1} \tag{4.9}$$

Thus C_H acts as a capacitor in series with the parallel combination of C_{sc} and C_{ss}.

The resistances and the capacities of the electrode, for simplicity, are described by an "equivalent circuit" which is an array of resistors and capacitors arranged to give the same impedance as the arrangement of resistors and capacitors in the sample itself. Thus, for example, the

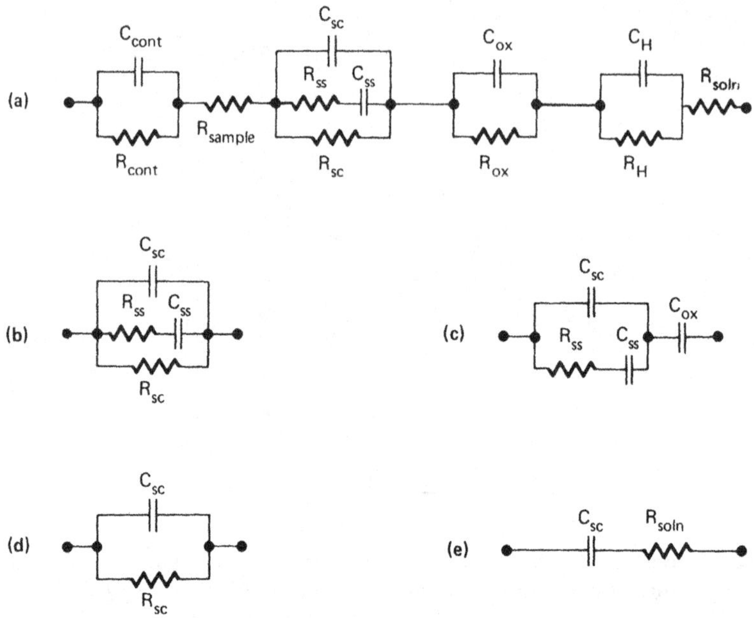

Figure 4.3. Equivalent circuits and their simplification. Circuit (a) includes most of the possible contributions to the impedance, where the subscript "cont" refers to the contact to the semiconductor, "sc" to the space charge, "ss" to surface states, "ox" to a passivating layer (a second solid phase) at the surface, "*H*" to Helmholtz, and "soln" to solution. Diagrams (b), (c), (d), and (e) represent cases where various of these contributions can be neglected because effectively $R = 0$ or ∞ or because $C = 0$ or ∞.

equivalent circuit in one of the more complex forms may be as indicated in Figure 4.3a. Here the subscript "cont" refers to the electrical contact to the semiconductor. The other sources of resistance and capacitance are fairly obvious. Hopefully this equivalent circuit can be substantially simplified. For example, the experiment should be designed so that the solution, the bulk of the sample, and the metal contact to the sample are of negligible impedance. Normally the Helmholtz capacity is so large it can be neglected (Section 4.1.3.2 below). If the surface is clean, there is no contribution C_{ox} or R_{ox} from a surface oxide film. Then Figure 4.3b may be a reasonable equivalent circuit.

In the equivalent circuit of Figure 4.3b there are three parallel impedances. The impedance of a capacitor is $(j\omega C)^{-1}$ of course, so the impedance of R_{ss} and C_{ss} in series is given by $R_{ss} + (j\omega C_{ss})^{-1}$; the other two are simple impedances. By the parallel impedance formula the impedance Z of the circuit is then

$$\frac{1}{Z} = \frac{1}{R_{sc}} + j\omega C_{sc} + \frac{1}{R_{ss} + (j\omega C_{ss})^{-1}} \tag{4.10}$$

where j is, of course, $(-1)^{1/2}$. Now an impedance bridge measures two numbers, the real part of the effective impedance (or admittance) and the imaginary part. That is, in a measurement of admittance Y one obtains two numbers R_{eff} and C_{eff} given by

$$Y_{\text{eff}} = 1/Z_{\text{eff}} = R_{\text{eff}}^{-1} + j\omega C_{\text{eff}} \tag{4.11}$$

If Equation (4.11) is equated to Equation (4.10), equating separately the real and imaginary parts, it is found that the effective capacity C_{eff} measured during an admittance measurement is given by

$$C_{\text{eff}} = C_{\text{sc}} + C_{\text{ss}}/(1 + \omega^2 R_{\text{ss}}^2 C_{\text{ss}}^2) \tag{4.12}$$

and the one measurement of C_{eff}, using the methods described below, will not separate C_{ss} from C_{sc}, but provide an unmanageable combination. The measurement will give a frequency-dependent effective capacity, which is an indication of possible equivalent circuit problems. This analysis of Figure 4.2b illustrates why it is so important to simplify the experimental situation so that a simple equivalent circuit can be used. If, for example, the frequency is high enough so that charge cannot be exchanged with the surface states in the period of a cycle, then effectively R_{ss} is relatively high, and the second term of Equation (4.12) can be neglected. Or, if the surface state capacity can be considered zero (see discussion of Section 4.1.3.3) similarly the second term is negligible, and the capacity measurement gives the space charge capacity, a meaningful number. In both these cases, the equivalent circuit is then the parallel circuit, Figure 4.3d. Then $C_{\text{eff}} = C_{\text{sc}}$ as desired, independent of frequency in the high-frequency regime.

The admittance Y_{eff} is measured by maintaining a constant ac voltage on the sample. Then if one measures the ac current flowing through the sample

$$I = V/Z = YV \tag{4.13}$$

and the measured current is proportional to the admittance. The 90° out-of-phase component of the current, in accordance with Equation (4.11) is proportional to C_{eff}. The 90° out-of-phase component is determined with phase-selective lock-in amplifier or with a bridge that is designed to give capacity values with a parallel equivalent circuit.

If one convinces himself that he has a series equivalent circuit, for example that the dominant impedances in the circuit are C_{sc} and R_{soln}, as in the equivalent circuit of Figure 4.3e, then the equivalent circuit has the impedance

$$Z = R_{\text{soln}} - j/\omega C_{\text{sc}} \tag{4.14}$$

and for simplicity of measurement and interpretation, one will hold the current constant and measure the voltage across the sample:

$$V = IZ \tag{4.15}$$

and directly measure C_{sc} again as the voltage component 90° out of phase with the current, using a phase-selective lock-in amplifier.

To summarize the above analysis, the object, and indeed the necessity, for the experimenter is to arrange the experimental conditions so that the equivalent circuit simplifies down to an understandable circuit, for only then can the parameters be analyzed. It does not necessarily have to be as simple as Figure 4.3d or 4.3e, but those are certainly desirable cases. One has only two parameters that are measurable, the real and the imaginary part of the impedance. If there are more than two parameters that need to be determined, the difficulty is obvious. To aid in the simplification of the equivalent circuit, one has at one's disposal the frequency of the ac signal, the conductance of the solution, and so on. For a satisfactory measurement of the parameters of interest, usually C_{sc}, one must vary the other parameters and ensure that they are not affecting the measurement.

The physical measurement of the impedance, to determine the capacity, is done using a manual capacity bridge, an automatic capacity bridge, a lock-in amplifier, or some other form of a phase-sensitive detector. Now the current–voltage characteristics of the electrode are not linear—they vary exponentially with qV/kT [Equation (3.34)]. Thus the ac voltage must be kept low, much less than kT/q, viz., much less than 0.025 V, to avoid nonlinearity, to avoid harmonics in the measured signal. This requirement precludes the use of some automatic bridges, but other than this requirement, their use is convenient and accurate. The use of phase-sensitive detectors or lock-in amplifiers that provide the out-of-phase component of the impedance (the capacity) on a continuous basis is becoming increasingly popular. Then V_S can be cycled, as in voltammetry (Section 4.2.1), and the capacity continuously monitored. In all these techniques, naturally, it is necessary for the scientist to determine whether the equivalent circuit for the system under study is a parallel or series circuit, and whether the bridge or other means of detecting the capacity is set up for constant current (series equivalent circuit) or constant voltage (parallel equivalent circuit) measurement. It should be noted that more complicated equivalent circuits than shown in Figures 4.3d or 4.3e can be analyzed,[1] but the measurements are difficult and the circuit must be well defined.

4.1.3. Analysis

In this section we emphasize the interpretation of capacity measurement to provide the fundamental parameters of the semiconductor electrode.

4.1.3.1. Space Charge Capacity. Normally, in semiconductor electrochemistry the parameter of interest obtained from the capacity measurement is V_s, which can be obtained most easily when the electrode potential is adjusted so the space charge layer is a depletion layer. The relation between the C_{sc} and V_s is obtained from the solution of Poisson's equation. The relation could be obtained directly from Equation 2.25, but it is common in the literature to include, in the derivation of the capacity, the contribution of electrons in the conduction band, which, as will be seen below, affects the calculated value of V_s by only kT/e (0.025 V). Although the correction is usually negligible, we will derive the C_{sc}/V_s relationship for a depletion layer including the correction.

From Poisson's equation, Equation (2.8), we have

$$d^2V/dx^2 = -(N_{sc}q/\kappa\epsilon_0)[\exp(-qV/kT) - 1] \tag{4.16}$$

where we assume N_{sc} positive charges in the depletion region, and we are assuming $n = N_{sc}\exp(-eV/kT)$ electrons in the conduction band. On integrating with $dV/dx = 0$ when $V = 0$, we have

$$(dV/dx)^2 = (2qN_{sc}/\kappa\epsilon_0)\{V + (kT/q)[\exp(-qV/kT) - 1]\} \tag{4.17}$$

At the surface $V = V_s$, and the factor $\exp(-qV_s/kT)$ is negligible for reasonable values of V_s. From Gauss' law we have

$$(dV/dx)_{V=V_s} = Q_{sc}/\kappa\epsilon_0 A \tag{4.18}$$

where Q_{sc} is the total charge in the space charge layer and A the sample area. By Equation (4.4), we simply differentiate the combination of Equation (4.17) and (4.18) with respect to voltage V_s and obtain C_{sc}:

$$\begin{aligned} C_{sc}^{-2} &= (2/qN_{sc}\kappa\epsilon_0 A^2)(V_s - kT/q) \\ &= (2/qN_{sc}\kappa\epsilon_0 A^2)(V_m - V_{fb} - kT/q) \end{aligned} \tag{4.19}$$

a much-used equation relating the measured space charge capacity to the surface barrier V_s. The second form of Equation (4.19) is obtained using Equation (2.36). Figure 4.4 shows a typical plot of C_{sc}^{-2} versus V, where V is the applied voltage. The example used[2] is that of ZnO. When the capacity is thus plotted against electrode potential V_m as per Equation (4.19), the diagram is often termed a "Mott–Schottky plot."

Two important parameters are determined by interpreting C/V data by Equation (4.19), the flatband potential V_{fb}, and the density of charge in the space charge region, N_{sc}.

The flatband potential (which could be corrected by kT/q if the measurement were accurate enough) is determined by extrapolation, as in Figure 4.4, from the depletion layer region of voltages to the intercept. Because V_m (say from Figure 4.4) must be related to a reference electrode, also V_{fb} must be so related. In theoretical discussions V_{fb} is usually given

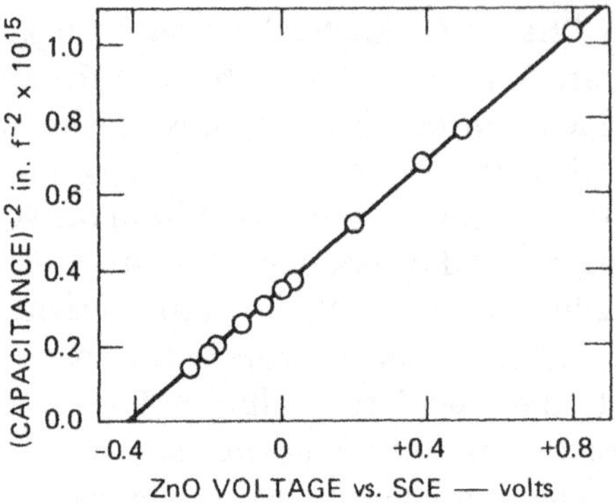

Figure 4.4. Mott–Schottky plot of C^{-2} versus electrode potential. The slope, with $A = 6 \times 10^{-6}$ m^2 and $\kappa = 8.5$, gives $N_{sc} = 6 \times 10^{24}$ m^{-3}, the intercept gives $V_{fb} = -0.41$ V(SCE) $= -0.17$ V(NHE).

relative to the hydrogen electrode; in discussions of experimental results it is often given relative to the saturated calomel electrode (SCE).

The density of charge N_{sc} in the space charge region, as measured by the slope of the Mott–Schottky plot in a deliberately doped crystal is, of course, the donor or acceptor density, but, as will be discussed in Section 5.1, for a less-than-ideal crystal it can include trapped electrons or holes, or deep energy levels that are charged. The flat band potential V_{fb} was shown in Section 2.4 [Equation (2.37)] to be of importance in determining the conduction or valence band edges of a semiconductor on the hydrogen reference scale of energies.

Formulas for the capacity as a function of V_s for the accumulation layer and the inversion layer have also been worked out.[R15,R16] Dewald[3] gives an equation for the capacity of an n-type semiconductor with completely ionized donors, that includes the transition between an accumulation layer and a depletion layer:

$$C_{sc}^{-1} = (2kT/q^2N_D\kappa\epsilon_0A^2)^{1/2}[\exp(-qV_s/kT) + (qV_s/kT) - 1]^{1/2} \times [|\exp(-qV_s/kT) - 1|] \tag{4.20}$$

This becomes the Mott–Schottky form for $V_s > kT/q$, and yields for negative V_s (an accumulation layer)

$$C_{sc} = (q^2N_D\kappa\epsilon_0A^2/2kT)^{1/2}\exp(-qV_s/2kT) \tag{4.21}$$

and C_{sc} increases as V_s becomes more negative (the accumulation layer becomes stronger). On the other hand as V_s becomes very positive and the depletion layer gives way to a strong inversion layer, it can be calculated,[4] combining Equation (2.30) (with V_s large and positive), Gauss' law ($dV/dx = Q/\kappa\epsilon_0 A$) and Equation (4.1):

$$C_{sc} = (q^2 \kappa\epsilon_0 A^2 p_b / 2kT)^{1/2} \exp(qV_s/2kT) \quad (4.22)$$

If the surface is always at equilibrium, these formulas lead to a double-valued curve for capacity as a function of V_s, as the surface progresses from an accumulation layer through a depletion layer to an inversion layer. The C/V_s relation is sketched in Figure 4.5. With an accumulation layer the capacity will be very high, as the distance separating the mobile charge at the semiconductor surface from the ions in solution is small. As the depletion layer is formed (V_s becomes positive), the capacity will decrease, in accordance with Equation (4.19), because the distance between the mobile carriers in the majority carrier band of the semiconductor and the conducting ions in solution becomes larger [d in Equation (4.2) becomes larger]. However, when an inversion layer forms, minority carriers are now at the surface. With a strong inversion layer, such that most of the charge dQ, associated with the change in voltage dV, is located near the surface in the minority carrier band, then d is again small and C is high. Thus from the beginning of an inversion layer to a strong inversion layer, the capacity increases again. A sketch of the expected behavior is given in Figure 4.5 designated as the "equilibrium" curve. Such a C/V

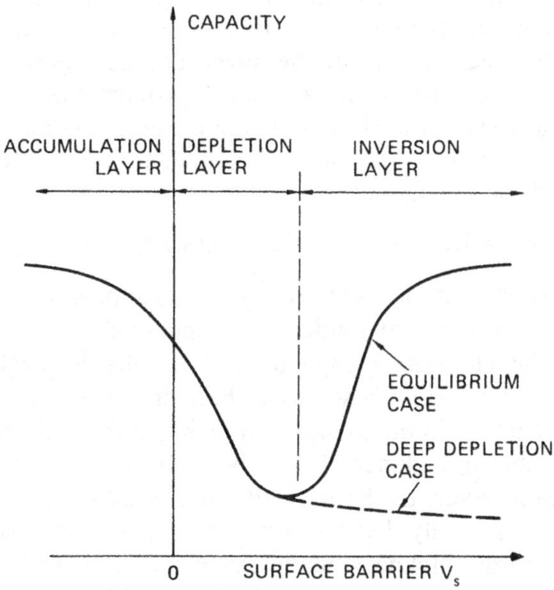

Figure 4.5. Variation of space charge capacity with surface barrier V_s on an n-type semiconductor with an accumulation or an inversion layer. Mobile carriers are at the surface in inversion so the capacity is high. If minority carriers do not accumulate at the surface when there is an inversion-type barrier, the "deep depletion" curve is followed.

relation is often observed with narrow bandgap semiconductors such as germanium, where minority carrier equilibrium is easily reached.

If equilibrium with minority carriers is not reached, as is very common with wide bandgap semiconductors where there is a negligible density of holes and a negligible generation rate of holes, the dotted curve of Figure 4.5 is typically observed, indicated as "deep depletion" (see Section 2.2.4.2). The depletion layer behavior, as described by Equation (4.19), continues to high voltages. Any minority carriers that are generated are consumed in surface reactions and are not accumulated at the surface, so an equilibrium inversion layer cannot be formed. The curve of Figure 4.4 for ZnO is an example of a deep depletion curve. ZnO, with its 3.2-eV bandgap, has few thermally produced holes, and these are always consumed at the surface by some reaction—if no other reactants are present they will react with the surface lattice oxygen, corroding the ZnO.[5]

4.1.3.2. The Helmholtz Capacity. In Section 2.2.3, the Helmholtz double layer for semiconductors and metals was discussed. The thickness of the Helmholtz region d_H on metals, where it can be measured by capacity, is of the order of a few angstroms. This is the effective distance between the charge on the solid surface and the distance-of-closest-approach of the nonadsorbed ions that provide the countercharge. In semiconductor work, until a more detailed theory can be developed, or until more accurate measurements can be obtained, it is quite sufficient to consider that the Helmholtz double layer is of similar thickness, usually negligible compared to the thickness x_0 of the space charge region. It should be kept in mind that capacitative impedance is proportional to double layer thickness, and as long as the Helmholtz double layer thickness d_H is much less than the space charge thickness, the impedance Z_H of the Helmholtz double layer will be negligible:

$$Z_H \propto 1/C_H = d_H/A\kappa\epsilon_0 \ll x_0/A\kappa'\epsilon_0 = 1/C_{sc} \tag{4.23}$$

where κ and κ' represent the dielectric constants appropriate for the Helmholtz region and for the semiconductor, respectively.

Although the Helmholtz capacity can usually be neglected in the equivalent circuit because it is so large, there are several cases where the other capacities in the circuit are large enough that the Helmholtz capacity gives a comparable impedance. If one is dealing with a strong accumulation or inversion layer, as theoretically discussed above with Equation (4.21), and experimentally discussed in the high-current cases of Section 6.6, the space charge thickness becomes small and the capacity becomes large. Also, if the doping is extremely high (10^{25}–10^{27} m^{-3}), the depletion layer capacity can be very large. For example, we will be discussing cases in Chapter 8 where the movement of ions to the surface of an oxide layer leads to a large space charge capacity and the Helmholtz capacity again becomes an important contribution. Finally it is possible, although there seem to have been no cases of such high C_{ss} observed to date, that the surface state capacity C_{ss} can be of the order of C_H and then, from Equation (4.9), C_H becomes important. Typically in semiconductor electrode studies, C_H is a source of error, not a source of interest.

4.1.3.3. The Surface State Capacity. As mentioned in Section 4.1.2 while discussing Equation (4.5), the surface state capacity C_{ss} is quite different in origin from C_{sc} or C_H. Here we can have charge storage: the charge that is stored in surface states from the conduction band upon application of a voltage dV can be returned to the conduction band upon removal of the voltage dV. But here we do not have anything resembling a parallel plate capacitor. There is no distance d associated with the surface state capacity.

Let there be a distribution of surface states in the bandgap of the semiconductor, and let the density be $N_{ss}/\text{m}^2\ \text{eV}$, where N_{ss} is a function of energy, $N_{ss}(E)$. At the Fermi energy there are $N_{ss}(E_F)$ states $\text{m}^{-2}\ \text{eV}^{-1}$. The scheme is shown in Figure 4.6. Now if the surface V_s is raised by dV_s, this means the Fermi energy will be lower by dV_s relative to the surface states, and an additional $N_{ss}(E_F)dV_s$ surface states will move above E_F and become unoccupied. Thus $dQ_{ss} = qAN_{ss}dV_s$ will be the corresponding change in surface state charge. That is, from Equation (4.5),

$$C_{ss} = dQ_{ss}/dV_{ss} = qAN_{ss}(E_F) \tag{4.24}$$

Thus, the measurement of the surface state capacity measures the density of surface states at the Fermi energy, and a plot of C_{ss} versus V_s gives the surface state density as a function of energy in the bandgap.

Unless we are measuring surface states, it is desirable to establish conditions where C_{ss} can be neglected. Now the transfer of the charge to the surface states is activated. As indicated in Figure 4.6, the electrons going to or from the surface states must cross a potential barrier V_s or $E_{cs} - E_c$. Such electron transfer may be very slow if the surface barrier is high, and so when measuring the capacity with an ac voltage of high enough frequency the surface states may not be able to change their charge during the period of the ac voltage. Then, as discussed in Section 4.1.2, the contribution of the surface states to the capacity may be neglected and the equivalent circuit simplified.

Another case where it is possible to neglect the surface state capacity may be when the charge is not "stored." Commonly at the solid/electrolyte interface, the charge in the surface states is not stored, in the sense that to give capacitative impedance the stored charge must be returned to the band from which it came. As will be discussed in Section 6.1, it is often found with semiconductors that if the oxidized form of an ion in

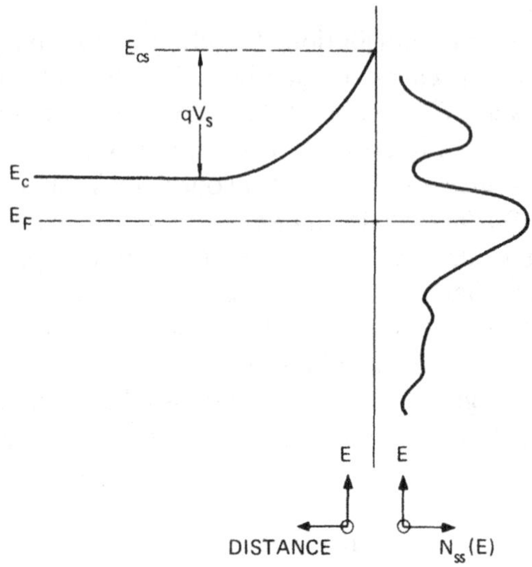

Figure 4.6. Surface states shown distributed as a function of energy to illustrate the origin of surface state capacity.

solution can capture electrons, the reduced form cannot inject them; and if the reduced form can inject electrons, the oxidized form cannot capture them. Thus the electron transfer is unidirectional only. Then it is not surprising, by analogy, that often surface states also do not show any surface state capacity. If a carrier is captured by an intrinsic surface state, it results in a broken bond, so the resulting species often dissolves into the solution or forms a new phase. For example, as will be discussed in Section 5.1.1.2, although there are definitely various types of charged surface states on germanium, as determined by techniques such as recombination and Helmholtz voltage shifts, they are not easily observed by capacity measurements.[6] However, as discussed in Section 6.4.2, the surface states associated with some special forms of the germanium surface state (Ge· or Ge–*X*, with *X* a halide) are apparently observed by capacity measurements.

On the other hand, interface states (as defined in Section 2.2.1) will show a surface state capacity. If there is an oxide or other passivating layer on the surface of the semiconductor, especially if the passivating layer is insulating, chemical transformation cannot occur, so the capture of charge by surface states is reversible—the charge is "stored." Then the interface state capacity becomes of great importance in interpretation of the results. And with many electrodes, an undesired second phase may be at the surface, with interface states in between. The classic case is silicon, covered with a thick silica layer. This system has been studied extensively, as reviewed by Goetzberger *et al.*,[7] because of its great importance in silicon device technology. In the studies described by Goetzberger, a metal layer rather than an electrolyte makes contact to the outside of the oxide, but the principles are the same. With silicon covered by silicon dioxide, it can usually be assumed that the silica is of effectively infinite resistance. Then the equivalent circuit of Figure 4.3c applies. This is analyzed by measuring the impedance first at high frequency, then at low frequency. At very high frequency, the impedance is taken to be that of the two capacitors C_{sc} and C_{ox} in series, because the impedance of the capacitors becomes much lower than the impedance of the resistor, so R_{ss} is assumed infinite, and therefore C_{ss} can be neglected. The oxide thickness is generally known, so C_{ox} can be calculated from Equation (4.2), and thus C_{sc} is determined. At very low frequency, the capacitative impedances are much greater than the resistive impedance R_{ss}. In practice, low frequency measurements are made using a ramp technique so the frequency can be effectively as low as 10^{-3} Hz, for example. In the ramp technique the voltage is increased at a constant rate, so $dV/dt = b$, when b is a constant. When R_{ss} is assumed of negligible impedance, the capacity of the circuit, C_t is obtained from the formulas for adding series and parallel capacitors:

$$C_t^{-1} = (C_{sc} + C_{ss})^{-1} + C_{ox}^{-1} \tag{4.25}$$

and, from $Q = CV$ as given in Equation (4.2), we have by differentiation

$$dQ/dt = I = C_t dV/dt = bC_t \tag{4.26}$$

so C_t is obtained by measuring the charging current, I. Then, from Equations (4.26) and (4.25), knowing C_{sc} and C_{ox} from the high-frequency measurements, the value of the surface state capacity, C_{ss}, is obtained from (4.25), and N_{ss} from Equation (4.24).

The analysis of the silicon/silica surface by this high-frequency/low-frequency approach, often termed the "quasistatic" method,[8] has been a powerful tool in learning how to produce lower surface state densities on silicon devices. Its use with an electrochemical contact, instead of a metallic contact to the outside of the oxide, is relatively new, but promises to provide new knowledge with respect to oxide layers on electrodes.[63]

4.1.4. Complex Mott–Schottky Plots

Although simple Mott–Schottky plots such as illustrated in Figure 4.4 are often observed, at times there are deviations from this simple behavior, and in this section we discuss some of the common deviations and their probable origin. Deviations include an incorrect flatband potential (the intercept V_{fb}), an incorrect slope for the curve, or a slope that varies with voltage, and a frequency-dependent capacity. We will discuss changes in flatband potential in detail here, as it is usually an experimental problem. We will mention the other problems, but postpone detailed discussions to Chapter 5, where they are related to, and are used to analyze, complex properties of the electrode.

First let us consider the various sources of flat band shifts, shifts in V_{fb}. Referring to the band picture as shown in Figure 2.6, and noting that V_{fb} is the (extrapolated) value of V_m when $V_s = 0$, we can simplify the band picture for this specific case to the form indicated in Figure 4.7. Here we specify, for simplicity, a hydrogen reference electrode, and specify the hydrogen couple as the redox couple in the solution. As defined in Section 2.3, the energy (on the hydrogen scale) of E_{cs} when $V_H = 0$ is given by E^o_{cs}. E^o_{cs} is sketched in Figure 4.7 and is clearly negative by this definition, as illustrated in the picture. Then directly from Figure 4.7 [or from Equations (2.31) and (2.37)] we have

$$qV_{fb} = \mu - E^o_{cs} + qV_H \tag{4.27}$$

Now the only variable in this idealized formulation for the intercept of the Mott–Schottky plot is V_H. Thus variations in V_H [e.g., due to pH

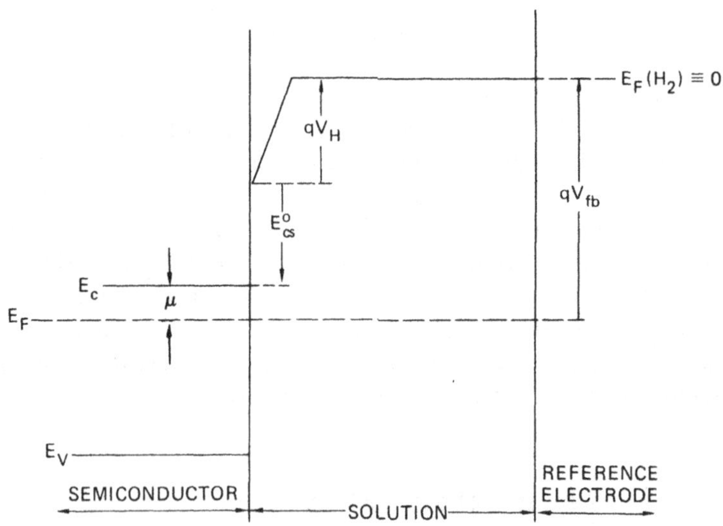

Figure 4.7. Sketch of a band model with the bands in the semiconductor flat, where the hydrogen reference electrode is used to define the zero. The parameters indicated with arrows pointing up are positive as sketched. With the bands flat as indicated, V_{fb} is measured as the electrode potential.

changes by Equation (2.20)] will appear directly as variations in the measured parameter V_{fb}. However, the double layer voltage V_H is not unique.

More generally, we can consider any extraneous voltages in the circuit may lead to a shift in V_{fb}. Any double layer voltage will, similarly to V_H, shift the voltage axis (e.g., in Figure 4.4) and thus shift the measured V_{fb}. Any other unknown voltage sources located outside the semiconductor space charge region such as an IR drop or a battery will shift the voltage axis by the amount of the voltage, and appear as a shift in V_{fb}. A common source for such an unknown voltage is a separate phase, a film, on the surface.

The best illustration where a film causes flat band shifts—and these have been studied in detail—is again silica on silicon.[7] We will briefly review the techniques used and the interpretations of flat band shifts on oxidized silicon. The technique that indicates most clearly the origin of flat band shifts for this case, is the use of cyclic sweeps of the voltage when measuring the capacity. Suppose during a sweep from negative to positive electrode potential, a certain flat band potential is observed. Now, as the voltage moves toward the most positive end of the sweep, suppose the charge distribution within the oxide changes. The double layer voltage within the oxide thus changes. Then returning on the reverse sweep, one will find a different V_{fb}. As the electrode potential moves back to the most negative end of the sweep, the charge distribution in the oxide will be restored to its original condition. A cyclic plot of capacity/voltage will show hysteresis that is due to such double layer changes. In the silicon/silica case it can easily be concluded that if the charge is moving from and to the silicon (electrons from the silicon are being injected and extracted from the oxide as the voltage goes from one extreme to the other), the hysteresis will be of one sign. The capacity, as the electrode potential sweeps more positive, will be lower than the capacity as the electrode potential is sweeping negative. If the charge is due to ions moving back and forth in the oxide, the hysteresis will be of the other sign.

Another possible origin of shifts in the flat band potential that may occur for very special cases is a nonnegligible C_H, as has been pointed out by De Gryse *et al.*[9] As discussed above, if the semiconductor is heavily doped, leading to a small value of the space charge layer thickness, then by Equation (4.19) the space charge capacity C_{sc} will be high. Then C_H may not be neglected. They analyze the expected capacity/voltage relationship for this case and conclude

$$C^{-2} = C_H^{-2} + (2/qN_{sc}\kappa\epsilon A_0^2)(V_s - kT/q) \tag{4.28}$$

By this analysis the slope of the Mott–Schottky plot is still determined by the space charge density N_{cs} in the solid, but the flat band potential has shifted by an amount ($\kappa\epsilon_0 qN_{sc}/2C_H^2$). This shift is, however, small—unless one is dealing with a solid of very high dielectric constant (e.g., TiO_2, with a dielectric constant in the c direction of 173) or an extremely high doping density (with $\kappa = 5$ it requires $N_{sc} \sim 2 \times 10^{26}$ m^{-3} to shift the flat band potential by about 0.1 V).

Thus, normally one concludes that unreproducible flat band potentials arise from shifts in V_H due to double layers caused by adsorption or surface states; arise from double layers in surface films; or arise from inaccuracies in eliminating voltage drops elsewhere in the circuit.

Changes in the slope of the Mott–Schottky plot usually indicate changes of the bulk properties (near the surface) of the semiconductor. They usually arise from changes in N_{sc} with time or with distance from the surface. Such changes can be associated with ion movement under the high electric fields of the space charge region (Section 7.3.4), trapping of holes and/or electrons in the space charge region, or a built-in variation in ion density with distance [in Equation (4.19) we assume N_{sc} independent of x]. Such variations are very useful in determining the bulk properties of semiconductors, and in Section 5.2 such measurements will be discussed.

A frequency-dependent capacity is another signal that Equation (4.19) may not be valid. Frequency-dependent capacity measurements while making Mott–Schottky plots can arise from one of three causes: (a) a poorly defined equivalent circuit, (b) trapping and detrapping of electrons with time constants near the period of the ac signals used, and (c) conceivably ion movement, particularly at low frequencies. We have discussed equivalent circuit problems in Section 4.1.2. The trapping effects can, in a sense, be considered to be equivalent circuit problems [e.g., Equation (4.12)] but a common, but inexcusable problem, is assuming a parallel circuit when a series circuit should apply, and having frequency problems because of this simple error.

In semiconductor electrochemistry, probably the most common cause (other than equivalent circuit errors) of a frequency-dependent capacity is trapping in the space charge region. Slow trapping at surface states [Equation (4.12)] is common when a passivating film protects the surface states from the solution, but, as mentioned above in Section 4.1.3.3, if the surface is clean, surface states will not in general store charge (Section 5.1). Ion movement in the passivating layer can be another source of frequency dependence. Tench and Yeager[10] suggest that ion movement in the semiconductor can cause frequency effects, particularly in heavily doped materials where the space charge region is very thin. Toshima and Uchida[11] report another source of frequency dispersion involving adsorbed ions, especially slow surface states associated with adsorbed ions. This is a special form of (b) above, where the adsorption of ions changes the energy level, and hence the time constant of the surface state.

In analyzing the origin of the frequency dependence of the capacity, it has been pointed out by Van Meirhaeghe *et al.*[12] that the frequency dependence of the capacity must be accompanied by a frequency-dependent parallel resistance, according to the Kramers–Kronig relations. If a model under test predicts the observed frequency dependence for the capacity, it must predict the frequency dependence for the resistance that the Kramers–Kronig relations demand. If the experimental frequency dependence of the resistance differs from this predicted relation, then the experimental observations must be in error.

The observation of frequency dependence in capacity has not proved a means of giving definite information about the solid, in general, except that it helps determine which equivalent circuit (Figure 4.3) is applicable. Experimenters, when observing frequency-dependent capacity, have speculated, for example, that it arises from bulk trapping,[13] bulk inhomogeneities,[14] and, of course, many authors have suggested surface states. But it is much easier to get definitive information about these phenomena by other techniques, so frequency-dependent capacity has not been used much as a probe.

4.1.5. Determination of Band Edges

The experimental determination of E_{cs} (or, in the case of p-type semiconductors, E_{vs}) follows directly from the analysis above of the Mott–Schottky plots. The determination was discussed at the end of Chapter 2, and for completeness Equation (2.37), developed there, is reproduced:

$$E_{cs} = \mu - qV_{fb} \tag{4.29}$$

Here μ is obtained by solid state physics measurements of the bulk properties of the semiconductor, derived directly from Equation (1.2). As μ is generally only the order of 0.1 to 0.2 eV, just the measure of V_{fb} provides a good esimate of E_{cs}. Similarly E_{vs} for p-type is given approximately by V_{fb}, again because E_F is within 0.1 to 0.2 eV from the valence band. For example, we see that if V_{fb}, the potential of the electrode relative to the hydrogen electrode, is negative, the Fermi energy, and hence the appropriate band edge, is above zero on the hydrogen scale.

The value of E^o_{cs} is the value of E_{cs} when the pH is adjusted to the point of zero charge, pzc, and can be measured at this pH, or can be obtained by extrapolation according to Equation (2.20) when (as is usually the case) Equation (2.20) can be considered valid. (See also Section 5.1.1.)

4.2. Measurements of the Current/Voltage Characteristics

4.2.1. General Techniques; Voltammetry

When current passes through an electronically conducting solid to an electrolyte, a chemical reaction must occur at the interface, as discussed at the start of Chapter 2. Electrons moving into the solution reduce species in solution; electrons moving into the electrode lead to oxidation of some species in solution. Thus the composition of the solution will change. In order to maintain, as much as possible, a constant composition for the solution next to the sample (the working electrode) when large currents pass, the counter electrode is relegated to a separate compartment. The configuration of Figure 4.8 is preferred over that of Figure 4.1 when currents of reasonable magnitude are passing. The reference electrode is inserted in a "Luggin capillary" which is close to the sample, so that the potential measured by the reference electrode is very close to that of the potential of the solution at the outer Helmholtz plane. A frit, or some other means of separating the compartments to prevent mixing of the solutions without offering appreciable electrical resistance, is placed between the cell containing the sample (the working electrode) and the counterelectrode. Thus any new species formed at the counterelectrode do not reach the sample. Of course, species produced at the sample are unavoidable.

In many cases we wish to program voltage changes between the sample and the solution, and measure the resulting current changes. Now if appreciable current is flowing through the counterelectrode, we cannot anticipate that the voltage drop between the counterelectrode and the solution will remain constant. Thus control of the "battery voltage," as indicated in Figure 4.8, does not control the voltage of interest, V, as measured between the sample and the reference electrode. As discussed earlier, the voltage V is called the "electrode potential." An electronic circuit, the "potentiostat," has been designed to program the potential V. It is a four-terminal circuit. The voltage that one desires to appear between the reference electrode and the sample is provided as an input voltage. The circuit has a feedback mechanism that adjusts the current through the counterelectrode so that the voltage between the sample and the reference electrode becomes the desired voltage. Thus, in discussions below we describe current/voltage curves where the voltage (the electrode potential of the sample relative to the reference electrode) is changed according to a prescribed program (e.g., dV/dt = const), and the current is measured. This is normally done using a potentiostat.

A common technique in the study of electrode reactions is "voltammetry," where the electrode potential is varied past the E^o of a redox

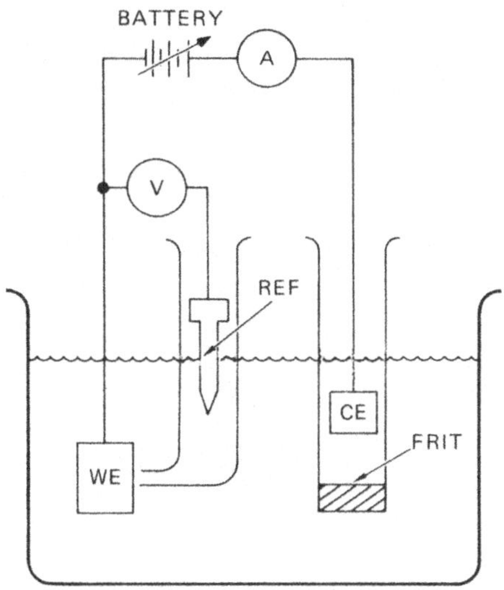

Figure 4.8. An illustration of the precautions that may be used when large currents flow. A Luggin capillary enables the solution potential to be monitored near the sample (WE, the working electrode). The counterelectrode, CE, is separated from the sample by a bridge or a frit to avoid contamination of the solution surrounding the sample.

couple of interest, in order to detect the reducing or oxidizing agent of that couple by the resulting current onset. By observing the current flow, the concentration of the electroactive species can be estimated. Most often it is used in the form of "cyclic voltammetry," where the voltage is cycled, with $dV/dt = \pm b$. The voltage sweep is "triangular": it is increased at a constant rate, then decreased at the same rate. The method is most used with metals or degenerate semiconductors. Its utility depends to some degree on the dominance of concentration polarization, as discussed in Section 3.1.2. If a reducing agent M is present in the solution, electron injection from the reducing agent will begin when the electrode potential reaches some critical voltage near E^o. After the ions in the neighborhood of the electrode, including adsorbed ions, have given up their electrons, others must diffuse in to maintain the current flow. Now if the concentration is low, the current will be limited to a low value. Thus the current will be in the form of a pulse or a peak, corresponding to the number of ions that are in the neighborhood or are adsorbed when the critical voltage is reached. With a series of reducing agents of varying critical voltages, a series of peaks, or step functions called "waves," will be observed on the current/voltage plots as the voltage is steadily increased.

Figure 4.9 is intended to illustrate the technique using the band model.

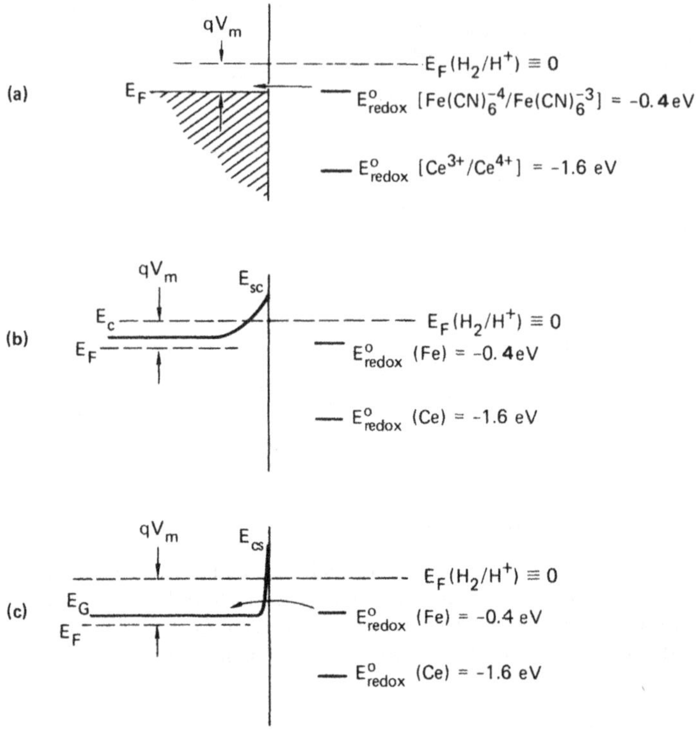

Figure 4.9. The origin of current waves in cyclic voltammetry. As described in the text, when E_F moves below E^o_{redox} in cases (a), a metal, and (c), a tunnelable surface layer, current can flow as shown by the arrows. In case (b), a modestly doped semiconductor, anodic current cannot flow until the voltage is high enough for breakdown.

Figure 4.9a illustrates the metal/electrolyte system; (b) and (c) illustrate a moderately and a heavily doped semiconductor, respectively. In all cases a voltage is applied, so the systems as shown are not at equilibrium. Also, in all cases let us assume only the reducing agent has been added to the solution. With the metal/solution case, as the electrode potential is made more positive, E_F is lowered in the diagram. Electron injection from the reducing agent of a couple will begin just before the potential when $E_F = E^o_{\text{redox}}$. This can be shown as follows: As discussed in Section 3.2, the net current is actually zero when $E_F = E^o_{\text{redox}}$ only in the special case where the concentration of reducing agent equals the concentration of oxidizing agent. At $E_F = E^o_{\text{redox}}$, $J = J_0$ from the reducing agent, and $J = -J_0$ from the oxidizing agent, where J_0 is the exchange current. If the concentration of oxidizing agent is essentially zero, then the current comes only from the reducing agent and is J_0 when $E_F = E^o_{\text{redox}}$. Thus, (neglecting concentration polarization) a practical amount of current from the reducing agent begins to flow when the Fermi energy in the metal is a little higher than E^o_{redox}, and has increased to J_0 when $E_F = E^o_{redox}$, and increases very rapidly as E_F drops below E^o_{redox}, because a high density of allowed empty levels (those above E_F) can accept electrons from the reducing agent. In other words, as the electrode potential V_m is made more positive in Figure 4.9a, current flows from the ferrocyanide, $Fe(CN)_6^{-4}$ ion when V_m ($= E_F/q$) reaches about + 0.4 V relative to the reference electrode, as shown by the arrow in the figure. The electrons tunnel through the Helmholtz layer. As the electrode potential continues to sweep more positive, the electrons from ions near the electrode are all transferred, and the current becomes diffusion limited, and perhaps drops to low value. Thus, one observes a peak in the voltammetry curve. Similarly when the electrode potential reaches approximately +1.6 V, another peak is observed, this time associated with the cerous ion in solution. On the return cycle, the ceric ion formed may give a peak of cathodic current as the voltage sweeps negative from +1.6 V, and any ferricyanide ion formed may give a peak when the voltage passes +0.4 V.

With this picture in mind, it is understandable that cyclic voltammetry has become[R19] a highly sophisticated tool in metal electrochemistry. It is used to analyze the species in solution; to analyze their interaction with the surface; and to analyze the other surface reactions, such as corrosion of the electrode, adsorption on the electrode, and so on. The potential of the current peak has been exhaustively analyzed as a function of voltage sweep rate, and can be interpreted in terms of the number of electrons involved in the rate-limiting step, and the degree of irreversibility of the reaction.[18]

Cyclic voltammetry does not offer quite such a useful tool for most semiconductor electrodes, but it has provided valuable contributions.[15,16] With a semiconductor electrode, the concentration of ions in solution is seldom measurable. Figure 4.9b illustrates the difficulty. This figure shows the band diagram for a typical depletion layer with the electrode potential V_m approximately 0.4 V relative to the hydrogen reference electrode. It is observed that when the value of V_m is 0.4 V, there will be no electron injection from the ferrocyanide, because the ferrocyanide level is no nearer the allowed empty energy levels, located at E_{cs}, than it was at $V_m = 0$ V. Primarily cyclic voltammetry is useful when the voltage drop V_m appears over an accumulation layer. Then the system resembles a metal (see the discussion of degenerate semiconductors at the end of Sections 2.2.4.2 and 3.4.3). In this case it can be expected that the current will begin near E^o_{redox}, just as with a metal electrode.

Another case where cyclic voltammetry on semiconductors will give results similar to a metal, is that where tunneling through the space charge region occurs, as illustrated in Figure 4.9c. If the thickness of the space charge region is less than say 30 Å, due to heavy doping [Equation (2.23)], then tunneling of electrons is possible. In this case, (a) current flows freely through the surface barrier, and (b) the Fermi energy is close to, or in, one of the bands, as estimated by applying Equation (1.1). Thus the semiconductor is almost, or is, degenerate, and acts like a metal[17] in many respects. And, in Figure 4.9c, as the level E^o_{redox} crosses the Fermi energy, one can expect electron transfer through the barrier (shown by the arrow) similar to the electron transfer through the Helmholtz barrier in the case of metals shown in Figure 4.9a.

Because of the limitations of current flow with semiconductor electrodes, as illustrated by Figure 4.9b, many techniques, such as the chronopotentiometric techniques[18] that are of great value with metal electrodes, are not found as yet to be of similar value with semiconductor electrodes, although use has been made of them[19]

Despite the above qualifications, general current/voltage measurements on semiconductor electrodes are, of course, the direct way in which electrode reactions are followed. Measurements of the current–voltage characteristics, either using straightforward dc measurements or using cyclic voltammetry, can be, and have been used to determine the properties of semiconductors (band edges) and the properties of solutions (energy levels, reorganization energies). As discussed in Section 6.5.2, for example, the detailed theory[20,21] of tunneling through a barrier has been applied[20] to determine "experimentally" energy levels and the reorganization energy λ. Another type of measurement, this one on a nondegenerate semiconductor, is the determination of band edges by studying the current for a series of redox couples as a function of their E^o. In accordance with the model of Figure 4.9b, electron injection into n-type semiconductors will not occur unless the energy level of the reducing agent is above E_{cs}. Similarly, electron capture will not occur unless the energy level of the oxidizing agent is suitably located with respect to E_{cs}. In Chapter 6 we will describe studies that were made of a series of redox species on ZnO to see at what value of E^o_{redox} current will flow.[22,23] From such a series, estimates of E_{cs} were made. The results and analysis will be discussed further in Sections 6.5 and 6.2, respectively.

4.2.2. Rotating Electrodes

In this section we discuss briefly the rotating-disk electrode (RDE) and the rotating-ring–disk electrode (RRDE). The RDE has been a powerful tool in metal electrochemistry, as reviewed by Riddiford,[24] Opekar and Beran,[25] Bruckenstein and Miller,[26] and in the book by Pleskov.[R20]

The electrode is made[27] in the form of a plate, a disk, with the normal to the flat area pointed down. It is rotated around the axis of the disk. Because of the viscous forces, the solvent flows along the surface of the plate from the center to the outer rim, flowing in a nonturbulent way. Fresh solution is drawn up to feed the flow as required. If the current is diffusion limited, the flow and the introduction of fresh electrolyte to replace the flowing electrolyte will bring ions to the surface faster than diffusion will. Analysis of the rate at which ions will be brought to the surface as a result of this viscous flow provides a simple formula, and the current is found to vary with the square root of the rotation frequency. Because the analysis yields a simple formulation, many important parameters of metal electrochemistry can be accurately obtained using rotating-disk electrodes.

With semiconductor electrodes, however, the rotating-disk electrode as such has been little used. The problem is that with a semiconductor electrode the electrode current is seldom diffusion limited. Normally, it is electron transfer limited for nondegenerate semiconductors, many orders of magnitude below that expected for diffusion-limited flow.

The rotating-ring–disk electrode (the RRDE), however, is a variation on the rotating-disk electrode that has proved very useful in semiconductor work as a method of monitoring the chemical changes due to electrode reactions. In the RRDE, the electrode (be it metal or semiconductor) is again in the form of a disk. In addition to the disk electrode, however, one has a ring of metal separated from the outer edge of the disk by a thin insulating layer. The disk and the encircling ring are on the same plane, normal to the axis of rotation. Beyond the metal is a second insulator, so only the thin ring is exposed to the solution. A sketch of the construction is shown in Figure 4.10. As the electrode rotates, the viscous forces cause the solution to flow along the face of the disk, and then to flow over to the ring, and any species formed on the semiconductor disk can be detected by voltammetry using the metal ring electrode.

This RRDE configuration has proved particularly useful in detecting photoinduced corrosion of semiconductor electrodes, where the consumption or the production of ions from solution at the disk semiconductor is quickly sensed using the ring. Results from such measurements will be discussed in Section 7.4. Normally in such photo experiments the light is chopped, to provide added assurance that only photocorrosion is being monitored, and to provide added sensitivity. The ring current of interest will, of course, follow the frequency of the chopper. Alternately a sinusoidal variation of the rotation frequency can be used.[28] A variation of the RRDE is the channel electrode,[29] where the solution flows past the electrodes.

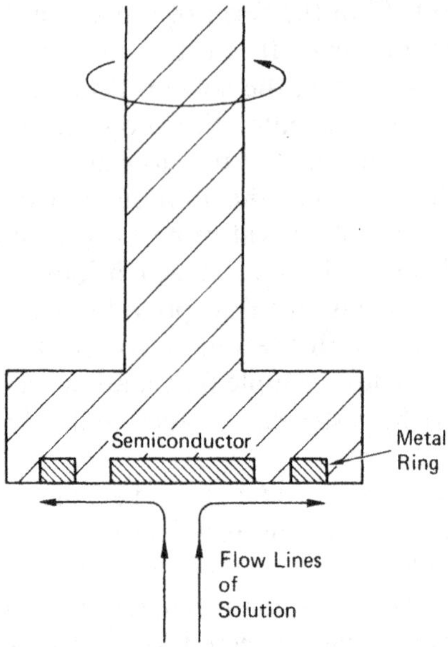

Figure 4.10. Schematic of rotating ring–disk electrode.

4.2.3. Illumination

The study of the reactions of photoproduced carriers is of great importance with semiconductor electrodes for two reasons. First, it is of potential technical importance in photocatalysis, photolysis, and solar energy conversion, as will be discussed in Chapter 9. Second, it has proved a useful tool in characterizing the carrier/ion reactions because of the ability to create the reacting carriers at will. Because of this latter ability, a large fraction of our knowledge in semiconductor electrochemistry has arisen from studies of photocurrents. Most of the techniques are obvious and will be indicated with the results. For example, the results on carrier injection from optically excited dye molecules in solution will be described in Section 6.5. To provide an example of the use of photoproduced carriers, in the present section we will discuss the measurement of diffusion length.

When minority carriers are photoproduced using photons of band gap energy, the average distance the minority carrier diffuses before deexcitation (recombination) is termed the diffusion length: L_p if the minority carriers are holes; L_n if electrons. This diffusion length can be determined[17,30] by monitoring the current due to the minority carriers. Electrochemical measurement turns out to be a convenient way of measuring the diffusion length. The diffusion length of minority carriers in a semiconductor is one of the important bulk properties of a crystal. Also, the rate at which photoproduced minority carriers come to the surface in an electrochemical experiment depends on the diffusion length, as we will show below, so the diffusion length is of interest in such electrochemical processes as solar energy conversion.

If one assumes the illumination is penetrating, the formulation for diffusion length measurement is particularly simple.[30] Specifically, if the wavelengths used are such that the diffusion length and the space charge thickness are both short compared to the reciprocal of the absorption coefficient α, then the rate of minority carrier generation per unit volume is independent of distance, and proportional to the light intensity:

$$dp/dt = bI_0 \tag{4.30}$$

where holes are indicated as the minority carrier of interest, so L_p is the diffusion length of interest. In the voltage range to give a depletion layer of thickness x_0, any minority carrier reaching x_0 will be swept to the surface by the electric field of the depletion layer. But by the definition of the diffusion length, all the holes (on the average) produced within L_p of the plane x_0 will reach x_0. As x_0 increases with voltage, the photocurrent increases with voltage.[31] The total current is from the holes generated in the volume $x_0 + L_p$:

$$J_p = bI_0(L_p + x_0) = bI_0(L_p + A\kappa\epsilon_0/C_{sc}) \tag{4.31}$$

where the second form derives from Equation (4.2). Then a plot of the current density J_p as a function of C_{sc}^{-1} gives the diffusion length L_p from the intercept at $J_p = 0$. As described above, C_{sc} can be readily measured and L_p determined.

A more detailed formula accounting for absorption of the light has been developed by Gärtner,[32] Butler,[33] Ginley and Butler[34] and leads to the relation

$$J_p = aI_0[1 - (1 + \alpha L_p)^{-1}\exp(-\alpha x_0)] \tag{4.32}$$

where a is a constant. This expression is based on a model for photocurrent where monochromatic radiation is used. For very low α, Equation (4.32) approaches (4.31). One determines L_p by varying the wavelength.

Another measurement of photoproduced current that gives considerable information about the reactions of carriers with ions at a semiconductor surface is the "current doubling" measurement. Under certain circumstances the reactions of minority carriers at the semiconductor surface lead to free radicals, and the presence of the free radicals is detected by unusually high currents. By observing the current as a function of parameters of interest, the behavior of the free radicals as a function of these parameters can be monitored. As the experimental technique cannot easily be separated from the theory of the electrode reaction, the description of the method will be presented at the same time as the observations using the technique, in Section 6.3.1.

4.3. Other Techniques

Many techniques common to other fields, particularly semiconductor physics, and physical or organic chemistry, can be useful in studies of the semiconductor electrode behavior. The most obvious example is simple chemical analysis of the solution following an electrode reaction. Or, conductance or Hall effect measurements to determine the properties of the semiconductor under study.[35] It is not our intent to review relatively standard techniques normally used to measure the bulk properties of one phase or the other. On the other hand, there are rather exotic techniques. These again we will not review in detail. An example would be Mössbauer analysis[36] where the results of the referenced paper are of interest to us because of the application to surface layers. Other, even more rarefied techniques that may be useful would be polaromicrotribometry,[37] where surface films are detected by friction methods, or an intriguing technique using liquid crystals to detect nonuniformities in current flow into a semiconductor.[38] Because semiconductor electrochemistry is a relatively new field, many techniques which appear to have potential value have not been utilized to date. We are emphasizing in this section those techniques that have been used, and have proven valuable in illustrating the basic features of semiconductor electrodes. Valuable supplementary reading will be found in the review of Pleskov.[35]

4.3.1. Techniques for V_s Measurement

We will review two techniques for V_s measurement other than the capacity measurement discussed earlier, namely, electroreflectance and the introduction of a generous supply of minority carriers (by, for example, illumination).

In the electroreflectance measurement bandgap radiation (photon energy greater than E_g) is reflected from the electrode. The reflectance varies with electric field near the surface due to the Franz–Keldysh effect, and the electric field at the surface can be conveniently varied by the applied voltage, in accordance with Equation (2.27) (where dV/dx is the field). A greater reflectance obtains at a higher electric field; a lower reflectance at a lower electric field. Experimentally a small ac voltage is applied between the semiconductor and the counterelectrode. This ac voltage varies the electric field at the semiconductor surface and so the reflectance shows an ac component. Now, if the bands are bent up by the dc voltage (V_s positive), the bands will be bent further when the ac voltage applied to the electrode is in its positive half cycle, and the reflectivity will be in phase with the ac voltage. However, if the bands are bent down by the dc voltage, the bands will flatten on the positive swing of the ac voltage, and the reflectivity will be out of phase with the ac voltage. When we have flat bands, the ac component of the reflectance will go to zero, as there is no field at the surface of the sample. The phase change and the zero at flat band offer an easy measurement, enabling one to determine whether V_s is positive, zero, or negative, and to determine directly the flat band potential V_{fb}. Aspnes and Frova[39] and Gobrecht and Thull[40,41] review studies at the semiconductor/electrolyte interface, and recently Vos and Krusemeyer[42] and Pond[43] have reviewed more recent work and the problems inherent in the approach. The technique has problems, probably related in many cases to the difficulty of moving a band to the real flat band configuration. As the flat band configuration is approached, but still not reached, high currents may flow and the voltage drop may occur elsewhere in the circuit. Then a further change in applied voltage causes no change in the band bending and this appears as an erroneous, or at best an insensitive, measure of the position of the flat band potential. Other problems are undoubtedly related to the effect of the reactions of the photoproduced minority carrier at the surface, because bandgap illumination causes a flood of minority carriers.

On the plus side, however, Vos and Krusemeyer do report that for TiO_2, if certain wavelength illumination is used, the sign reversal in electroreflectance does agree with the value of V_{fb} as obtained by extrapolation of the Mott–Schottky plot of capacitance.

Another technique[44] to measure the flat band potential V_{fb} is to provide a large density of minority carriers at the surface, much larger than can be consumed by a surface reaction or by recombination. If initially a depletion region exists at the surface, the minority carriers will be drawn to the surface and cause V_s to approach zero. If. for example, the sample is n type and V_s is positive, holes are introduced. They will move to the surface and the surface charge N_s (originally negative to produce a positive V_s) will become less negative because of the positively charged holes. Then, by Equation (2.25), the band bending, V_s, will decrease. As the light intensity increases, the band bending will approach zero, the flat band case. If, on the other hand, the bands bend down (V_s negative) initially resulting from a net positive charge at the surface, electrons that are photoproduced will be attracted to the surface, and again the band bending will approach zero. Normally the excess carriers are produced by light, and the change in band bending is monitored by the change in electrode potential. If one can convince oneself that under high illumination the bands are flat, i.e., $V_s = 0$, then the change in electrode potential ΔV_m gives the value of V_s in the dark:

$$\Delta V_m = \Delta V_s = V_s \tag{4.33}$$

However, if the surface recombination of minority carriers is rapid (perhaps due to a covering oxide), or if the minority carriers are consumed in corrosion or another electrode reaction, the V_s may saturate, approaching a value other than zero, and the results will be in error. Thus this technique is very questionable if the results are not supported by other evidence.

An interesting variation on this nonequilibrium method to measure V_{fb} was reported by Pleskov,[14] who, rather than light, obtained the minority carriers by locating a p–n junction near the surface. Then he biased the p–n junction in the forward direction and caused minority carriers to pour into the surface region and flatten the bands. Again the electrode potential change of the surface region, relative to a reference electrode in the solution, enabled the determination of V_s before the introduction of the minority carriers.

4.3.2. Techniques to Determine Surface Species or Phases

In this short section two techniques will be mentioned briefly, namely, ellipsometry and attenuated total internal reflectance. Both of these techniques have been reviewed many times.[R4] For example the complete volume of *Surface Science*, Volume 56, is devoted to ellipsometry, so for a more detailed discussion, the papers therein and the references from these papers will give a comprehensive background of the use of ellipsometry. The objective here is only to list the techniques and to discuss them sufficiently for later reference in the book.

Ellipsometry is basically a technique used to observe films on a flat surface. Plane polarized light, with the plane of polarization rotated 45° from the plane of incidence, is reflected from a surface. The two components of the polarization, a component polarized perpendicular and a component polarized parallel to the surface, are then equal in amplitude in the incident beam. Now if there is a film on the surface, the two components of polarization in the reflected beam will not be equal either in phase or amplitude. The difference in phase between the two reflected components is termed Δ, and the ratio of two amplitudes is termed ψ. By a suitable arrangement of polarizers, compensators, and analyzers, the values of Δ and ψ can be determined accurately. The thickness and the index of refraction of the film can be determined with the Drude theory.

Where ellipsometry detects the thickness and index of refraction of surface phases, attenuated total reflection (ATR) detects the composition of surface phases and adsorbates of monolayer or multilayer concentration. In attenuated total reflectance, or total internal reflectance spectroscopy, infrared light enters the edge of a wafer. It is internally reflected many times from the surface, and exits from the other edge, so that a species present on the surface of the wafer will have many chances to absorb light, and its absorption spectrum will be seen. The crystal is made in the form of a parallelepiped. The light is introduced at one end of the sample, at a low enough angle that when it reaches one of the flat surfaces it is beyond the critical angle for total reflection. If the parallelepiped is thin enough, the light can be reflected many times from the surface before it emerges. Now when light is thus reflected totally, electromagnetic theory tells us that the wave has an exponential tail extending the order of a micron beyond the surface. Thus adsorbates, or thin phases at the surface, have ample opportunity to absorb the light and make their presence known.

Harrick[(45)] has pioneered the use of the ATR technique, particularly at the solid/gas interface. Hansen *et al.*,[(19)] Möllers and Memming,[(46)] and Yeager's group[(47,48)] have developed the technique for use with semiconductor electrodes. One can study free electron absorption by this method, but it is of more general value as a tool to study composition and composition changes in the solution, or at the surface of the electrode.

4.3.3. Techniques to Study Electrode Reactions

The "transitor technique" determines whether an electron transfer reaction occurs via the conduction band or the valence band, that is, whether electrons or holes are primarily responsible for the current across the interface. A schematic of the circuit is shown in Figure 4.11. Germanium is indicated as the semiconductor, and actually it was the first[(49)] semiconductor used with this configuration.

With the configuration of Figure 4.11, with an n-type material facing the electrolyte, there are both conduction band electrons and valence band electrons available to reduce an ion in solution. A p–n junction is provided near the surface and the junction is reverse biased. If valence band electrons are consumed (holes are injected), and the hole diffusion length is long enough, the hole will proceed into the p-type region. Thus, any current associated with the valence band processes will be measured as current through the contact to the p-type material. Any current associated with conduction band processes will be measured through the n-type material. Thus the ammeters A_p and A_n distinguish the current associated with valence band and conduction band processes, respectively. A high energy level such as E_{ox} in Figure 1.6 could only capture

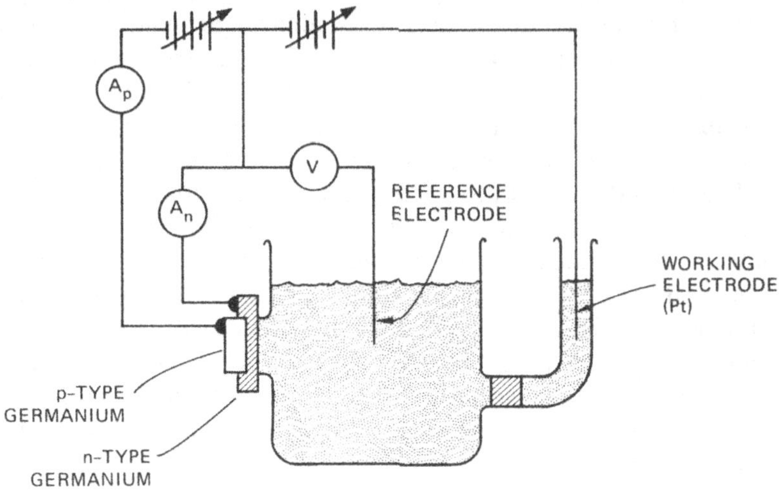

Figure 4.11. The "transistor technique," where the electrode has a p–n junction near the surface so the valence band and conduction band processes can be distinguished.

conduction band electrons. Valence electron capture (hole injection) would indicate, on the other hand, a low energy E_{ox}.

Another technique, the measurement of electrical noise, has been used to study electron and hole transfer, particularly to test whether in a particular two-electron reaction the two electrons are transferred almost simultaneously, or whether one electron is transferred to form an intermediate species and then the other electron is transferred later. A parameter termed the correlation coefficient will describe how many electrons are transferred per event, and its frequency dependence will determine approximately the average length of time between the two electron transfer steps. The technique has been used by Cardon and Gomes.[(50–52)] Unfortunately the results can be unclear because of other contributions to the noise, such as $1/f$ noise at low frequencies, and trapping phenomena at high frequency.[(52)] For example, in studies of the reduction of $Fe(CN)_6^{-3}$ to $Fe\,(CN)_6^{-4}$, which obviously must be a one-electron reduction, they found by noise measurements that the above problems led to erroneously high measured values of the number of electrons per event, and the minimum value they found, at an intermediate frequency, was 1.3 electrons per event. However, for the reduction of I_3^- a value of unity was found by Tyagai and Lukyanchikova[(53)] on CdS, and for the oxidation of I^- on ZnO (by holes) a value of unity was found by Gomes and Cardon.[(51)]

The study of luminescence associated with electron transfer to or from electrodes can provide additional information regarding chemical and electronic processes occurring at an electrode. The luminescence can appear either in the solution or in the solid.

Luminescence in the solution has been observed in cases where the conduction band of a solid is high enough that electron transfer to the excited state of an ion in solution occurs, and the luminescence is due to radiative deexcitation.[(54)] Alternately, the electrode reaction can produce a species in the solution capable of reacting with a species already present. Then the luminescence in the solution can arise due to a chemical reaction in solution.[(55)] As a third possible sequence of events, a reaction in solution causing luminescence can lead to electron transfer.[(56)] Such processes can easily be distinguished and provide direct information regarding the steps in the electrode reaction.

Luminescence in the solid is generally associated with the presence of minority carriers. Three types of studies are of interest involving electrode luminescence. One is a study of the effect of electrode potential on normal fluorescence, where the fluorescence is characteristic of the solid induced by minority carriers, (the minority carriers being created by illumination). If the electrode potential is adjusted to sweep such minority carriers to the surface, the fluorescence will decrease. This application of luminescence offers an interesting way to measure the diffusion length of minority carriers,[(57,58)] a technique analogous to the capacity technique described above [Equation (4.31) or (4.32)]. The second case where a study of luminescence in the solid can lead to an understanding of electrode reactions[(21)] is where the solid luminesces, but the minority carriers are introduced by an electrochemical reaction.[(59)] This is one of the best ways to prove conclusively that an energy level in solution is active enough to inject minority carriers. The third interesting case of luminescence in the solid[(60)] is where the minority carriers are obtained by breakdown processes due to overly high voltages applied to the electrode. Breakdown can be of two types, avalanche breakdown, or direct tunneling through the energy gap. In either case, as discussed in Section 2.2.4.3, minority carriers are produced. The observation of luminescence in this case provides information regarding the electric field in the semiconductor due to the applied voltage,[(61)] which must of course, as discussed in Section 2.2.4.3, be high enough to cause breakdown.

Finally, it is of interest to mention the electrophoresis measurement, the measurement of zeta potential shifts on powders that arise due to illumination. In a sense this is not an electrochemical measurement, but in another sense it is, because the influence of the illumination is to create electrons and holes which come to the surface and react. Hauffe and Volz[(62)] studied this effect on various semiconductors and were able to deduce information about the electron and hole reactions that must be occurring. The measurements and interpretations will be discussed further in Section 7.3.2.

5

The Properties of the Electrode and Their Effect on Electrochemical Measurements

In this chapter experimental measurements are reviewed that illustrate how the properties of the electrode have been determined using electrochemical techniques, and how the properties of the electrode are found to affect electrochemical reactions at the surface. In Section 5.1 we discuss the properties of the "perfect crystal" where completely ionized donors or acceptors, surface states, and even surface phases are considered attributes of a perfect crystal. Studies of such properties of the perfect crystal by electrochemical measurements[R14,R23.1] are made usually in an indifferent electrolyte (negligible current) and have been a straightforward application of the methods of Sections 2.2.4, 2.4, and 4.1. But, as described in Section 2.4, some surface parameters of the solid, such as any voltage across surface phases or Helmholtz voltages due to adsorbates on the solid, must be known to interpret the measurements. Thus, first we discuss the charge on the solid, and the adsorbates on the solid, that give rise to the Helmholtz double layer. Then the behavior of the space charge double layer is discussed, with two objectives—first, to indicate how the solid is studied using electrochemical techniques; second, to indicate how the properties of the defect-free solid influence current flow to or from an active electrolyte. Experimental observations are emphasized where the behavior is determined more by the properties of the solid than by the properties of the species in solution; the behavior of ions in solution is discussed in detail in Chapter 6.

In Section 5.2 we discuss the behavior of crystal defects, where deep levels in the bulk semiconductor, dislocations, grain boundaries, and variation with distance or time of the donor or acceptor density, are considered defects. Again, two objectives will be recognized: to review how the defects have been studied using electrochemical techniques, and to review how the defects are found to affect current flow between the electrode and an active electrolyte.

Finally, in Section 5.3 we present a summary of results on the band edges of semiconductors, as measured electrochemically, for use as a reference. This section will serve, hopefully, a dual purpose; not only providing the basis for grouping semiconductor materials in accordance with their properties, but also providing literature references for less studied materials that are not referred to elsewhere in the book.

5.1. The Behavior of the Perfect Crystal

5.1.1. The Helmholtz Double Layer: The Surface Charges on the Electrode

5.1.1.1. The Dependence of V_H on pH. In Section 2.3 the methods of theoretical estimation of the energy of band edges for a semiconductor were developed, and by Equation (2.31) it was shown that any changes in the Helmholtz double layer voltage V_H result in equal changes in the measured value of the band edge energy. Thus, the first requirement in characterizing the solid is to characterize the Helmholtz double layer. In Section 2.2.3 in turn, we have showed that V_H varies with the pH of the solution. Here we want to illustrate with a few examples how well the theoretical relation [Equation (2.20)] for the Helmholtz double layer as a function of pH is followed, and discuss deviations for the perfect crystal. This will indicate the degree of confidence expected for extrapolations of measured band edges to the pH of zero Helmholtz voltage, as will sometimes be done in the listings of Section 5.3.

First, we derive more explicitly the relation between the flat band voltage and the pH. From Equation (2.31) we have

$$E_{cs} = E^o_{cs} + qV_H \tag{5.1}$$

or with Equation (2.36)

$$qV_{fb} = E^o_{cs} - \mu + qV_H \tag{5.2}$$

and, including Equation (2.20) (which implies only protons and hydroxyl ions are adsorbed),

$$qV_{fb} = E^o_{cs} - \mu + B - 2.3kT(p\text{H}) \tag{5.3}$$

Equation (5.3) can be better expressed as

$$qV_{fb} = E^o_{cs} - \mu + 2.3kT(\text{pzc} - p\text{H}) \tag{5.4}$$

when we recall that $V_H = 0$ at the point of zero charge, pzc.

As was discussed in Section 2.2.3, the Helmholtz double layer is expected to depend on pH in accordance with Equation (2.20) for insulators and semiconductors, but not for clean metals. For clean metals, the value of V_H is determined by electron exchange between the solution and the solid, in general, whereas in insulators the value of V_H is determined by adsorption/desorption of ions, usually protons or hydroxyl ions in aqueous solution. The reasons why semiconductors should follow the behavior of insulators rather than metals were discussed in Section 2.2.3. The point is that if experimentally a relation such as Equation (5.3) is found valid, and strongly adsorbing or complexing ions (such as I_3^- or CN^-) are avoided, then the model that the Helmholtz layer is determined by adsorption/desorption of hydroxyl ions can be assumed valid.

Experimentally, in most cases for nondegenerate semiconductors, the flat band potential does indeed follow Equation (5.3). Lohmann[2] determined Equation (5.3) to be valid for ZnO; Fujishima *et al.*[3] and Dutoit *et al.*[4] for TiO_2; Madou *et al.*[5] for GaP; Laflere *et al.*[6] for GaAs; Kennedy and Frese[7] for Fe_2O_3; Halouani *et al.*[8] for $Ga_{1-x}Al_xAs$ and $GaAs_{1-x}P_x$; Möllers and Memming[9] for SnO_2; and Boddy[10] for Ge. Many authors have shown that V_H is independent of redox ions in solution as expected by the theory of Section 2.2.3. Bernard and Handler[11] have shown that V_H is independent also of conductivity type for GaAs, another indication that V_H does not depend on electron exchange. They showed this by determining that the difference in flat band potential between *p*- and *n*-type materials is about equal to the difference in Fermi energies. We can therefore expect that for these materials (except germanium, as discussed below) the flat band potential will follow Equation (5.4) if the pzc = IEP for the solutions studied (see Section 2.1.1).

More recently Ginley and Butler[12] have shown that with CdS the flat band potential is determined both by the pH and the sulfide ion concentration in solution. Thus both H^+ and HS^- are presumably adsorbed. Actually they conclude that H^+ and HS^- are the potential-determining ions, the conclusion stemming from an analysis of the variation of V_{fb} with the concentration of the various species in solution. To say that H^+ and HS^- are the potential-determining ions is thermodynamically equivalent to saying that S^{2-} is the potential-determining ion. They find the IEP on CdS occurs at $[S^{2-}] = 4 \times 10^{-9}\,M$.

Presumably, the reason for the simple behavior of the Helmholtz layer on ionic solids arises because of the nature of their adsorption sites. Very few substances can displace water from the adsorption sites because water is bonded typically about 10 times as strongly as most other common adsorbates.[R13] The V_H-determining adsorption of protons or hydroxyl ions occurs so energetically on ionic solids because on the surface of ionic semiconductors, or on the surface of the oxide layers that normally cover covalent semiconductors, there are acidic and basic sites, as discussed in Section 1.1.2.1. The Lewis acid sites strongly attract the electron pair on the hydroxyl ion; the Lewis base sites strongly attract the pair of unoccupied orbitals on the proton. Thus the equilibrium of Equation (1.6) (and a corresponding equation for OH^- adsorption on cations), determining whether more H^+ or OH^- is adsorbed, determines the charge on the solid surface[R4] and the resulting V_H.

5.1.1.2. The Dependence of V_H on Surface States on Covalent Crystals. On a highly covalent semiconductor such as germanium, the usual pH dependence of V_H is observed, but the behavior is complicated by adsorption with covalent bonding.[R10] Specifically, water molecules adsorbing onto germanium are reported to dissociate into radicals, OH· and H· radicals, and these form covalent bonds with the germanium surface.

The surface properties of the germanium electrode differ depending on whether a cathodic or an anodic current is flowing, or has been flowing, across the interface. This is concluded from the fact that the minimum capacitance (Figure 4.5) occurs at a different potential, depending on history. There are shifts of up to about 0.4 eV in the flat band potential V_{fb}. Brattain and Boddy[13] first suggested that these shifts on germanium were associated with adsorbed dipoles. Gerisher, Maurer, and Mindt[14] analyzed the behavior in more detail and suggested that the shift was due to a current-stimulated change in the surface layer. After a cathodic treatment (the germanium biased negative with respect to the solution), the adsorbed layer is hydrogen atoms; after an anodic treatment, the adsorbed layer is hydroxyl groups. The reactions leading to these layers are

$$\text{Ge:OH} + e^- + \text{H}^+ = \text{Ge}\cdot + \text{H}_2\text{O} \tag{5.5}$$

$$\text{Ge}\cdot + \text{H}^+ = \text{Ge:H} + h^+ \tag{5.6}$$

where a cathodic current causes both equations to go to the right, an anodic current causes both equations to go to the left. There are thus two relatively stable surface layers, Ge:OH and Ge:H, with an intermediate condition with dangling bonds on the germanium. The adsorbate dominating an open circuit depends on the recent history of current flow.

The same behavior, the adsorption either of H· or OH· depending on current flow, has been reported by Madou and his co-workers[5] on GaP; and by Gerischer and Mattes[15] on GaAs.

Although the model of Equations (5.5) and (5.6) explains many of the observed features of the experiment (such as the observation of the Ge· as a recombination center, as will be discussed in Section 6.4.2), the two important features of the measured V_H are not explained well. First, the shift of 0.4 V in Helmholtz voltage is not really to be expected. Each of the surface groups, Ge:H, Ge·, and Ge:OH, is electrically neutral. Neither reaction (5.5) nor (5.6) leads to a change in surface charge, if we assume the H^+ ions and the water molecules come from the solution. Second, it is somewhat puzzling how the germanium surface could follow Equation (5.3), that is, have a pH dependent V_H as discussed in the preceding subsection, and yet exhibit the strongly covalent behavior of Equations (5.5) and (5.6). Equation (5.3) suggests an ionic solid, and Equations (5.5) and (5.6) suggest a clean covalent surface. A possible explanation for this behavior is that the germanium is covered with a thin hydrous oxide layer, and the surface state (the Ge:OH or Ge:H) is at the interface between the hydrous oxide and the covalent semiconductor (and so it is better termed on interface state). Then, as the equivalent of Equations (5.5) and (5.6) proceeds, the protons or hydroxyls are "interface states" probably associated with Ge:H or Ge:OH bands, become oxidized or reduced to form the covalent bonds indicated. Under these conditions, with little ion movement, the charge remains near the surface and the 0.4-V or so shift in V_H is explainable. The adsorption of H^+ and OH^- will, of course, occur at the outer surface of the oxide and lead to the pH effects. Such a structure is to be expected, for, in aqueous solution, any of the semiconductors, Ge, GaAs, or GaP, would be expected to have a thin oxide layer.

5.1.1.3. The Dependence of V_H on Surface States on Ionic Crystals. In Section 5.1.1.1 above, it was made clear that on ionic crystals, V_H depends on intrinsic surface states in their role as adsorption centers, particularly for H^+ and OH^- adsoprtion. But, intrinsic surface states could also exchange electrons with redox ions in solution, and affect V_H that way. Thus at this point it is necessary to explore the tendency of intrinsic surface states (in either the bare or hydroxylated form) to become charged and thereby affect V_H.

Charge exchange with intrinsic surface states on ionic semiconductors is often postulated by workers in the field to explain abnormal experimental results, including shifts in V_H (measured as shifts in V_{fb}). The influence of such surface states and of surface phases on the Helmholtz potential is, actually, not well understood. However, any surface state, even an intrinsic surface state, that can be highly charged (the order of 10^{17} electronic charges/m^2 or more) should affect the Helmholtz voltage. This expectation stems from Equation (4.1). Because surface states are at the solid/solution interface, and because we expect a Helmholtz capacity of the order of 0.1 F/m^2, a surface state charge of 10^{17} q/m^2 will, from Equation (4.1), lead to a Helmholtz voltage change of 0.16 V. Thus to a first approximation, to evaluate whether the surface states influence V_H, we need only to determine whether they can be charged to the above level.

Intrinsic surface states may be present in densities up to 10^{19}/m^2, and thus by the above argument, they should affect V_H, but such a role has not often been observed. As discussed in Section 1.1.2.1, one may generally expect donor surface states near the valence band (neutral when occupied) and acceptor states near the conduction band (neutral when unoccupied). If the E_{redox} of a redox couple in solution is above the acceptor state, one expects the acceptor states to become charged. If the E_{redox} of a redox couple in solution is below the donor state, one expects the donor states to become charged. In either case, one then expects the Helmholtz potential (as measured by V_{fb}) to be dominated by the electron exchange between the ions in solution, and the surface states. If the electron exchange between the bulk semiconductor and the surface states is the rate-limiting step in any overall current flow, of course, the semiconductor should still have a transfer coefficient of zero, as discussed in Section 3.4.3. However, V_H will be determined by surface state effects, not ion adsorption. Such surface state control of V_H on ionic semiconductors, where nonadsorbing redox couples exchange electrons with intrinsic surface states, has not apparently been observed. Normally pH is the only controlling variable.

The reason surface state effects on ionic crystals are not found to control V_H may simply be that the sites that could offer intrinsic surface state in the band gap are stabilized by bonding to ions from solution. These intrinsic surface states are really the same sites that provide the acid or basic character of the surface.[16] Then a surface cation—one that at the solid/vacuum interface provides the intrinsic acceptor state for capturing an electron—may prefer in solution to share two electrons with an adsorbing OH^- ion.

$$M_L^{2+} + OH^- \rightarrow [M—OH]^+ \tag{5.7}$$

The acceptor surface states are expected to shift toward the conduction band by such an acid–base reaction. In other words, once the OH^- is adsorbed, the surface cation no longer attracts an unpaired electron.[17] Similarly, the donor level associated with surface lattice oxygen, the donor intrinsic surface state, will move toward, or into, the valence band when the lattice oxygen shares its electrons with an adsorbing proton.

$$O_L^{2-} + H^+ \rightarrow [O—H]^- \tag{5.8}$$

This possibility is discussed further in Section 6.4. As will be reviewed in Sections 6.1 and 6.4, researchers have, in several cases, inferred the presence of intrinsic surface states at the surface of ionic solids, based on observations other than V_H shifts. In most cases intrinsic surface states are postulated, based on unusual current/voltage characteristics or on measurements of corrosion. In the case of intrinsic surface states that are involved in corrosion (Chapter 8), the lack of Helmholtz potential effects by their interaction with redox couples is understandable. Here, if an intrinsic surface state becomes charged, the corresponding ion dissolves into solution or forms a new phase. Thus the charge does not remain on the solid to affect V_H.

On the other hand "normal" surface states, not involved in corrosion after capturing carriers, have been reported. For example, as will be discussed further in Chapter 6, Kohl and Bard[18] suggest surface states (e.g., at 0.5 eV below E_{cs} for ZnO) on ZnO, CdS, GaP, and TiO_2. Tyagai and Kolbasov[19] suggest surface states on CdS. The lack of observed V_H effects, as a function of redox couple in solution, places an upper limit (as evaluated above) on the surface state density: it must be the order of 10^{17} m^{-2} or less in concentration.

On the other hand, the effect on the Helmholtz potential of adsorbed species (as opposed to intrinsic surface states) is well documented. Three distinct cases can be distinguished: first, the adsorption of ions of the host crystal; second, the adsorption of foreign ions because of ion exchange phenomena; and third, the adsorption of foreign ions as a precursor to precipitation.

The case of Section 5.1.1.1, the adsorption of OH^- and H^+ on oxides, can be considered an example of the first case, the adsorption of ions of the host crystal. Such a suggestion is based on the realization that the distinction between a hydrous oxide and an anhydrous oxide is often blurred in aqueous solution. As described in Section 5.1.1.1, such adsorption usually dominates in controlling V_H on oxides. A similar example of the adsorption of ions of the host crystal dominating V_H is the adsorption of HS^- and H^+ on CdS.[12]

The adsorption of foreign ions by ion exchange has been reviewed in Section 2.1.1.3. There the concept of the IEP was introduced to cover just this case, where the pH that gives $V_H = 0$ varies because of foreign ion adsorption. In that section electrophoresis was cited as a good way to determine the IEP. However, changes in the IEP can be monitored also by changes in the flat band potential. Examples of studies where the flat band potential (as opposed to electrophoresis) was used to determine the change in V_H due to ion adsorption are provided by the work of Uchida *et al.*[20] They showed that in aqueous solution the flat band potential of SnO_2 depends on alkaline earth ion concentration when at high concentration. They also showed[21] that in molten sodium chloroaluminate, the flat band potential of SnO_2 depends on the chloride concentration with V_{fb} varying as the log of the chloride concentration.

The adsorption of ions as a precursor to precipitation has been studied by James and Healy,[22] and can be considered another form of adsorption. Here hydrolyzable metal ions of valence $+n$ are shown to adsorb when the pH rises, when the ions accumulate enough OH^- ligands to assume the form $M(OH)^{+1}_{n-1}$.

$$M^{n+} + OH^-_{n-1} \rightarrow M(OH)^+_{n-1} \tag{5.9}$$

Such behavior occurs just before the pH becomes high enough to precipitate the ion as the neutral hydroxide $M(OH)_n$. In this case specifically, there will be a very rapid dependence of V_H on pH as the pH goes through the narrow region in which adsorption occurs, and into the higher values where a surface layer of the metal hydroxide forms.

5.1.1.4. Double Layers Associated with Surface Phases. Little will be said in this chapter about the formation, or electrical properties, of separate phases at the surface of semiconductor electrodes. We are interested here only in their influence on the apparent V_H. The presence of separate surface phases is a recurring problem in semiconductor electrochemistry (and, of course, metal electrochemistry). Efimov and Erusalimchik in their review of silicon and germanium[R24] indicate that even in HF, measurements of the silicon surface are difficult because of the inevitable presence of an oxide layer. Madou *et al.*[5] note the formation of Ga_2O_3 or $Ga(OH)_3$ insoluble layers on GaP while making cyclic voltammetry measurements, and Harvey[23] notes a similar problem with GaAs. In the interpretation of measurements, both on semiconductor electrodes and on metal electrodes, such layers must be recognized before a realistic model can be developed. Further discussion of this important characteristic of the electrode will be postponed and described in detail in Chapter 8, after the measurements on "clean" electrodes are reviewed in the interim chapters. Here we simply wish to indicate that the formation of such surface phases can be expected to substantially alter the Helmholtz potential.

When a surface phase develops, three reasons for a change in V_H can be visualized. If the surface phase is an insulator, the pzc will be different, in general, from that of the host material, so V_H will change. If the surface phase is conducting, electron exchange with redox couples in solution will determine V_H in a way similar to the case of a metal or of surface states. And finally, interface states at the solid/solid interface or space charge in the surface phase, may induce a substantial voltage drop across the thin surface phase that is not a real Helmholtz voltage, but is very hard to distinguish from a Helmholtz voltage.

Although our knowledge of the influence of surface phases on V_H is meager and poorly documented, recognition of these surface phase effects is important in describing the semiconductor. For good experimental data to compare to theoretical models of band edges for various semiconductors (Section 2.3), (a) we should have IEP = pzc; (b) we must have no double-layer voltage present across a surface phase; and (c) the measurements must be made at the pzc, or it must be safe to extrapolate the results to the pzc. The point is that if there is any danger of a surface phase, then (b) and (c) both are probably invalidated. When an oxide phase is present, the possibility of a residual voltage across the oxide phase at flat band cannot be precluded even at the pzc. Then the real band edge E_{cs}^o of GaAs with an oxide layer would not be measured, but the band edge plus this residual voltage.

5.1.2. The Space Charge Region of the Perfect Crystal

In this short section we describe the influence of the ideal space charge region on electrochemical measurements. The section is short because most of the results are described elsewhere.

Observations of the space region of the perfect crystal in an indifferent electrolyte (negligible current flow) follow almost by definition the theoretical models of Section 2.2. For example, the Mott–Schottky plot of the capacity/voltage relationship is found to give a fairly accurate value for the charge in the space charge region [Equation (4.19)], when partially ionized flaws or partially ionized bulk impurities are not contributing.[1] The density of majority carriers at the surface follows, as far as can be determined, the expectations of the space charge theory as given by Equation (2.10).

The flow of current with an active electrolyte, on the other hand, does not always follow simply from the models of the space charge region. The problems have been mentioned earlier. Of particular interest are the influence of flaws in the solid on the reverse diode characteristics of the semiconductor electrode, observations of tunneling through the space charge region, and the observed elimination of space charge effects when the semiconductor is made degenerate.

In the often observed case, where J_0 [Equation (3.33)] is very low, the space charge barrier height, V_s, present under open circuit conditions has no theoretical meaning. V_s at open circuit simply represents the barrier reached when the forward current of the electrode equals the (unpredictable) reverse current that arises at flaws. The case was discussed in more detail in Section 3.3.3.2 while introducing Equation (3.34).

The irreproducibility of reverse currents, and hence the lack of significance of the barrier height at open circuit conditions, has been pointed out several times.[R10,24,25,26] However, it is still always tempting to adopt the Gerischer model (the equilibrium model) in its entirety, and to assume $E_F = E_{redox}$ (equilibrium) at open circuit. Such temptation should be resisted unless one has a very narrow bandgap semiconductor where the thermal generation of minority carriers is significant at ambient temperatures. One should normally assume that V_s under open circuit conditions has no theoretical significance.

The "diode" characteristics in the dark of the perfect semiconductor/electrolyte interface (Section 3.3.3.2), although not as good as theory predicts, can be excellent because of the low reverse current when no light and few flaws are present. For example, Morrison and Freund[27,28] studied the reverse current of ZnO, using an etching procedure that apparently eliminated the effect of dislocations from the ZnO surface. In this etching procedure,[29] first a nonpreferential etch was used to remove superficial damage. Then a highly preferential etch was used such that the exposed face was not etched, but all other faces were. The result was that the remaining dislocations were etched deeply, while the planar portions of the exposed surface were not etched. Thus a relatively flawless surface could be exposed to the solution with presumably little current from the dislocations because they were at the end of long resistive pores. Under these conditions, anodic voltages greater than 100 V could be applied to the surface with very low reverse currents (tenths of milliamps/ m^2). However, as discussed Section 5.2, when dislocations intersected the surface relatively huge reverse currents passed.[28,27]

The tunneling of electrons or holes through the space charge barrier is another phenomenon of interest with the "perfect" semiconductor that does not follow immediately from the simple fluctuating energy level model of electron transfer. Tunneling may occur through the space charge region of a heavily doped semiconductor. In studies to date, the results indicate that tunneling can occur in either direction between a band in the solid and a localized surface energy level. The surface level can be due to an ion in solution, an adsorbed species, or can be an intrinsic surface state. No cases have been reported of tunneling from the valence band to the conduction band of the semiconductor, followed by the minority carrier moving to the surface, except perhaps one of the first cases analyzed—that of Boddy,[10] who described tunneling from levels on the TiO_2 surface at 3.2 eV below the conduction band (the position of the valence band), and showed agreement between his results and a simple model. The resulting holes (or holes trapped on a surface state) were found to oxidize water to oxygen. It is possible that if he had assumed intrinsic surface states somewhat above the valence band associated with surface oxygen ions, which would be more in line with current thinking for oxide semiconductors, he would also have been able to show reasonable agreement.

In this section we primarily want to mention tunneling as an important characteristic of the perfect crystal. The detailed theories and the current/ voltage characteristics will be reviewed in Section 6.5.2, because tunneling has been used more to study ions in solution and current flow than it has been (in semiconductor electrochemistry) to study characteristics of the solid.

However, because some of the flaws described in Section 5.2 can lead to tunneling effects, we will at least indicate here the characteristics of the solid that permit tunneling. First, if tunneling into or out of a band is to make an important contribution to the current, compared to processes where the carriers reach the energy of the band edge, then the energy of the tunneling electron should be at least, say, 0.1 eV away from the band edge. Second, for tunneling to occur, the thickness of the surface barrier must be less than say 30 Å at the energy of the tunneling electron. Tunneling current decreases exponentially with increasing barrier width, and 30 Å is a reasonable barrier width. Thus, if the field at the surface of the semiconductor is more than 0.1 V/30 Å, or 3×10^7 V/m, significant tunneling is at least possible. An example band diagram for tunneling is indicated in Figure 5.1. From Equations (4.18), (2.24), and (2.23), the electric field is given by

$$(dV/dx)_{V=V_s} = (2qN_{sc}V_s/\kappa\epsilon_0)^{1/2} \tag{5.10}$$

and so for tunneling to be reasonable with a surface barrier of say 0.4 V, assuming a dielectric constant for the semiconductor of 5, we conclude that N_{sc} must be at least the order of 10^{24} m^{-3} (in order that the barrier be thin enough). The sample must be quite highly doped.

Tunneling at the surface of *d*-band semiconductors, a class of semiconductors that is becoming of interest because of their corrosion resistance (Section 7.3) is always a likelihood because of the low mobility of carriers in the materials. The low mobility arises because the *d* bands are narrow, as discussed in Section 1.1.3.1. Tunneling is common, not directly because of the low mobility, but because the low mobility requires a highly doped material before one has a useful electrode. If, for example, the carrier mobility is 10^{-7} m^2/Vs, and one wishes (to avoid ohmic losses in the semiconductor) to have a conductivity higher than 10^{-1} mho/m, then application of Equation (1.8) dictates that one must dope the semiconductor to a level of the order of 6×10^{24} m^{-3}, well in the tunneling range.

Tunneling at the surface of semiconductors that conduct by polarons will sometimes be observed for similar reasons. Polarons occur in semiconductors with a very high dielectric constant, such as TiO_2 or titanates.

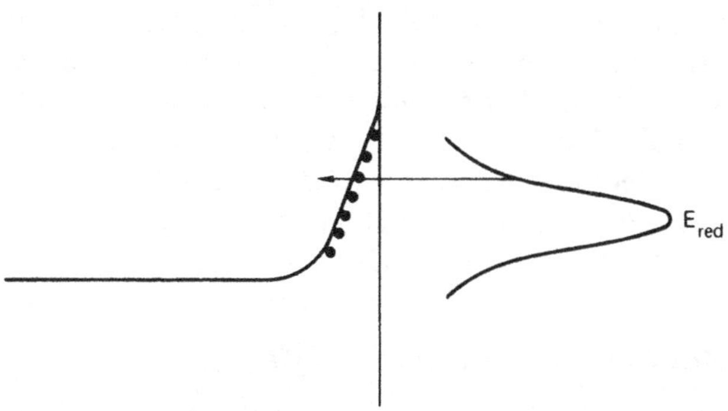

Figure 5.1. Tunneling from a reducing agent in solution, through a narrow space charge layer, into the conduction band of a semiconductor. The barrier must be less than say 30 Å in thickness for appreciable tunneling current.

In such materials the "conduction band electron" with its negative charge polarizes the lattice around itself in the same way that an ion polarizes solvents (Section 1.2.4). Then for the electron to move, the polarized region must move with it. The electron plus its associated polarization is called a polaron. Because ions must be moved a little in order to move the polaron, the mobility of the polaron is low, and for high conductivity a heavy doping of the semiconductor is required. Again, tunneling effects must often be considered.

The polaron in semiconductor electrochemistry has not been given a great deal of attention. The energy level of the electron can be much lower than the "real" conduction band in such a system, and it can show fluctuations just as the energy level associated with ions in solution does. Although there does not seem to have been a good analysis of electron transfer between polarons and ions in solution, the case may resemble electron transfer between two ions in solution, and hence differ quantitatively, and to some extent qualitatively, from the fluctuating energy level model developed in Chapter 3.

In the basic discussions of Chapter 3, polaron effects in charge transfer to or from ions in solution were not mentioned. The omission is simply due to the lack of theoretical development in the literature. As described in the preceding paragraph, charge transfer involving a small polaron may well be equivalent to ion–ion transfer, modified because of the fields in the space charge region and in the Helmholtz region. Certainly the problem of polaron exchange is open for good theoretical analysis. In our discussion, in Chapter 6, of electron capture by TiO_2, the probable polaron nature of the conduction electrons (and probably the holes) must be considered as a possible explanation for many unexpected results.

Another form of interesting behavior of the "space charge region" in a defect-free semiconductor electrode—one that we will mention many times again—is the behavior of a degenerate semiconductor. As discussed sufficiently in Sections 1.1.3 and 3.4.2, this is the case where the bands are bent sufficiently (enough voltage is applied or the semiconductor is doped enough) that the Fermi energy at the surface is either above the conduction band edge or below the valence band edge. A degenerate semiconductor electrode behaves in a quasimetallic way, with high currents flowing, and the transfer coefficient moving toward the value 1/2 characteristic of metals (Section 3.2).

5.2. The Behavior of Electrode Defects

5.2.1. Introduction

In this section we will discuss the behavior of bulk flaws in the electrode, and, as in the last section, we are interested from two points of view, first how electrochemistry provides methods for studying flaws, and second, how flaws affect the electrochemical behavior of semiconductors and oxide-covered metals. The flaws of interest are variations with distance or time of donors or acceptors, deep levels (traps), dislocations, grain boundaries, and interface states between a semiconductor and its oxide.

Because the behavior of perfect crystals is reasonably well understood, the measurement of deviations from this ideal behavior can provide substantial information about flaws in the semiconductor. And the electrochemical approach offers a convenient way of analyzing flaws. A very important asset of the electrochemical approach is the ability to make a measurement on a surface, remove the sample, treat it to change its properties, and reimmerse it to study the changes. Naturally, the electrochemical techniques are primarily useful for studying flaws in the bulk near the surface (in the space charge region). There are two general ways to study flaws. One is to examine the semiconductor using an indifferent electrolyte where, at least in principle, no current flows and the information arises from capacity/voltage measurements (Mott–Schottky plots). The second is to examine the semiconductor with active ions in solution, where the information is gleaned by the observation of current flow. In Section 5.2.2 we discuss the former, in Section 5.2.3 the latter.

The effect of flaws on the current flow through the electrode is particularly important. Point defects in the space charge region can result in tunneling of current through to the electrolyte, and result in complex time- or voltage-dependent characteristics. The presence of dislocations or grain boundaries in the material can lead similarly to extra electrode current, and at times to apparent breakdown across the space charge layer.

In cases where the current must pass through oxide films, defects in the oxide can completely control the current/voltage characteristic. Energy levels in the oxide can lead to enhanced tunneling through the oxide, and dislocations or grain boundaries can lead to excess currents. In fact, in the extreme case of an oxide that is so badly formed that it is noncoherent, such flaws allow the electrolyte to reach the substrate and one obtains current directly from the substrate to the solution. In principle, the behavior of the surface film can be expressed in terms of the behavior of the various types of flaws, point defects, dislocations, and grain boundaries, but a complicated combination of the flaws is always present in such a film. Thus measurements of surface layers will not be emphasized in this section: the object here will be to examine the effects of the various flaws, with each as the sole flaw present in the solid. The use of single crystal electrodes is the only possibility for isolating flaws. Thus effects on semiconductor crystals must be examined. The general behavior of surface films with their multiplicity of defects will be discussed in more detail later in Chapter 8, where the present section will be used as background.

5.2.2. Deviations of Mott–Schottky Plots Due to Bulk Flaws

In this section we are interested in abnormalities in the slope of the Mott–Schottky plot, which [from Equation (4.19)] are immediately suggestive of change in N_{sc}. Two major difficulties other than slope problems are commonly observed in Mott–Schottky plots, and were discussed in Section 4.1.4. One is a shift in the flat band potential V_{fb} with no other anomalies. Usually this is associated with a Helmholtz layer variation or other stray voltage drops occurring in the circuit, across oxide layers, contacts, etc. The other is frequency dependence of the capacity, which is usually either associated with electron exchange with surface states or other energy levels that have time constants near the frequency used in the capacity measurement, or associated with complicated or poorly interpreted equivalent circuits. In this section we will only discuss irregularities in the slope of the Mott–Schottky plot, first considering cases resulting from a nonuniform spatial distribution of ions or a variation of ion density with time, then considering cases resulting from the trapping of carriers as a function of position or time. It is worth noting that such irregularities are entirely undesirable if one is studying electrochemical processes, but can be of great interest when one is studying the solid.

5.2.2.1. Nonuniform Doping and Deep Levels. Nonlinear Mott–Schottky plots that are not time dependent usually arise because the space charge N_{sc} varies either with distance or with voltage, or both. We distinguish between completely ionized donors or acceptors where N_{sc} is a function of distance, and incompletely ionized donors and acceptors that the Fermi energy passes through at some voltage, so they become charged, changing N_{sc}. These are not easily distinguished on a Mott–Schottky plot where the voltage is the variable used to vary the position of the edge of the depletion layer.

Consider first the behavior expected if it can be assumed the donors are completely ionized, but their density varies with distance from the surface. If the space charge region is thus nonuniformly doped, then clearly a straight line relation in the Mott–Schottky plot cannot be expected, for Equation (4.19) is based on a constant N_{sc} with distance. The doping profile $N_{sc}(x)$, can, in principle, be determined by analyzing the Mott–Schottky plot. Butler[30] has used the approximation that the slope of the Mott–Schottky plot at any given x_0 (the width of the space charge region) gives $N_{sc}(x_0)$. At each voltage he determined N_{sc} from the Mott–Schottky plot, determined x_0 by the optical method described in Section 4.2.3, and thus could plot N_{sc} versus x. Ambridge[31] suggests a similar method.

Another case where the density N_{sc} may vary with distance only, and not time or voltage, is the case when a separate phase or surface film is on the surface. For example Gerischer and Gobrecht[32] studied CdS and CdSe in an electrolyte with dissolved sulfides or polysulfides, and found that the sulfide from the electrolyte ion-exchanges with the anion in the solid under illumination. Not only is CdSe changed, forming a surface layer of CdS, but also CdS undergoes ion exchange. The resulting surface layer is amorphous. The Mott–Schottky plots, again estimating N_{sc} from the slope, indicate a large defect density near the surface.

In general, however, we can expect any observed variations in the Mott–Schottky slope to be associated with deep incompletely ionized donors or acceptors, rather than with shallow, ionized donors or acceptors. This is because we seldom have shallow levels with concentration varying with distance. The only case where we can expect shallow donors or acceptors to vary in concentration with distance is the well-defined case when impurities, that are known to provide shallow traps, are intentionally introduced, say by diffusion, from the surface. Problems with deep traps are more common, because unintentionally introduced deep traps are common. We can, for example, have deep traps whose concentration is independent of distance where the deep traps are caused by impurities or defects that are in the original crystal before the crystal is sliced. And, we can have deep traps with a distance-dependent concentration for many reasons. Diffusion of unknown species in from the surface, mechanical damage to the surface, a separate phase at the surface, and surface states or surface impurities, all can be expected to provide deep traps with the concentration varying with distance from the surface in some manner. Thus below we emphasize these more important cases, where incompletely ionized impurities and imperfections (deep traps) are present with simple spatial configurations. These cases will illustrate how deep traps may affect the slope of the Mott–Schottky plot.

Consider first the behavior expected if a deep level, of concentration independent of distance, is crossed by the Fermi energy as the bands bend. In the derivation of Equation (4.19), we assumed that all the donors or acceptors in the space charge region have the same charge independent of voltage. We assumed they were always ionized, or perhaps always neutral. But consider an n-type sample with not only shallow donors N_D that are always ionized, but also deep levels N_t that become ionized with increasing V_s because they cross the Fermi energy. For illustration, assume the set of deep levels N_t is located at an energy say 0.2 eV below the Fermi energy in the bulk. When the bands are flat, the levels N_t will be essentially completely occupied. But when a surface barrier of 0.2 eV is present, the Fermi energy passes through the N_t levels right at the surface so they become ionized. And as V_s increases, the depth to which the N_t levels are ionized will increase. Thus between V_{fb} and $V_{fb} + 0.2$

V, the slope of the Mott–Schottky plot will be that associated with N_D, but beyond that, the slope will be much lower as the charge near the surface increases to $N_D + N_t$.

This case, where deep energy levels are monoenergetic but are distributed uniformly through the space charge region, has been more quantitatively analyzed by Kennedy and Frese,[7] and as discussed above, it leads to a Mott–Schottky plot with two slopes. The value for the flat band voltage V_{fb} will be meaningless (more or less) unless (a) one is sure that the deep levels are uncharged when the bands are flat, and (b) one can make measurements close to V_{fb}.

The other particularly simple case of deep traps, a case even simpler than the one discussed above with the traps distributed uniformly through the sample, is the case of monoenergetic interface states.

If there is an extremely high density of deep, monoenergetic interface states at a certain energy, the Fermi energy will be pinned at that energy and the capacity will be independent of voltage. Figure 5.2 illustrates the situation. In Figure 5.2a, a distance d is indicated between the interface states and the outer Helmholtz plane where the countercharge resides.

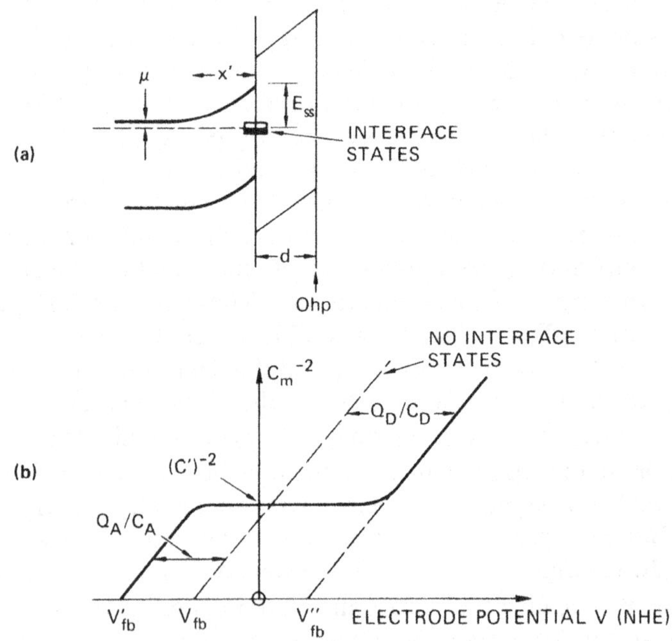

Figure 5.2. The influence of monoenergetic interface states (or surface states) on the Mott–Schottky plot. (a) The band model showing the interface states with no net charge; (b) the expected Mott–Schottky plot, showing how the curve differs with the interface states present, with V_{fb} shifted to a higher or lower electrode potential depending on the charge on the states.

The distance d may be as small as the thickness of the Helmholtz double layer, but if a passivating layer is present, d will normally be much larger. However, we will assume d is substantially less than the thickness of the space charge layer x'. In any case, if there are Q of these interface states per unit area, and if Q_D are donors, and Q_A acceptors, then the interface state charge is zero when there are Q_D electrons in the surface states (viz., when the surface charge equals the density of donor-type interface states). Now suppose the electrode potential is made negative, so E_F moves above the interface states. The interface states will become fully occupied, and so the states will have a negative charge Q_A, and the voltage due to the interface states across the distance d is $-Q_A/C_d$, where C_d is the capacity across the surface layer of thickness d. The expression for the voltage comes directly from the parallel plate capacity formula, Equation (4.2). Thus the apparent flat band potential V_{fb} is shifted from the "correct" value V_{fb} to $V'_{fb} = V_{fb} - Q_A/C_d$ as indicated in Figure 5.2b. The flat band shift results from the same arguments as used to derive Equation (4.27), where the voltage $-Q_A/C_d$ here is equivalent to V_H in that expression. If more applied negative voltage is applied to move E_F further above E_{ss}, the voltage will appear across the space charge region (assuming d is small) because the surface state charge cannot become more negative. Then the linear Mott–Schottky plot will be observed as indicated by the left branch of Figure 5.2b, with a slope given by N_D of the semiconductor.

If the electrode potential is made positive, so that E_F shifts below E_{ss} so the interface states are completely unoccupied, then the voltage due to the surface states across the distance d is $+Q_D/C_d$, and as indicated in Figure 5.2b, the apparent flat band potential indicated as V''_{fb} will be correspondingly shifted.

If the monoenergetic interface states are neither completely unoccupied or completely occupied, then by the definition of the Fermi energy, Equation (1.1), the Fermi energy is in the immediate vicinity (energy-wise) of the surface state band. The Fermi energy is pinned. The bands are bent by a definite amount, shown in Figure 5.2a as $E_{ss} - \mu$, and corresponding to this band bending, there is a definite space charge capacity, C', given by Equation (4.19) as

$$(C')^{-2} = (2/qN_D\kappa\epsilon_0 A^2)(E_{ss} - \mu - kT/q) \tag{5.11}$$

Then, as indicated in Figure 5.2b the capacity is C', independent of the electrode potential, over a large voltage span. In this middle range of electrode potential most of any change in applied voltage occurs across the distance d, simply inducing the charging of the surface state capacity. The span of voltage ΔV over which the capacity is constant is given clearly by

$$\Delta V = Q_D/C_d + Q_A/C_d \tag{5.12}$$

The above is a very simple situation, as we have assumed all the energy levels that pin the Fermi energy are located on a plane, and there is effectively only one energy for the levels, E_{ss}. However, it provides a very useful example of (a) Fermi energy pinning; (b) changes of slope of the Mott–Schottky plot, due to voltage that occurs as an energy level crosses the Fermi energy; and (c) changes of the intercept due to double layer voltages that develop outside the space charge region.

An important case where this model can be usefully applied is that where the interface states are surface states at the solid/liquid interface, and $C_d = C_H$, and d is the thickness of the Helmholtz double layer. This case was discussed in Section 5.1.1. The example is seldom realized. The fact that it is seldom realized is the best evidence we have that high surface state densities are seldom active at the solid/liquid interface. If they were active, V_{fb} would change for a "perfect" material as indicated in Figure 5.2b. As was discussed in Section 5.1.1, for ionic semiconductors the observed Helmholtz voltage does not apparently change with applied voltage as the model would require. However, one can interpret adsorption/desorption of H· and OH· on covalent semiconductors such as germanium [Equations (5.5) and (5.6)] by the model of Figure 5.2.

The next degree of complexity in energy level distribution is the case where the energy levels are interface states on a plane, as in Figure 5.2, but are distributed in energy. Frese and Morrison[33] have made Mott–Schottky plots for oxidized GaAs, with the object of studying surface states at the GaAs/native oxide interface. In this case the distance d of Figure 5.2 becomes the oxide thickness. The object of the experiment in this case is to measure the density of interface states in the bandgap region as a function of energy, using an analysis similar to that associated with Figure 5.2. We will not describe the details here, but simply point out that in Figure 5.2b the slope of the flattened region is related closely to the density of energy levels. If the central region is actually of zero slope, ($dC^{-2}/dV = 0$), then there must be an infinite density of states. The higher the slope in the central region, the lower the density of surface states. Thus, during a given measurement the capacity gives the position of the Fermi energy [E_{ss} in Equation (5.8)], and the slope of dC^{-2}/dV gives the density of interface states at that Fermi energy position. Then by sweeping the voltage (or more accurately, sweeping the space charge capacity) such that the Fermi energy moves through the band gap, the density of interface states as a function of energy can be determined. The method derives from MOS (metal–oxide–semiconductor) work of Terman.[34]

The most complex case, that of bulk energy levels distributed nonuniformly both in space and in energy, also occurs in nature but is too difficult to interpret. For example, Pham and Hueller[35] have studied silicon with defects produced by ion implantation (with no annealing). They find essentially a constant high capacity, with a value suggesting that the Fermi energy is pinned in the gap. The interpretation of the data is very difficult, for the states have a strong influence not only on capacity but also on minority carrier current flow. However, as Pham and Hueller point out, the technique of capacity measurement is certainly very sensitive to the defects produced by ion implantation.

Charged dislocations in the surface region should also have this same sort of influence on the Mott–Schottky plot because levels due to dislocations are distributed in distance and probably[36] in energy. This case will be discussed below because in the studies of dislocations available there has always been a further complication—trapping effects that lead to time dependence.

5.2.2.2. Ion Movement and Trapping. Time dependence of the capacity is another common feature of Mott–Schottky plots leading to nonlinear curves. Of course, we are most interested in time-dependent variations with time constants the order of the times involved in the experimental measurement, say between 10^{-1} and 10^{4} sec. One source of such time constants is movement of ions in the space charge region due to the electric field. A variation in the charge N_{sc} can occur because ions move either in or out. A second source is slow ionization of traps in the space charge region. First, we will discuss ion movement and then electron or hole trapping effects.

The high electric fields at a semiconductor surface are often high enough to cause field-aided diffusion of charged ions in the solid, and thus change the doping (N_{sc}) of the semiconductor. From Equation (1.11), neglecting the diffusion term, we have for the flow of impurities or defects in the solid

$$J = -qN\mu(d\phi/dx) \tag{5.13}$$

Combining the derivative of Equation (2.22) with Equation (2.23) to eliminate x_0 and evaluating at $x = 0$, we obtain

$$-\left.\frac{dV}{dx}\right|_{x=0} = \left(\frac{2qN_{sc}V_s}{\kappa\epsilon_0}\right)^{1/2} = \left.\frac{d\phi}{dx}\right|_{x=0} \tag{5.14}$$

and with say $N_{sc} = 10^{24}\ \text{m}^{-3}$ and $\kappa = 3$ we have approximately

$$\left.\frac{d\phi}{dx}\right|_{x=0} = 10^{8}V_s^{1/2}\ \text{V/m} \tag{5.15}$$

for a depletion layer. With an accumulation layer we find from Equation (2.26), using the same κ and N_{sc}:

$$\left.\frac{d\phi}{dx}\right|_{x=0} = 10^8 \exp\left(-\frac{qV_s}{2kT}\right) \tag{5.16}$$

Thus with V_s a high positive or negative number, a high field results. With these high fields at the surface, it is quite possible for fast-diffusing ions to move into or out of the electrode and change the characteristics. For example, the mobility of hydrogen ions in zinc oxide, as extrapolated from Thomas and Lander's measurements,[37] is 2×10^{-20} m^2/V sec. This would lead from Equation (5.16) in (5.13), to a flow of hydrogen into the sample[27] of about 5×10^{19} ions/m^2 sec if we assume the electrode surface is hydrated, so has a high equivalent surface density (10^{27}/m^3), and if we assume V_s in the accumulation layer is 0.5 V. From such numbers, one can expect significant ion movement into and out of semiconducting electrodes to be common.

An analytic solution describing how the charge density in the space charge region changes as a function of time and distance from the surface could, in principle, be found, but in practice the analysis is too difficult. One would have to start with Poisson's equation, Equation (2.6), and include a variable space charge term (variable with time and distance) due to the moving ions. Also, the rate at which ions are activated from the solution into the solid, thus providing the source of ions to include in the diffusion equation, must be assumed. Such an analysis does not seem to have been attempted.

Experimentally such ion movement in semiconducting oxides has often been observed. As a first example, we simply compare the case here to that of anodic oxidation of metals. In the growing metal oxides, fields of the above magnitude cause transfer of ions (either the cations or oxygen) to induce continuing oxide growth. During anodic oxidation, foreign ions are also found to be mobile due to the fields applied. For example, Mackintosh *et al.*[38] find As, Sb, Bi, Se, and Te can move in an anodic alumina layer over aluminum during the anodic oxidation process. Kornfeil[39] discusses proton movement in an MnO_2 layer over platinum. Dyer and Leach[40] discuss the movement of hydrogen into TiO_2 and Nb_2O_5 layers under cathodic potentials, and the optical changes accompanying this movement.

Ions also move in semiconductor space charge regions. Haneman and Steenbeeke[41] report Cu and Ni movement in the space charge region of TiO_2, due to the fields normally applied for electrochemical measurements. In this case the ion movement is in the *c* direction, the direction of easy diffusion. Harris and Wilson[42] and Butler[43] also find effects suggestive of donor movement in TiO_2. Other results could perhaps be attributed to such injection in TiO_2.[43] For example, Noufi *et al.*[44] observe, in the reduction of many ions, a nonzero transfer coefficient for TiO_2, suggestive of metalliclike behavior in the sense that a significant fraction of the applied voltage appears across the Helmholtz region. Although not suggested by the authors, such behavior would be consistent with positive ion diffusion into the accumulation layer region leading effectively to surface-state-like behavior. Such diffusion in TiO_2 could cause erroneous interpretation of cathodic observations, and possible anodic. Sayers and Armstrong[45] report that oxygen ions diffuse into TiO_2 during illumination to restore stoichiometry to a reduced crystal.

Other cases of diffusion of ions include Mohapatra's[46] work on the introduction of H^+ or Li^+ into WO_3 by a cathodic bias. In this case the object was to introduce color centers for electrochromic effects. Yoneyama *et al.*[47] introduced an ion, believed to be protons, into GaP, as observed by a resulting photocurrent peak at less than band gap energy. The photocurrent is presumably associated with excitation from, or to, a defect level. Gottesfeld and McIntrye[48] report that the anodic oxide on iridium permits easy, reversible electrochromic changes due to proton movement.

Such field-aided diffusion can be rate-limited by the injection of the species into the solid, by the mobility of the species in the bulk semiconductor, or by diffusion along flaws. Yoneyama *et al.*[47] deposited a metal on the surface of their GaP to induce adsorption of the hydrogen atoms (formed by the cathodic reduction process) and thus were able to substantially increase the surface concentration. In this case, presumably the hydrogen injection was rate limiting until the metal catalyst was desposited. If diffusion is the limiting factor, such field-aided diffusion of ions can be increased when linear or planar flaws (dislocations or grain boundaries) are present in the semiconductor or the surface film.[49] Because around a dislocation pipe there are regions of dilated lattice, diffusion and mobility of foreign ions usually has a lower activation energy than in regions of perfect crystal. Whether diffusion along dislocations dominates depends of course on the dislocation density and the temperature.

The other dominant cause of time dependent Mott–Schottky plots is the trapping or detrapping of holes and electrons. Such trapping can result in either time dependence of N_{sc} or voltage (distance) dependence of N_{sc}. N_{sc} changes with trapping because N_{sc} in Equation (4.19) is not the density of impurities in the space charge region, but the density of immobile charge, and trapped electrons and holes contribute to immobile charge. Thus, when carriers become trapped in levels where they should not be under equilibrium conditions, the slow charge emission as the trap occupancy moves toward equilibrium causes the time dependence of N_{sc}. Such nonequilibrium conditions are usually brought about by two situations. One is illumination, where the presence of minority carriers in nonequilibrium amounts changes the occupancy of the traps, and during subsequent dark periods the time-dependent effects are observed as the sample struggles to regain equilibrium. Alternately, a change in trap occupancy can occur due to voltage changes. Such processes are very common with the electrolyte/semiconductor interface. Figure 5.3 illustrates such a case, where an n-type semiconductor electrode is shown at equilibrium with most levels above the Fermi energy unoccupied in Figure 5.3a. The trap levels are indicated by circles in the bandgap at an energy E_t below the conduction band edge. Levels occupied by an electron are indicated with $-$, unoccupied levels with $+$. If a cathodic pulse is applied, making the semiconductor more negative relative to the counterelectrode, the bands bend down, as in Figure 5.3b. Because the bands are bent down, there are far more electrons in the conduction band near the surface, and the trap levels are now below the Fermi energy. Thus, both from a kinetic and an equilibrium point of view, the traps fill with electrons rapidly. When the cathodic voltage pulse is removed (Figure 5.3c) the trapped electrons are not in equilibrium, as they are occupied, but are now above the Fermi energy. However, for the electrons to return to the conduction band, there is an activation energy $E_c - E_t$ to be overcome. The return to the equilibrium condition of Figure 5.3a can be very slow; the exact time constant depends on the energy difference $E_c - E_t$. Worse, a spectrum of trapping level energies is usually present in these cases, so one has a spectrum of time constants.

Of course, as the traps empty, the effective charge N_{sc} in the space

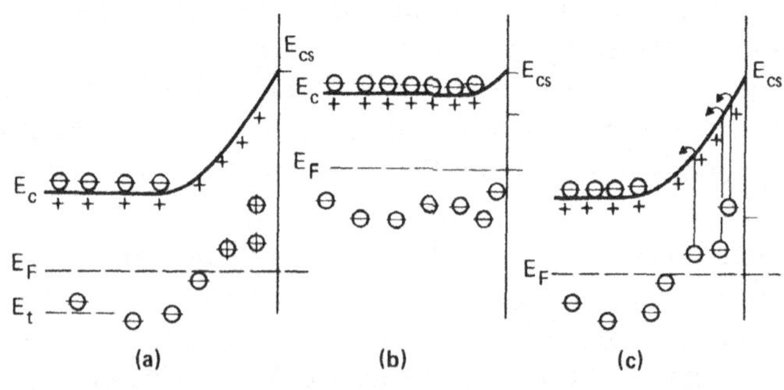

Figure 5.3. Charge trapping in the space charge layer. In (a) the sample is at equilibrium, with the Fermi energy essentially separating occupied from unoccupied levels. In (b) a cathodic (semiconductor negative) pulse is applied, permitting electrons to reach the deep trap levels near the surface. In (c) the charge on these deep trap levels is reemitted to the conduction band, but slowly due to the large activation energy.

charge region that must be used for the analysis of the electrode behavior is varying with time, so the capacity varies with time.

The two stimuli described above that lead to time variation in N_{sc} will usually lead also to distance variations in the density of trapped electrons, and therefore to a voltage dependence of the Mott–Schottky slope. For example, if in Figure 5.3b the bands are not made completely flat by the voltage pulse (and seldom can bands be made flat at a semiconductor surface), the traps near the surface will not become occupied. Then N_{sc} will vary with distance. Similar considerations apply for traps filled or emptied by minority carriers produced by illumination. Near the surface of a semiconductor with a depletion layer, there will be a different density of minority carriers than is present in the bulk. Again, the trapping will be distance dependent. With these considerations in mind, it is apparent that without a carefully designed experiment, the trapping and detrapping of charge in the space charge region can be extremely complex in any real electrochemical system, especially when the electrode has a spectrum of defects.

Techniques have been used to study such trapping in solid state physics. An old technique, termed thermally stimulated currents, TSC,[50,51] has been used to study bulk trapping of the type mentioned. In this type of study one illuminates the sample at very low temperature, such that the trapping levels shift from their equilibrium occupation but the temperature is too low (the rates are too slow) for equilibrium to be regained. Then the temperature is slowly raised, and as the temperature reaches a value such that a given trap can empty, the resulting carriers in the bands are observed by conductance. One can, in principle, analyze for the distribution of trap energies and densities. Another, but similar, technique that measures trapping in surface states has been developed more recently. The method is termed DLTS, deep level transient spectroscopy. In this case modern electronic techniques are applied[52] to monitor the capacity of the space charge region, and the capacity (see Section 4.3) can easily be interpreted in terms of charge emission from surface states. A cathodic pulse (for *n*-type), such as indicated in Figure 5.3 is applied at low temperature, filling the surface states. Again the temperature is slowly raised so charge emission from the traps can occur and the capacity, as a function of temperature, analyzed to give the trap distribution. This technique has been applied,[52,53] using the MOS (metal–oxide–semiconductor) configuration rather than being applied to the EOS (electrolyte–oxide–semiconductor) configuration. However, the principle is identical, except that in the EOS case there will be a more modest accessible temperature range.

In both cases described above the system is amenable to theoretical interpretation because there are no traps in a space charge region. In the

TSC measurement the traps are in the bulk semiconductor and surface effects can be neglected, and in the DLTS measurement the traps are all in the plane of the interface. So the traps in both cases are always at the same electrical potential. If the traps are spread out through the space charge region, and are also of varying energy, the analysis may be more difficult[54] because the trapping and detrapping of electrons occurs in a region of varying potential. As described above, the trapping is then particularly nonuniform. To use the techniques described above, the surface barrier must be eliminated—the bands must be made flat—so that the trap occupancy at time zero can be assumed to be independent of distance from the surface. Johnson[55] has used this technique to study traps near the surface in GaAs.

5.2.2.3. The Effect on Mott–Schottky Plots of Dislocations and Grain Boundaries. With the concepts of Sections 5.2.2.1 and 5.2.2.2 in mind, one can qualitatively interpret Mott–Schottky plots to indicate the properties of flaws in a semiconductor. An example of such a study is the study of dislocations at the ZnO surface.[27,28] In this study the dislocations were introduced by abrading the surface, and, when desired, removed by etching the abraded region away. Dislocations introduced by abrading are "dislocation loops" that begin and end at the same surface, penetrating usually a distance the order of the grain size of the abrasive used.[56]

In Figure 5.4, four curves are shown that will help illustrate the effect of trapping at dislocations on the Mott–Schottky plot. With the sample as perfectly etched as possible, curve (c) and curve (d) are obtained, with curve (c) taken in the dark, and curve (d) under illumination. During curve (d) photoproduced holes flow to the surface, but, as described in Section 7.3, they simply corrode the electrode and go into solution as Zn^{2+}, not affecting the Mott–Schottky plot at all. There is never a high enough hole density retained in the solid to represent a significant charge density. Thus curves (c) and (d) are identical. With dislocations present entirely different results are obtained, as observed in curves (a) and (b) for one crystal face, and curves (e) and (f) for another. Curve (a) represents measurements taken in the dark, and it is observed that the slope is very high, indicating a very low N_{sc} [Equation (4.19)]. Electrons are trapped at the dislocations during the period when the bands are relatively flat, as in Figure 5.3b, and the net positive charge in the space charge region is lowered by the density of the trapped electrons. The measurements, particularly those at high positive potential, are taken rapidly to minimize the time-dependent release of carriers as illustrated in Figure 5.3c, but time dependence is observed so the density of electron traps can only be estimated qualitatively. In curve (b) of Figure 5.4 are shown results when the sample is illuminated so holes are present in the space charge region. In this case, the slope is low, indicating that N_{sc} is very high. The interpretation is that the positive charge due to donors in the ZnO [of density

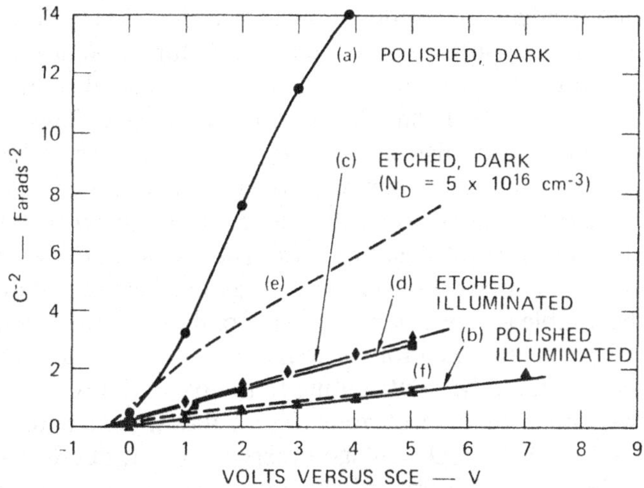

Figure 5.4 Determination of net charge density in the space charge region.

easily calculable using curve (c)] is supplemented by trapped holes. Thus it can be concluded, from these results, that dislocations in ZnO act both as electron and hole traps. To obtain trap densities and energy levels, more sophisticated studies would be necessary such as described above, with some means to simplify the behavior (by using a lower temperature and DLTS measurements, by using rapid measurements to avoid time dependence, and by making a careful determination of when the bands are flat to avoid distance dependence). However, the results do provide a nice illustration of the general behavior of trapping, as well as pointing out the active role that dislocations can play in trapping.

Grain boundaries, being in a sense a gross array of dislocations, can also provide trapping states. The states at grain boundaries have much in common with interface states, the surface states between a semiconductor and a passivating layer.

The observations of Vanden Kerchove *et al.*[57] should also be mentioned, although to date the observations are not easily interpreted. They observe that the flat band potential, V_{fb}, of TiO_2 depends on whether the sample is "well etched." This phrase presumably distinguishes whether the surface damage (dislocations), which stems from cutting and grinding is or is not removed.

5.2.3. Current Flow Associated with Bulk Flaws

5.2.3.1. Dark Current Due to Flaws. In this section we discuss current flow in the dark, in either direction, where flaws in the semiconductor affect the current flow. As pointed out in Sections 3.3.3.2 and 5.1.2, early studies of wide bandgap electrodes showed that the reverse diode current between an electrode and a redox couple in solution is usually associated with flaws. Such current due to flaws leads to nonequilibrium exchange current (Section 3.3.3.2). It should be noted that bulk flaws, being recombination centers, can also be sources of reverse current because of space charge generation of minority carriers [Equation (1.5) and its discussion]. This generation is primarily expected in narrow bandgap ($\lesssim 2$ eV) materials. We will not describe this process, a process well documented in solid state diode theory, but will restrict the present discussion to effects, observable on wide bandgap semiconductors, that seem related to the electrolyte/semiconductor system.Here we wish to examine the various flaws that could be contributing to the observed reverse current, both the low-voltage reverse current due to flaws, designated as J_a in Equation (3.34), and the high-voltage reverse current, due to flaws, that can resemble breakdown. Forward current to active ions in solution can also be associated with flaws. Normally the forward direction of the current to a redox couple is considered independent of flaws in the semiconductor (except as the flaw density constributes to n_s or p_s). However, as discussed in Section 3.4.2, when the reaction is highly exothermic (for example E_{ox} is much below E_{cs} on an n-type material), the forward current by theory should be low, and defects may play a role in inducing a high forward current.

Thus in this section we will consider the role of defects in increasing both the reverse and the forward currents of the semiconductor electrode.

The concepts originally presented,[(24)] that the small reverse current on an n-type semiconductor may be associated with "imperfections and impurities on the surface of the sample," have been only slightly clarified in later studies. Vanden Berghe *et al.*[(58)] observe that the reverse current is independent of pH, electrolyte, or N_2 bubbling on a CdS crystal, and conclude that "effects on the CdS anode are connected with the presence of imperfections in surface and subsurface regions." Nozik[(R25)] discusses the reverse current as it affects solar cell performance, noting its irreversible characteristic. However, there have been very few positive experiments to identify the source of the reverse current into a wide bandgap semiconductor. As discussed in earlier reviews,[(R4,R10)] the current may be due to point defects at the surface, or to dislocations. In general, studies have been made on single crystal material with no grain boundaries, but the effect of grain boundaries should resemble a high density of dislocations.

We will first discuss the possible influence of point defects, then discuss the influence of dislocations. At low voltages and moderate concentrations ($\lesssim 10^{24}$ m^{-3}) point defects in the bulk do not appear to have the necessary characteristics to affect either the dark reverse or forward current. Such an effect would require transfer of carriers between ions in solution and localized energy levels near the surface, followed or preceded by thermal activation of the carrier to the majority carrier band, and such transfer seems improbable. As indicated in Chapter 3, the rate of direct carrier exchange between any fluctuating energy level and a band is expected to be much higher than the rate of transfer between a nonfluctuating isolated bulk energy level and a band.

On the other hand, if a very high density of point defects is present, then by a mechanism equivalent to impurity band conduction, carriers could be transported through the bulk material as described for dislocations below. There is evidence of a significant effect of a high density of point defects in increasing the reverse current in the low-voltage region from the work of Hueller and Pham.[(59)] They intentionally introduced a high density of defects by ion implantation, and observed a high reverse current at the silicon–electrolyte interface.

With a high reverse voltage applied, indirect effects can lead to significant current associated with point defects. There are two mechanisms. One is the ionization of traps associated with the defects, as discussed in reference to Figure 5.3, that can lead to extremely high surface fields, much higher than when only shallow donors or acceptors are contributing. Such ionized traps will lead to premature breakdown. The second is tunneling through deep levels near the surface. Schmickler[(60)] has analyzed this possibility, terming it "resonance tunneling," where the energy level from the flaw provides a "path" for the electron or hole to tunnel from the solution to the majority carrier band of the semiconductor. Alternately, such defect levels in a surface film can provide a path for an electron or hole to tunnel through the film into a semiconductor.

Interface states at the interface between a semiconductor and a covering film is a class of flaws related to point defects. Such surface states can pin the Fermi energy so that any applied voltage appears across the film, and the resulting high fields can lead to early breakdown of the film.

However, as mentioned, to date there is little experimental evidence (except that of Hueller and Pham, discribed above) where the controlled introduction of point defects has been examined, and the resulting increase in reverse current analyzed.

Linear and planar defects, dislocations and grain boundaries, can be expected by the following arguments to permit a substantial current flow in either direction, even at low voltages. Such flaws produce a line (or a plane) of defect sites leading to the interface. This produces a continuous pathway of partially occupied orbitals from the surface to the bulk, associated with the dangling bonds at the dislocation. Thus in germanium, dislocations provide a line of energy levels[(36)] at 0.2 eV below E_c. With such a pathway, there is the possibility that current can be transferred short distances by electrons hopping from one to the next of these overlapping energy levels.

Consider, for example, the possible role of dislocations in permitting forward current to flow, especially carrier capture by an ion in solution whose energy level is deep in the band gap from the band edge (a highly exothermic carrier capture). By the simple fluctuating energy level model, such capture should not occur. The role of dislocations can be compared to the role of surface states in such "forbidden" carrier capture. The surface state role was discussed in Section 3.4.2, particularly with reference to Figure 3.5. Now dislocations will be a particularly effective substitute for surface states. As described above, the carrier can be captured by dislocation levels a short distance from the surface and be conducted along the dislocation to the surface. If, in particular, E_{ox} of the oxidizing agent is anywhere near the dislocation energy level, then most electrons within tunneling distance from the surface will be removed from the dislocation levels. This will leave a large density of electron traps available, and the cathodic current may be increased considerably.

Consider also the role of dislocations in increasing the reverse current; both the current associated with injection by ions in solution and the current associated with injection by intrinsic surface states, viz., corrosion. In either of these processes, conduction of carriers along the dislocation in the reverse direction will be even more effective than in the forward direction because the electric field aids the movement. One can expect a simple process for reverse current flow. Essentially the model is as follows. In the bulk of the material (or in the high field region near the surface where the carriers can tunnel), majority carriers are excited to the majority carrier band from the dislocation levels. The resulting "trapped minority carrier" on the dislocation can make its way to the surface under the influence of the space charge field. When the minority carrier arrives, it can either react with an ion in solution or induce corrosion. In the latter case, as discussed further in Section 7.4, the presence of minority carriers at the surface is equivalent to the removal of chemical bonds holding the surface atom to the crystal. When enough minority carriers arrive to release the surface atom from the crystal (break all its bonds), the atom leaves, the electrical circuit is completed, and the corresponding current is registered as reverse current.

At higher reverse voltage, the current associated with the dislocations can resemble Zener breakdown at high voltages, where majority carriers can tunnel from the dislocation levels to the majority carrier band. And, of course, in a thin passivating film over a metal or a semiconductor, dislocation pipes can, by the above mechanism, provide a conducting path through the film.

In addition to the above possibilities based on their linear character, dislocations can aid breakdown by the same processes that were mentioned in reference to point defects above, namely, offering, when ionized, a very high N_{sc}. The high N_{sc} can be due to impurities because in the neighborhood of the dislocation the lattice is dilated, so dislocations provide an easy path for impurity movement. Thus they can contribute to the high N_{sc} by allowing in-diffusion of impurities under high surface fields.

It is clear from the above discussions how critical the control of dislocations at semiconductor surfaces can be in determining current flow under extreme conditions. Unfortunately, most semiconductors cannot be prepared with a reasonably low density of dislocations present. Thus in many discussions in Chapter 6, the interpretation offered for many measurements will often be qualified because the same results could occur due to the presence of dislocations.

There have been some experimental studies showing the influence of dislocations, but unfortunately too few and often too ambiguous in interpretation, which is to be expected with such a complex flaw. Experimentally Tranchart *et al.*[(61)] find apparent breakdown associated with dislocations on GaAs. In this case the effect was identified as due to dislocations because they observed that etch pits develop at the dislocations when the current passes through the sample. In neutral *p*H, where the Ga_2O_3 produced by the current flow is insoluble, an oxide forms over the dislocations. The current is found to be associated with the electric field at the surface, as concluded because the voltage of the "breakdown" is found to decrease rapidly with increasing donor density. However, the "breakdown" occurs at about 1/10 the expected breakdown voltage as calculated by the model for bulk avalanche breakdown (Section 2.2.4.3) without dislocation effects. In Section 5.2.3.2 we discuss a case[(25)] where flaws are associated with photodecomposition of CdS, and suggest that the flaws could be dislocations. Vanden Kerchove *et al.*[(57)] find even the flat band potential V_{fb} of TiO_2 and titanates depends on proper etching (dislocation removal). A qualitative analysis of the effect of dislocations on the reverse current was made by Morrison and Freund[(27,28)] using ZnO. In these studies high anodic currents in an indifferent electrolyte were observed when the sample was mechanically polished to introduce dislocation loops. The currents increased so rapidly with voltage that the I/V curves resembled breakdown for voltages as low as 1 V (SCE). However, as the samples were etched to remove the damaged layer the "exchange current" at low voltages decreased and the "breakdown field" increased. With a well-etched sample, no breakdown was observed (with a doping of $10^{23}\ m^{-3}$) up to 100 V.

Thus the evidence is clear that dislocations can be a contributor to current flow to and from a semiconductor electrode under extreme conditions, but the measurements to date are not sufficiently sophisticated to prove a detailed mechanism.

5.2.3.2. Photocurrent as Affected by Bulk Flaws. Bulk flaws, intrinsic surface states, and interface states all compete with ions in the electrolyte for the capture of minority carriers produced by illumination. Thus if we are to analyze photoeffects, such as photocatalysis associated with minority carrier capture by ions, we must first explore what is known about photoeffects in the solid.

Bulk recombination at point defects has been thoroughly analyzed, the models satisfactorily proven, and the resulting decrease of the flow of minority carriers to the surface has been well documented.[62,R26] The concept of the diffusion length of minority carriers, L_p (holes) or L_n (electrons), was introduced in Section 4.2.3, and, as indicated there, it is the average distance a minority carrier can diffuse in the bulk semiconductor before recombining with a majority carrier. Thus on the average, only minority carriers created within a diffusion length of a given plane will reach it before recombining. The diffusion length is given by the expression

$$L_{p,n} = (D_{p,n}\,\tau)^{1/2} \tag{5.17}$$

where the subscript n is used if electrons are the minority carrier, p is used if holes are the minority carrier. The parameter D is the diffusion coefficient for the carrier, and τ is the bulk lifetime of the minority carrier.

Now bulk recombination, the process that controls τ, almost always occurs via flaws in the semiconductor. If a high density of flaws is present, flaws that can capture a minority carrier and then a majority carrier, then L_p or L_n will be very low.

Bulk recombination at dislocations in particular plays an important role in lowering the diffusion length or τ for a minority carrier, as was recognized many years ago[63] but satisfactorily analyzed only recently.[64] The role of dislocations in photoelectrochemistry is clear, for example, from a study of Figure 5.4. As discussed in connection with Figure 5.4, dislocations can act either as electron traps or as hole traps. Thus one would expect them to act as efficient recombination centers to lower τ and hence L_p. Measurements of L_p on ZnO by the techniques described in Section 4.2.3 showed[27,28] indeed that with an abraded surface, a surface with dislocation loops present, the value of L_p in the damaged region was far lower than with the same sample etched. These studies were repeated by Tench and Gerischer[65] and the observations were confirmed.

Other effects of bulk flaws on the electrochemical characteristics of a semiconductor during illumination arise indirectly because the flaws are important in controlling luminescence, photocurrent, and photovoltage. A few examples of the influence of flaws on these measurements have been reported in electrochemical studies. Luminescence is radiative recombination and removes photoproduced carriers before they can reach

Table 5.1. Flat Band Potentials of Oxide Semiconductors

Material	Bandgap	V_{fb}	pH	Ref.[a]	pzc	Ref.[a]	E^o_{vs}	E^o_{cs}	χ
p-CuO(?)	1.7(?)	0.5	7	*A*, R27					
n-Fe_2O_3	2	−0.1	13	*B*, *C*, *D*	8.6	*E*	2.2	0.2	4.7
		0.15	9	*A*					
n-CdO	2.1	0	13	*C*, *D*	12	*F*	2.2	0.1	4.6
p-NiO		1.2	3	*G*					
n-WO_3	2.6	0.1	1	*H*					
		−0.5	9	*A*, R27					
		0	13	*I*, *D*	0.4	*J*	3.4	0.8	5.2
n-V_2O_5	2.75	0.9	7	*A*, R27					
n-Bi_2O_3	2.8	0.2	9	*A*, R27					
n-PbO	2.8	0	9	*A*, R27					
n-TiO_2	3	−0.6	13	*K*, R27, *L*, *D*	5.8	*M*	2.7	−0.3	4.33
		−0.05	2.1	*N*, R27, *K*					
n-Ta_2O_5	3.1(?)	−1.2	13	*O*, *D*	2.9	*D*	2.5	−0.6	4.1
	4	−1.2	13	*C*					
n-ZnO	3.2	−0.85	13	*P*, *D*	8.8	*Q*	2.6	−0.6	4.2
		−0.1	4.8	*K*, R27					
		−0.4	12	*K*, R27					
n-Nb_2O_5	3.4	−0.75	13	*O*, *R*					
n-SnO_2	3.7	−0.45	13	*S*, *R*	4.3	*T*	3.8	0.1	4.49
		−0.3	13	*C*, R27					
		0.5	1	*U*					
n-ZrO_2	5.0	−1.8	13	*O*, *B*, *C*	6.7	*V*	2.5	−1.4	3.4

[a] The letters *A–V* correspond to the following references:

A. K. L. Hardee and A. J. Bard, *J. Electrochem. Soc.* **124,** 215 (1977).
B. M. A. Butler, D. S. Ginley, and M. Eibschutz, *J. Appl. Phys.* **48,** 1914 (1977).
C. H. H. Kung, H. S. Jarrett, A. W. Sleight and A. Ferretti, *J. Appl. Phys.* **48,** 2463 (1977).
D. M. A. Butler and D. S. Ginley, *J. Electrochem. Soc.* **125,** 228 (1978).
E. R. J. Atkinson, A. M. Posner, and J. P. Quick, *J. Phys. Chem.* **71,** 550 (1967).
F. G. A. Park, *Chem. Rev.* **65,** 177 (1965).
G. D. M. Tench and E. Yeager, *J. Electrochem. Soc.* **120,** 164 (1973).
H. W. Gissler and R. Memming, in *Semiconductor Liquid-Junction Solar Cells*, see Ref. R57.
I. M. A. Butler, *J. Appl. Phys.* **48,** 1914 (1977).
J. R. K. S. El Wakkad and H. A. Risk, *J. Phys. Chem.* **61,** 494 (1957).
K. W. P. Gomes and F. Cardon, in *Semiconductor Liquid-Junction Solar Cells,* see Ref. R57.
L. J. G. Mavroides, D. I. Tchernev, J. A. Kafalas, and D. G. Kolesar, *Mater. Res. Bull.* **10,** 1023 (1975).
M. Y. G. Berube and P. L. DeBruyn, *J. Colloid Interface Sci.* **27,** 305 (1968).
N. R. Memming and G. Schwandt, *Surf. Sci.* **4,** 109 (1966).
O. P. Clechet, J. Martin, R. Oliver, and C. Vallony, *C. R. Acad. Sci. Ser.,* **C282,** 887 (1976).
P. F. Lohman, *Ber. Bunsenges. Phys. Chem.* **70,** 428 (1966).
Q. L. Blok and P. L. DeBruyn, *J. Colloid Interface Sci.* **32,** 518 (1970)
R. A. J. Nozik, *Ann. Rev. Phys. Chem.* **29,** 189 (1978).
S. M. S. Wrighton, D. L. Morse, A. B. Ellis, D. S. Ginley, and H. B. Abrahamson, *J. Am. Chem. Soc.* **98,** 44 (1976).
T. S. Mukai, Y. Yokoyama, T. Wakamatsu, and N Narazaki, *Suiyokai-Shi* **17,** 49 (1970).
U. F. Möllers and R. Memming, *Ber. Bunsenges. Phys. Chem.* **76,** 469 (1972).
V. K. C. Ray and S. Khan, *Ind. J. Chem.* **13,** 577 (1975).

the surface to react. Thus, for example, Ellis and Kares[66] find, on CdS, that as the radiative recombination (through Te traps in CdS) increases, the photocurrent through the electrode decreases. Maltby *et al.*[67] have analyzed the effect of bulk trapping on the surface photovoltage. Heller *et al.*[68] report work by Bijvik on experimental photovoltage measurement at the semiconductor–electrolyte interface, and show the bulk levels affect the results. Gissler and Memming[69] observe photocurrent in WO_3 at a wavelength too long to excite electron–hole pairs and conclude the

Table 5.2. Flat Band Potentials of Group IV and III/V Semiconductors

Material	Bandgap	V_{fb}	pH	Refs.[a]
p-Si	1.1	−0.6	1	*A*, *B*
n-SiC	2.9	−1.35	1	*C*, R27
p-SiC		1.7	1	*D*
		1.34	14	*B*, *D*
n-InP	1.25	−0.8	14	*E*, *F*, R27
		−0.35	2.1	*G*, R27
		−0.65	9.2	*G*
p-InP		0.9	2.1	*G*, R27
		0.7	9.2	*G*
		−0.1	14	*D*
n-GaAs	1.43	−0.95	2.1	*G*, R27
		−1.35	9.2	*G*, R27
		−1.5	12	*G*
p-GaAs		0.4	2.1	*G*, R27
		0	9.2	*G*, R27
		0	14	*D*
		0.4	1	*D*
n-GaP	2.3	−1.9	14	*E*
		−1.1	2.1	*G*, R27
		−1.2	4.8	*G*, R27
p-GaP		0.8	2.1	*G*, R27
		0.7	4.8	*G*, R27
		0.2	14	*D*

[a] The letters *A–G* correspond to the following references:

A. R. M. Candea, M. Kostner, R. Goodman, and J. Hickok, *J. Appl. Phys.* **47**, 2724 (1976).
B. A. J. Nozik, *Ann. Rev. Phys. Chem.* **29**, 189 (1978).
C. M. Gleria and R. Memming, *J. Electroanal. Chem.* **65**, 163 (1975).
D. J. O'M. Bockris and K. Uosaki, *J. Electrochem. Soc.* **124**, 1348 (1977).
E. M. S. Wrighton, A. B. Bocarsly, J. M. Bolts, and A. B. Ellis, in *Semiconductor Liquid-Junction Solar Cells*, see Ref. R57.
F. A. B. Ellis, J. M. Bolts, S. W. Kaiser, and M. S. Wrighton, *J. Am. Chem. Soc.* **99**, 2848 (1977).
G. W. P. Gomes and F. Cardon, in *Semiconductor Liquid-Junction Solar Cells*, see Ref. R57.

Table 5.3. Flat Band Potentials of Chalcogenides

Material	Bandgap	V_{fb}[a]	pH	Refs.[c]	V_{fb}[b]	Refs.[c]
Bi_2S_3	1.2				−0.8	*A*
n-CdTe	1.5	−0.55	indep	*B*, R27	−1.4	*C*
n-CdSe	1.7	−0.45	indep	*B*	−1.2	*C*
		−0.17	4.8	*D*		
n-CdS	2.4	−0.65	indep	*B*	−1.2	*C*
		−0.55	4.8	*E*	−0.7: $[S^-] = 10^{-5}$	*F*, *G*
		−0.75	9.2	*E*	−1.1: $[S^-] = 1$	*F*, *G*
ZnSe	2.7	−2.5	7	*H*		

[a] In indifferent aqueous solution.
[b] With S^{2-}/S_2^{2-}, Se^{2-}/Se_2^{2-}, or Te^{2-}/Te_2^{2-}, as the electrolyte.
[c] The letters *A*–*H* correspond to the following references:

A. L. M. Peter, *J. Electroanal. Chem.* **98,** 49 (1979).
B. A. B. Ellis, S. W. Kaiser, J. M. Bolts, and M. S. Wrighton, *J. Am. Chem. Soc.* **99,** 2839 (1977).
C. M. S. Wrighton, A. B. Bocarsly, J. M. Bolts, and A. B. Ellis, In *Semiconductor Liquid-Junction Solar Cells*, see Ref. R57.
D. E. C. Dutoit, R. L. Van Meirhaeghe, F. Cardon, and W. P. Gomes, *Ber. Bunsenges. Phys. Chem.* **79,** 1206 (1975).
E. W. P. Gomes and F. Cardon, in *Semiconductor Liquid-Junction Solar Cells*, see Ref. R57.
F. T. Inoue, T. Watanabe, A. Fujishima, and K. Honda, in *Semiconductor Liquid-Junction Solar Cells*, see Ref. R57.
G. H. Minoura, T. Oki, and M. Tsuiki, *Chem. Lett.* 1279 (1976).
H. J. Gautron, P. Lemasson, F. Rabago, and R. Triboulet, *C.R. Acad. Sci. Ser. C* **287,** 349 (1978).

current arises from electron excitation from the valence band to a midgap impurity band. The holes move to the surface giving the usual electrochemical reaction (Section 6.3). It is difficult with this model to suggest how the electron conducts current. They propose the electron may either reach the back contact via the impurity band (which in this case is reasonable because the layer is only 5000 Å thick), or may thermally reach the conduction band of the WO_3.

Finally, the experiments of Vanden Berghe, Gomes, and Cardon[25] on illuminated CdS should be mentioned as evidence for the role of flaws in increasing the rate of photocorrosion. They found that photocurrent was present at any wavelength up to 800 nm, and concluded that the absorption of the light was at flaws in the space charge region. When this photocurrent passed, Cd^{2+} ions were found in the solution, indicating that the CdS was decomposed. It is possible that the decomposition occurred via dislocations, as discussed in Section 5.2.3.1. In that section it was proposed that holes in the dislocation were produced by thermal excitation or by breakdown, and then the "trapped" holes could move along the dislocation to the surface and there induce decomposition of the material. In the experiments under discussion here,[25] the holes in the dislocation are produced by illumination instead, but the model is otherwise identical. Alternately, the optical excitation could be from the valence band to the dislocation as in the preceding example, with free holes moving to the surface and the trapped electrons moving along the dislocation until they can be thermally excited. The latter is a less likely reaction, as an excitation energy of up to 1 eV is not easily acquired from thermal processes.

5.3. Observed Flat Band Potentials for Various Semiconductors

In Tables 5.1–5.4 are listed the measured flat band potentials for various semiconductors, the pH at which the potentials were measured,

Table 5.4. Flat Band Potentials of Mixed Oxides

Material	Bandgap	V_{fb}	pH	Refs.[a]	pzc	Ref.[a]	E^o_{vs}	E^o_{cs}	χ
$Hg_2Nb_2O_7$	1.8	0.32	13	*A*, R27					
$Hg_2Ta_2O_7$	1.8	0.42	13	*A*, R27					
n-$FeTiO_4$	2.2	0.45	14	*B*, R27					
n-Fe_2TiO_5	2.2	0.75	14	*B*, R27					
n-$FeTiO_3$	2.6	0.2	14	*B*, R27					
	2.1	−0.35	13	*C*, *D*	6.3	*D*	2.1	0	4.3
n-$CdFe_2O_4$	2.3	0	13	*E*					
n-$YFeO_3$	2.6	−0.2	14	*F*, R27					
$SrTiO_3$	3.2	−0.85	13	*C*, *D*	8.6	*D*	2.6	−0.6	3.7
		−1.15	13	*G*, R27					
		−1.0	13	*E*					
n-$BaTiO_3$	3.3	0.38	0	*H*, R27					
		−0.46	14	*H*, R27					
		−0.95	13	*C*, *I*	10	*D*	2.5	−0.75	
n-$CaTiO_3$	3.4	−0.5	0	*J*, R27					
n-$KTaO_3$	3.5	−0.5	2.5	*G*					
		−1.0	7	*K*					

[a] The letters *A*–*K* correspond to the following references:

A. H. H. Kung, H. S. Jarrett, A. W. Sleight, and A. Ferretti, *J. Appl. Phys.* **48**, 2463 (1977).
B. D. S. Ginley and M. A. Butler, *J. Appl. Phys.* **48**, 2019 (1977).
C. M. A. Butler and D. S. Ginley, *Chem. Phys. Lett.* **47**, 319 (1977).
D. M. A. Butler and D. S. Ginley, *J. Electrochem. Soc.* **125**, 228 (1978).
E. H. H. Kung, H. S. Jarrett, A. W. Sleight, and A. Ferretti, *J. Appl. Phys.* **48**, 2463 (1977).
F. M. A. Butler, D. S. Ginley, and M. Eibschutz, *J. Appl. Phys.* **48**, 1914 (1977).
G. J. M. Bolts and M. S. Wrighton, *J. Phys. Chem.* **80**, 2641 (1976).
H. J. H. Kennedy and K. W. Frese, Jr., *J. Electrochem. Soc.* **123**, 1683 (1976).
I. M. A. Butler, *J. Appl. Phys.* **48**, 1914 (1977).
J. D. I. Tchernev, Int. Conf. Photochem. Conv. and Storage Solar Energy, (Abstract 3), London, Ontario (1976).
K. P. J. Boddy, D. Kahng, and Y. S. Chen, *Electrochim. Acta* **13**, 1311 (1968).

and, in some cases, the conduction band or valence band edge at the pzc. The flat band potentials are vs. NHE (on the hydrogen reference scale). The listing will offer the reader a summary of data available on typical semiconductors and references to work on semiconductors that are otherwise not mentioned in this book.

In the tabulation, the flat band potential rather than E_{cs} or E_{vs} is listed, because that is the value most often given. In general, E_{cs} is approximately 0.1 or 0.2 eV above the flat band potential for *n*-type semiconductors, and E_{vs} is approximately 0.1 or 0.2 eV below the flat band potential for

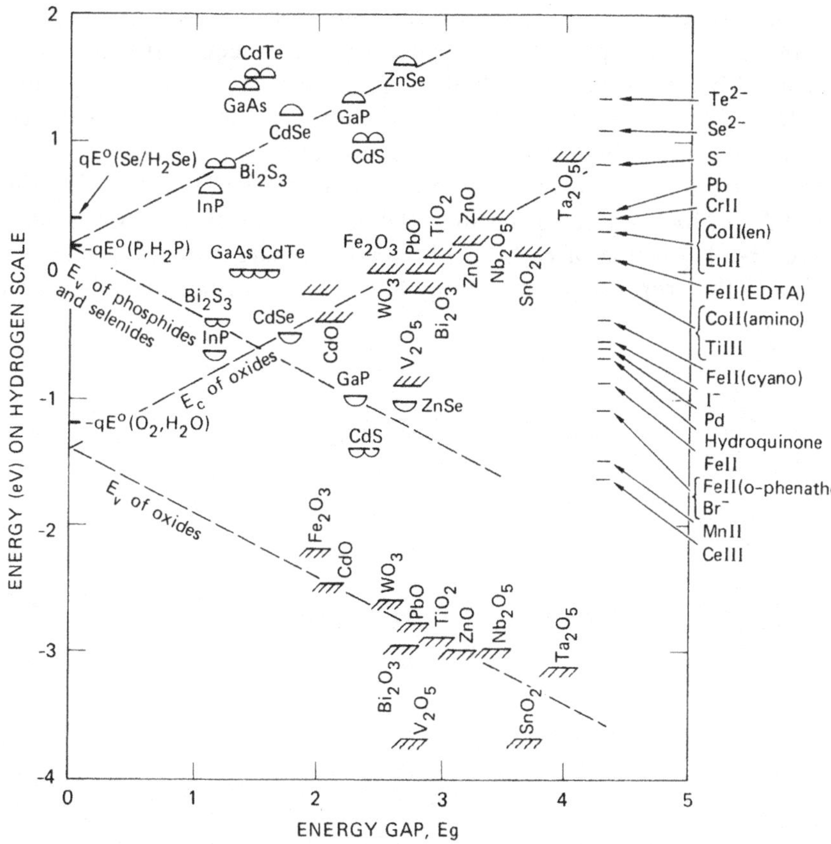

Figure 5.5. Summary of energy gaps and band edge energies (determined from V_{fb}, not corrected for μ) for several compound semiconductors. Energies are estimated with *p*H 7 for oxides, arsenides, and phosphides. Energies are estimated with polychalcogenides, present for sulfides, selenides, and tellurides. For comparison the estimated $E^o_{redox} = -qE^o$ is shown for a series of redox couples, identified by the reducing agent of the couple. Knowing the band edges and the E^o_{redox} for a semiconductor/redox electrolyte system, one can indicate whether carrier transfer is possible. The dashed lines, sketched in as $E_{v,c} = E_0 \pm \frac{1}{2}E_g$, are indicated as a memory aid, and have no obvious theoretical significance.

p-type semiconductors. Most of the data are accurate only to ± 0.1 or 0.2 eV in any case, because usually the Mott–Schottky plots from which V_{fb} is usually estimated are slightly curved and the exact intercept cannot be determined with accuracy better than the above probable error figure.

The references for this section (Section 5.3) are listed separately from the references for Sections 5.1 and 5.2. Harris and Wilson,[R27] Nozik,[R25] and Mavroides[R28] also list rather complete tables of V_{fb} for a series of semiconductors.

Finally, for convenience, in Figure 5.5 the estimated flat band potentials or the conduction and valence bands for the various semiconductors (except the mixed oxides) are plotted as a function of their band-gap. In this figure the band edge energies are normalized to pH 7 using Equation (5.3). An approximate empirical relation seems to hold and is emphasized by the solid lines. However, there is no theoretical justification for such a relationship, and the relationship should be considered a memory aid, not a fundamental connection between V_{fb} and E_g.

In Figure 5.5 is also listed a series of redox couples, showing the value of E^o_{redox} for easy comparison to the band edges of the various semiconductors. The value of E^o_{redox} is obtained from the standard redox potential, and thus represents the energy level of the species hydrolyzed as is found in 1 M acid.

6

Observations of Charge Transfer at an Inert Semiconductor Electrode

6.1. Introduction

The theoretical arguments of Chapter 3 suggest that semiconductors should have electrode properties that differ in many ways from those of metal electrodes. And the discussions of this chapter will show this to be true, as reviewed also in many articles analyzing their behavior.[R29,R4,R10,R30,R31] The behavior of semiconductors is different from metal electrodes in a way that offers unique capabilities for practical use, and different in a way that offers the possibility of providing rapid improvement in the basic understanding of redox processes at a solid/electrolyte interface. In the present chapter we describe many of the behavior patterns of the semiconductor electrode that apparently represent redox reactions under ideal conditions: (a) when the properties of the electrode do not change (no corrosion); (b) when there is (presumably) little change in the inner coordination sphere of the ion, and (c) when there is presumably no adsorption of the reactants or products. The behavior patterns are compared to the predictions of the fluctuating energy level model, and resulting modifications in the model are reviewed.

The unique properties of semiconductor electrodes are based, as indicated in earlier chapters, on a few dominating features. The first unique feature is that the band edges do not move relative to the energy levels of ions in solution. Then, if the band edges are favorable for electron transfer to a given redox couple, the transfer will occur independent of voltage (although the rate will change because of the dependence of surface carrier concentration on voltage). If the band edges are unfavorable for electron transfer to a given redox couple the transfer will not occur, independent of voltage (as long as the semiconductor is nondegenerate). This is quite different from the behavior of metal electrodes, where for all redox couples electron exchange is possible, because the energy at which occupied and unoccupied states are accessible (here the Fermi energy) does move relative to the energy levels of the ions in solution. Thus, as will be discussed below, electron injection into the conduction band of a solid from a species with E^o_{redox} well below the band edge will not occur, independent of applied voltage, unless it becomes degenerate and acts like a metal.[1,2] With a metal electrode, the oxidation or reduction of any species may require an overvoltage to bring the energy level involved in the rate-determining process toward the Fermi energy, but the reaction will always go.

A second unique feature of semiconductor electrodes in redox reactions is our ability to control the properties of the semiconductor, and to control the band in which the reacting carrier is to appear. It will be found that use of this feature aids immeasurably in the analysis of the steps of electrode reactions. For example, the transistor technique (Section 4.3.3) permits the analysis of reactions because one can determine whether the electron exchange prefers to proceed at a high energy level (conduction band) or at a low energy level (valence band). Another example involves the creation of reactive carriers by illumination of the semiconductor. The discussions below in Section 6.3.1 show what a powerful tool this can become in understanding the steps of a reaction. Such options are not available with a metal electrode.

Such unique characteristics lead to the ability with semiconductor electrodes to measure the characteristics of ions in solution and the steps of an electrochemical reaction with facility not attainable on a metal electrode. For example, parameters such as the reorganization energy λ, and such as the energy levels of ions in solution, have been measured. Two techniques have already been developed[3,4] with which to measure the reorganization energy λ that depend on the use of a semiconductor electrode. (See Sections 6.2.1 and 6.5.2.) It should be noted of course, that values of λ for various ions have been measured also using a metal electrode[5] but the procedure is much more difficult as it involves calculating the preexponential term in J_0. The measurement of the energy levels of a two-equivalent species has been done to date in only one case[6] but the ability to do so has been demonstrated, and hopefully will be utilized more in the future. The behavior of radicals at the surface of semiconductors has been studied by photoproducing the radicals[7,8,R31] taking advantage of their relative stability on a semiconducting surface over that on a metal surface. Many such observations will be discussed in this chapter, and a few particularly instructive examples of the use of semiconductor electrochemistry to identify chemical reaction steps, and to characterize the properties of ions, will be described in detail in Section 6.6 after the individual techniques and observations have been introduced in the earlier sections.

The simple theory described in Chapter 3 cannot describe in detail all the experimental behavior of redox reactions on semiconductors. One example of inadequacy that has been discussed earlier (Section 5.1.2) is the extremely low "exchange current," or "reverse current," J_0 [Equation (3.33)], which is usually so low that both components of a redox couple cannot actively exchange carriers with the same band of a nondegenerate semiconductor. If the oxidizing agent of a couple can readily accept electrons from a band, the reducing agent often cannot inject electrons (at least at a rate such that it can be detected over stray currents the order of 10^{-5} A/m^2). In fact, in one case at least, if the value of E_{redox} for a couple is right at the band edge of a nondegenerate semiconductor, no direct carrier exchange in either direction between that band and that couple can be detected. Table 6.1 shows this case, where various redox couples on ZnO are found either to inject or to extract electrons, depending on whether E^o_{redox} is higher or lower than E_{cs}, but where the cupric/cuprous couple with $E^o_{\text{redox}} \simeq E_{cs}$ neither injects nor extracts electrons. The cerium case is an anomaly that will be discussed further in Section 6.2.

Another apparent failure of the model that is yet to be completely resolved is the failure of the simple fluctuating energy level model for very strongly exothermic reactions. For example, if E_{ox} is far below E_{cs}, then, by Equation (3.45) or (3.52), for example, the current should be very low. But a low current is seldom observed. It seems in general that "nature can find a way" to make such exothermic processes go. Evidence will be presented in Section 6.2.1 that under some special conditions nature cannot find a way, but the evidence is not clear.

The above provides a sampling of the type of topics to be described in the present chapter, and hopefully also indicates the motivation for

Table 6.1. Electron Injection into and Electron Capture from the ZnO Conduction Band[a]

Couple	E^o, eV	Injects into	Captures from
Cr(III/II)	−0.4	Yes	No
Eu(III/II)	−0.4	Yes	No
V(III/II)	−0.3	Yes	No
Co(III/II)en	−0.3	Yes	No
Fe(III/II)EDTA	+0.1	Yes	?
Cu(II/I)	+0.15	No	No
Fe(III/II)cyano	+0.35	No	Yes
MnO_4^-/MnO_4^{-2}	+0.6	No	Yes
Ir(IV/III)hexachloro	+1	No	Yes
Ce(IV/III)	+1.6	No	No

[a] From Table 5.1, the conduction band of ZnO is at $E^o = -0.15 \pm 0.2$ eV at pH 7.

further progress along this line of research that is providing such a strong basic understanding of electrochemical redox processes.

6.2. Majority Carrier Capture

Most of the studies of majority carrier capture from inert semiconductors are studies of cathodic currents from *n*-type rather than anodic currents from *p*-type semiconductors. This is because anodic hole capture from *p*-type material usually results in corrosion of the material (Section 7.4), whereas electron capture from most *n*-type semiconductors, for example ZnO, TiO_2, GaAs, GaP, and so on, does not result in corrosion (reduction to the metal). The stability of the *n*-type material during cathodic processes is perhaps somewhat surprising, for on many metal electrodes (e.g., Pt, Fe, Ni, Cr, In, Ta, Ag) cathodic currents are used to reduce oxide layers back to the metal.[R9] The reasons why many semiconductors are stable in conduction band processes will be discussed in Section 7.4. The corrosion during anodic processes on *p*-type semiconductors, for example *p*-Si, *p*-Ge, *p*-GaAs, and Cu_2O, can be of two forms. When the majority carriers (holes) reach the surface, the electrode can either form a surface layer of an oxidized product, or go into solution (Section 7.4). We will say a few words about hole capture in this Introduction, but in the rest of Section 6.2 we will examine studies of electron capture from *n*-type material.

Studies of hole capture from *p*-type semiconductors have generally emphasized competitive reactions,[9] or *d*-band semiconductors. In the "competitive reaction" classification, workers have attempted to prevent corrosion by introducing redox couples to compete with corrosion for hole capture. It is found qualitatively in such studies, as it is found quantitatively for electron capture, that a high concentration of holes at the surface leads to a high capture rate.

The other common case of hole capture studies from *p*-type semiconductors is where transition metal oxides are used, where the corrosion of the oxide does not occur because the holes arise in a *d*-band, rather than in an oxygen-based *sp* band. A particularly noteworthy case is *p*-NiO. As described in Section 5.1.2, the *d* bands associated with the transition metal levels are narrow, with a low carrier mobility, so such materials must usually be very heavily doped to induce reasonable conductivity. Because of this heavy doping, the transition metal semiconductors are usually studied in the high-current region, often the current exchange involves tunneling, and a detailed discussion will be postponed to Section 6.6.2 where tunneling and high currents are emphasized.

In this section we review experimental observations of moderate current flow of electrons to oxidizing agents in solution and compare the observations to the theories of Chapter 3. We will not discuss the quasimetallic high-voltage case, where the semiconductor becomes degenerate (the Fermi energy moves into the majority carrier band), the voltage appears across the Helmholtz double layer, and the material starts to behave like a metal, passing very high currents. Such metallic behavior will be discussed in Section 6.6.2 below. Here we emphasize cases where the semiconductor does not become degenerate; cases where in principle the straightforward fluctuating energy level model for semiconductors, developed in Section 3.3.3, should apply. It will be found that most features of the fluctuating energy level model apply accurately, and will require little discussion. The most controversial part of the discussion is the case of highly exothermic electron capture processes. The theoretical introduction to this exothermic case is discussed in Section 3.4.2, with particular reference to the surface state model of Figure 3.5, and in Section 5.2.3.1, with particular reference to dislocations as the cause of the failure of the fluctuating energy level model. At the end of this section (Section 6.2), we will briefly summarize the electrode behavior with exothermic electron capture and the models that seem to be required to explain the experimental observations.

6.2.1. Direct Carrier Transfer to Ions in Solution

In this section we describe cases where there is no obvious reason to suggest that surface states or other defects participate in the electron capture reaction. First, we must discuss electron capture when there is no active ion present. If no active ions are present, if the solvent is not easily reduced or oxidized, and if the semiconductor is not corroded, a reasonably wide bandgap semiconductor will pass no appreciable current at any voltage where there is a depletion layer present. By "no appreciable current," currents typically 10^{-5}–10^{-6} A/m^2 are considered. This is several orders of magnitude lower than "no appreciable current" on a metal electrode. As discussed in Chapter 5, flaws in the sample (perhaps corrosion at such flaws?) may lead to currents this order of magnitude. As discussed in Section 2.2.1, we will consider the material a "perfectly polarizable semiconductor electrode" with respect to charge transfer polarization (Section 3.1.2) when the current falls below this order of magnitude. Thus, for example, Gleria and Memming[10] find that SiC is polarizable in water between +2 and −1 V vs. SCE.

When an oxidizing agent is now added, the cathodic current is found proportional to the density of electrons at the surface and, in general is found proportional to (first order in) the density of ions in solution. In other words, Equation (3.3) is generally found valid. Presumably if there were appreciable adsorption of the ions, such a proportionality would not be found, because for strong adsorption, the surface coverage is not proportional to the concentration in solution.[R4] The first-order relationship was verified for the reduction of a series of oxidizing agents on ZnO.[11,12] As an interesting example that could not be predicted by the simple theory, we have the reduction of I_2 on CdS and GaAs. As described by Cardon and Gomes,[13] the first-order dependence of cathodic current on iodine concentration with CdS and GaAs is observed (Gerischer and Mattes[14] and Gerischer and Meyer[15]), and indicates that the reduction occurs with the reduction of the molecule as the first step, rather than the first step being the dissociation of the iodine molecule. Such a first-order dependence on the concentration of oxidizing agent is not always observed. For example, Tyagai and Kolbasov[16] report that when iron or ferricyanide ions are reduced in the presence of the reducing agent at a CdS and CdSe electrode, the reduction rate varies as $c_{\text{ox}}^{0.7}c_{\text{red}}^{-0.2}$ rather than first order in the oxidizing agent and zero order in the reducing agent as is "normal" for semiconductor electrodes. These authors suggest that this system is an example of a case where the reducing agent can inject electrons into the semiconductor, and suggest that the current dependence can be explained in terms of surface state effects, where an equation similar to Equation (3.30) is obeyed. However, such a mechanism to explain the effect of the reducing agent would imply a high anodic current due to the reducing agent in solution under anodic bias, and such is not observed. An alternate explanation may be that the deposits of sulfur or selenium on the samples led to a quasimetallic behavior, similar to the behavior of semiconductor electrodes covered by "insulating" oxides as discussed in Chapter 8. However, without further study, the reason for such atypical behavior is difficult to explain in detail. In general, however, cathodic reduction on n-type semiconductors is "typical" in that a first-order dependence on c_{ox} and n_s is found.

From this discussion we expect Equation (3.27), modified by Equation (3.28), to apply, and the cathodic current to be given by

$$J_c = q\bar{c}n_s\sigma(N_t - n_t) = q\bar{c}n_s(\sigma[X]) \tag{6.1}$$

where $[X]$ is the effective surface density (in m^{-2}) of accessible oxidizing agent molecules. We will assume $[X]$ is proportional to the concentration of oxidizing agent in solution. As all the other factors in Equation (6.1) can be measured, $\sigma[X]$ can be experimentally determined.

Using Equation (3.62) (expressing E_{ox} as $E^o_{\text{redox}} + \lambda$), in combination with Equations (3.34) and (6.1), we find

$$\sigma[X] = \sigma d c_{\text{ox}}(kT/\pi\lambda)^{1/2} \exp[-(E_{cs} - E^o_{\text{redox}} - \lambda)^2/4\lambda kT] \tag{6.2}$$

where the σ is shown on both sides of Equation (6.2) to simplify later discussions.

The unknown parameter λ can be obtained experimentally by using Equation (6.2) and varying either E^o_{redox} or E_{cs}. One of the other expressions of Section 3.3.3.4 could be used, leading to a somewhat different preexponential in Equation (6.2). From Equation (6.2) it can be seen that the experimental value of $\sigma[X]$ should pass through a maximum, as either E^o_{redox} or E_{cs} is varied. On the one hand the parameter E^o_{redox} can be varied, simply by varying the chemical species. The scatter in the data using such a variation is expected to be considerable, but will improve if nonadsorbing ions are used, so the assumptions of Equation (3.62) are met, and if λ does not vary too greatly from ion to ion. On the other hand we can vary E_{cs}. A much more accurate value of λ for a particular species should be thus obtainable from equation (6.2), maintaining the ion in solution constant (using only one oxidizing agent) and varying E_{cs}. The easiest way to vary E_{cs} is by varying the pH of the solution [Equation (2.31)], but unfortunately when varying the pH, the species in solution may change (Section 1.2.4.5). In the following discussion we will review such experiments where E^o_{redox} or E_{cs} is varied and the parameters describing the solid or ions in solution (E_{cs} and λ) are estimated by analyzing the kinetics of electron capture.

Figure 6.1 shows results obtained[12] using the approach of varying E^o_{redox} to analyze Equation (6.2). The figure shows the cathodic current from ZnO associated with a series of one-equivalent oxidizing agents that offer a wide spectrum of E^o_{redox} ($= -E^o$). The figure illustrates the first-order dependence of the cathodic current on the surface electron density, n_s, as required in Equation (6.1). From Equation (2.10), n_s varies exponentially with qV_s/kT. Thus the slope of the curves in Figure 6.1 are the order of 60 mV/decade, in agreement with theory. In addition to the data shown, other ions were studied that showed negligible cathodic currents in this V_s range. Specifically Ce^{+4} (in nitric acid), the cupric ion and the chromic ion were all found to induce negligible cathodic current. With the value of J normalized to an oxidizing agent concentration c_{ox} of 10^{-2} M, the value of J/n_s, inserted in Equation (6.1), yields $\sigma[X]$ for each oxidizing agent at this concentration.

In Figure 6.2 the resulting relation between $\sigma[X]$ and the redox potential E^o ($= -E^o_{redox}$) is shown. The bell-shaped curve expected from Equation (6.2) appears to be in evidence. It should be noted that the interpretation of Figure 6.2 in the cited reference [12] is in error. However, the experimental data are of value.

As mentioned above, such qualitative agreement of Figure 6.2 and Equation (6.2) implies moderate accord for the two major assumptions. The first assumption is that specific adsorption of the ions can be neglected, so Equation (6.1) is able to provide a good description of $\sigma[X]$. Thus the reduction is presumably associated with tunneling between the solid and the nearby ions [perhaps ions within 15 Å of the surface, in

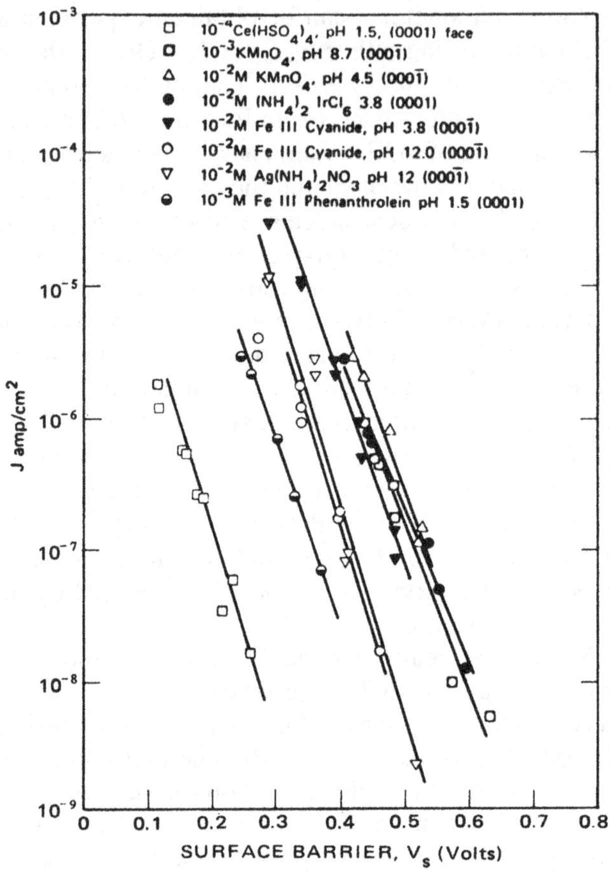

Figure 6.1 The variation in cathodic current with the surface barrier, for various one-equivalent oxidizing agents on ZnO.

accordance with Equation (3.62) and the accompanying discussion]. The other assumption—one that is undoubtedly weaker—is that variation of λ from ion to ion will not cause so much scatter in the points as to make the curve unrecognizable. As will be described below, other experimental work yields values of λ from 0.4 to 1.7 eV for the various species, so the assumption is undoubtedly poor.

Now Figure 6.2, interpreted using Equation (6.2), permits evaluation of the various parameters. For example the maximum value for $\sigma[X]$ occurs at a particular value of E^o, E^o_{max}, and should occur according to Equation (6.2) when

$$-E^o_{\text{max}} = (E^o_{\text{redox}})_{\text{max}} = E_{cs} - \lambda \tag{6.3}$$

The observation of E^o_{max} permits the estimation of a value for σ, and this estimation is insensitive to the exact value of λ used because the preexponential factor of Equation (6.2) is rather insensitive to λ. With the maximum of $\sigma[X]$ the order of 10^{-3}, d assumed to be 15 Å, and $c_{ox} = 0.01$ M or 6×10^{24} m^{-3}, we find from Equation (6.2) that σ is about 10^{-18} m^2, a reasonable value for a mean ionic capture cross section.

The parameter E_{cs} can be estimated from E^o_{max} in Figure 6.2 without knowing λ accurately. One can insert a theoretical value (or the mean of experimental values as discussed later for λ), of $\lambda \cong 0.8$ or so into Equation (6.3). This yields a value for E_{cs} close to the values reported in Section 5.3 for ZnO.

By comparing Figure 6.2 with Equation (6.2), a value for λ should also be obtainable. However, it turns out apparently that there is too much scatter in the values of λ for such an analysis to be made. For example, if one uses the slope of the curve as drawn, the curve originally sketched to pass through the experimental points, one obtains an average value of λ of 0.2 eV, much lower than predicted by theory or observed in other experiments.

Now the most important use of Figure 6.2 is not its use as a method for the evaluation of parameters, but as a check on the fluctuating energy level model. Other techniques for the evaluation of parameters are more

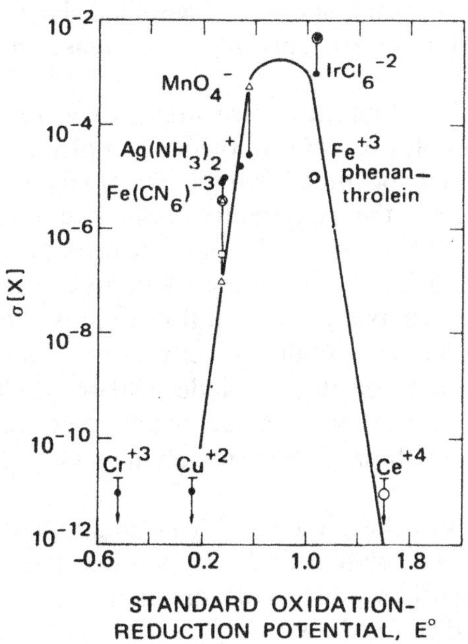

Figure 6.2 The variation in electron reactivity as a function of the redox potential (the energy level) for various one-equivalent oxidizing agents. The encircled symbols, indicating pH, read as follows: ○, 1.5; ●, 4; △, 9; and □, 12.

accurate (because of the problem here of the λ variation). The importance of Figure 6.2 is the observation that, in agreement with the fluctuating energy level model [Equation (6.1)], the current appears to decrease for E^o_{redox} more negative than the critical value $(E^o_{redox})_{max}$. In fact, for cerium the current is found to be negligible. The observation bears on the important question of whether the current ever decreases for exothermic reactions as predicted by the fluctuating energy level model. There are only a few observations in the literature suggesting that under ideal conditions (e.g., damage-free surface; cf. Section 3.4.2) the fluctuating energy level model does hold true, and there are an equal number of observations in the literature where the model apparently fails; high currents are observed for highly exothermic reactions. The present case is one of those that suggest that the fluctuating energy level model may be valid. Others are discussed later in this section.

The feature under discussion, the decrease in J_c for highly exothermic processes, is supported in this study by three measurements. Figure 6.2 shows two of the three, the ceric ion (in nitrate) and possibly the phenanthrolein complex of iron. A third observation was reported in the cited article,[12] namely, a value of $\sigma[X]$ of 10^{-8} for the sulfate complex of cerium. As the energy level of the complex would be higher than the uncomplexed cerium by an unknown amount (see Section 1.2.4.5 or 7.2), an experimental point is not shown in Figure 6.2. However, the observed low value of $\sigma[X]$ is consistent with a decrease in $\sigma[X]$ at $E^o_{redox} < (E^o_{redox})_{max}$.

It should be noted that these measurements were made on relatively dislocation-free samples of ZnO. In the studies of Figure 6.2 an etch was used on the ZnO (and specifically on the (0001) face where the important results were obtained) that apparently eliminated the effect of dislocations, as described in Section 5.2.3.1. And in accordance with the models of Sections 3.4.2 and 5.2.3.1, the presence of dislocations may be a common problem that permits high current flow to ions with very low energy levels. Most wide bandgap semiconductor crystals are difficult to grow with a low dislocation density, and indeed many studies are made, unfortunately, with lapping damage inadequately removed from the surface, and lapping damage will provide extremely high (say, 10^{16} m^{-2}) densities of dislocations.

Another means to check the validity of Equation (6.2) experimentally is by varying E_{cs}. A variation of E_{cs} by varying the Helmholtz voltage [cf. Equation (2.31)] is a much more direct way to evaluate λ than the variation of E^o_{redox} discussed above.

Vanden Berghe, Cardon, and Gomes[3] studied electron capture by potassium ferricyanide, the $Fe(CN)_6^{3-}$ ion, at the (0001) surface of ZnO, changing E_{cs} by a pH variation. They used an etching procedure suitable for removing dislocations. They found the highest cathodic current (at constant n_s) at pH 5, the lowest cathodic current at pH 12. They analyzed the results using Equations (6.1), (6.2), and (2.31), and concluded that the Gaussian for the oxidizing agent ferricyanide is as indicated in Figure 6.3, with E_{cs} for ZnO (0001) becoming equal to the E_{ox} of ferricyanide when the pH is between 5 and 9 (perhaps pH 6). In the analysis, λ is the parameter that must be fitted to match theory and experiment, and they concluded that $\lambda = 0.75$ eV, a value close to that theoretically predicted [Equation (1.47)].

These results also have bearing on the validity of the fluctuating energy level model for exothermic reactions. It will be noted in Figure 6.3 that the decrease in the capture rate at pH 12 is supporting evidence for the validity of the simple fluctuating energy level model, which predicts the capture cross section decreases if E_{ox} is below E_{cs}. Of course, the evidence is less direct than in the results of Figure 6.2 because the argument is somewhat circular: the relative position of E_{ox} and E_{cs} was calculated assuming the simple model to be valid, and thus assuming the only reason for σ to decrease as the pH rises (as E_{cs} rises) is the Gaussian distribution. It is conceivable that other factors (e.g., a slight change in hydrolysis) may lower the apparent capture cross section as the solution becomes more basic. However, the assumption of Vanden Berghe *et al.* that the change in E_{cs} is dominant is not unreasonable.

There is some inconsistency between the interpretation of the results

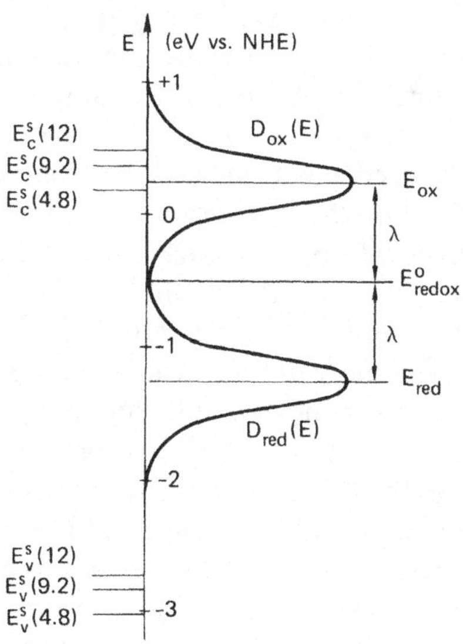

Figure 6.3 Energy scheme for ZnO in contact with an electrolyte containing equal concentrations of $Fe(CN)_6^{3-}$ and $Fe(CN)_6^{4-}$.

of Figure 6.2 and of Figure 6.3. The experimental values obtained are consistent: the value of $\sigma[X]$ (at 10^{-2} M) observed by Vanden Berghe *et al.*[3] is the order of 10^{-6}, within the range of values reported in Figure 6.2. However, if we accept the interpretation of Figure 6.3, the ferricyanide ion should be at the maximum of the $\sigma[X]$ vs. E^{o} curve, especially near pH 9. But in the curve of Figure 6.2, ferricyanide is displaced from the maximum by about 0.4 V. The apparent discrepancy may be due to scatter in Figure 6.2 arising from a variation in the capture cross section of the various ions. Certainly if it is accepted that ferricyanide at pH 6 should show the maximum $\sigma[X]$ observable, at constant σ, then the higher values observed for MnO_4^- and $IrCl_6^{2-}$ must arise owing to much higher values of σ. Alternatively, there is the possibility that the discrepancy arises in the interpretation of Figure 6.3. If a pH change causes an energy level shift of the ferricyanide ion, then the slight change in apparent σ with pH cannot be analyzed as was done in Figure 6.3, and cannot be interpreted as an indication that $E_{ox} \sim E_{cs}$ for ferricyanide on ZnO. Unfortunately, if this is the source of the discrepancy, the method of Vanden Berghe *et al.* for the calculation of λ is invalid.

6.2.2. Indirect Electron Transfer to Ions in Solution

The cases discussed above provide two examples (more will be described later) where the simple fluctuating energy level model seemed to apply directly, even when the reduction was highly exothermic. There are, on the other hand, many examples where excited states, surface states, or flaws in the solid can be shown to be involved, or are suspected of being involved in the electron transfer, so the simple model is inadequate. We will turn our attention to a few examples of such cases, cases of highly exothermic electron capture.

Grabner[17] monitored electron capture from the ZnO conduction band by rubrene radicals, where the rubrene radicals were generated at a nearby platinum electrode by reducing rubrene cations. He concludes in this article that electrons are not captured by the R^+/R level, 0.8 eV below the ZnO conduction band, but a cathodic current is first observed only (Figure 6.4a) when the ZnO has a degenerate accumulation layer, and V_H is sufficient to lower the excited state of the rubrene cation—the $R^+/{}^3R^*$ level (Figure 6.4b)—to the ZnO conduction band edge. The electron capture, of course, leads to an excited rubrene molecule, and Grabner speculates that the quenching of luminescence from this excited molecule is catalyzed by the degenerate ZnO. This model, which suggests indirect electron capture through excited states may, however, require some revision to fit more recent results.[126] Another example of electron capture by oxidizing agents with E_{ox} well below E_{cs}, where the excited state of

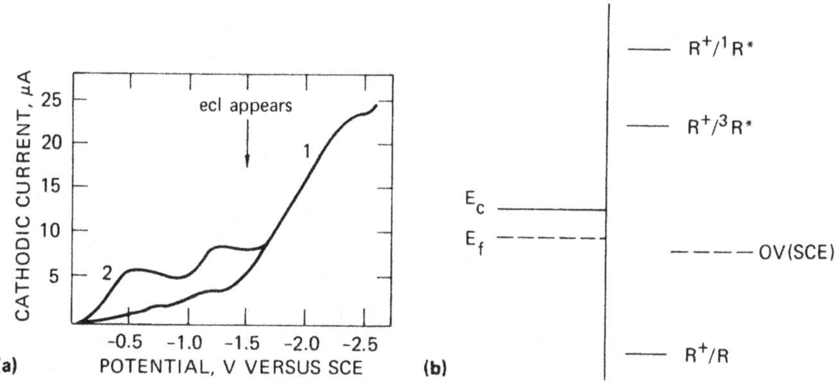

Figure 6.4. (a) Single-sweep voltammetry of 3 mM rubrene in benzonitrile (0.1 *M* TEAP) at a ZnO electrode. Voltage scan: 50 mV sec. Electrode area: 0.23 cm^2. Curve 1: normal scan. Curve 2: scan after generation of rubrene radicals cations at a Pt electrode in front of the ZnO electrode at 1.0 V versus SCE for 3 min. (b) Energy levels of ZnO and ionized rubrene. The observation that the ZnO must be accumulated to move electrons to the surface [curve 2 of (a)] suggests that the electrons move to the excited states of the rubrene rather than the ground state.

the ions is utilized as discussed in Section 3.4.2 (Figure 3.5), is a process observed by Gleria and Memming[18] in the electron capture by the bipyridyl complex of ruthenium, $Ru(bipy)_3^{3+}$. The process is identifiable because the electron transfer from the excited to the ground state of ruthenium is radiative, and the resulting luminescence is observed.

A second mechanism for electron capture by oxidizing agents with E_{ox} well below E_{cs} is by tunneling. If the donor density is sufficiently high, the surface barrier becomes very thin and tunneling through it can occur, as illustrated in Figure 5.1. However, that figure shows tunneling from the solution into the conduction band; tunneling from the conduction band to the solution, of interest here, can also occur. Tunneling from the conduction band to an ion in solution is expected to be very unusual. One expects tunneling to occur only if the oxidizing agent has an energy level at a very particular energy. It has to equal the energy of the bulk conduction band edge while the surface barrier is still bent up enough to offer a narrow barrier for tunneling. The desired configuration is shown in Figure 6.5b, where ion (i) has the right energy level. For a highly anodic (the semiconductor positive) voltage, Figure 6.5a, this same energy level of ion (i) in solution will be too high for the electrons from the conduction band to tunnel to, and in any case, tunneling is impossible [even to ion (ii)] because the barrier is too wide. If the voltage becomes more cathodic (less positive), as in Figure 6.5c, the energy level of ion (i) is too low for tunneling, so the current to ion (i) decreases as the voltage becomes more

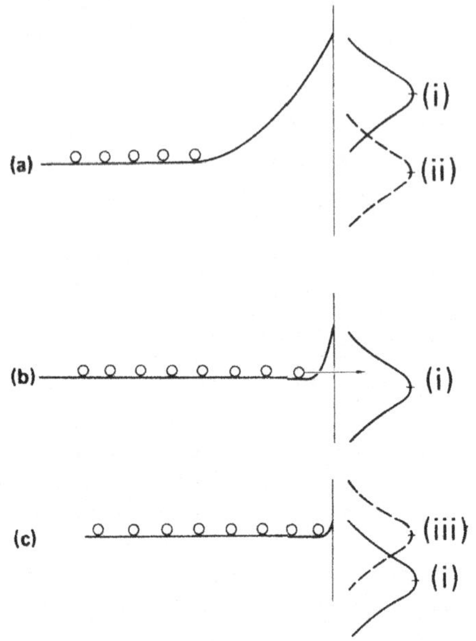

Figure 6.5. Required conditions for cathodic tunneling from a conduction band to an oxidizing agent in solution. The current to level (i) will go through a maximum because in (a) the electrode potential is too positive for tunneling, in (c) it is too negative.

cathodic. Tunneling to ion (iii) is possible, but such tunneling is inextricably mixed up with normal cathodic currents over the top of the barrier, and cannot be distinguished from this normal cathodic process. Thus only ions in a limited energy span can be positively identified as leading to tunneling—identified by a maximum in current as the electrode is made more negative.

Pettinger *et al.*[19] observed such electron transfer by tunneling to $IrCl_6^{2-}$ ions from ZnO, with the peak occuring at 0.4 eV vs. SCE. An analysis of the position of the $IrCl_6^{2-}$ level[R4] indicates it is in the right energy range for such a peak in cathodic current to be observed, and the observation thus provides confidence that the tunneling models are qualitatively correct.

Other tunneling observations thay may indicate majority carrier tunneling to the surface are less direct. Uchida *et al.*[20] observed completely reversible electrode behavior on highly doped (4×10^{26} m^{-3}) SnO_2 in molten LiCl/KCl eutectic at 450°C. They ascribed the ease of reduction of a series of ions in solution to the ease of tunneling, with their highly doped electrode. Laitinen[R32] also reviews tunneling at the SnO_2 electrode and notes reversible behavior when heavily doped (tunneling); irreversible when lightly doped (no tunneling).

A third apparent reason for the failure of the fluctuating energy level model when the level E^o_{redox} is well below E_{cs} is the assistance of electron capture by surface states or other energy levels in the solid, as indicated in Figure 3.5 and the accompanying discussion. A simple numerical analysis of such a phenomenon was done near the end of Section 3.3.3.4, and it was shown that with a highly exothermic reaction, electron capture via surface states could easily be more rapid than direct transfer to ions in solution. The surface state mechanism has been suggested, to explain some of their observations, by Bard and co-workers.[21,22]

Frank and Bard,[22] Kohl and Bard,[21] and Yeh and Hackerman[23] have provided a particularly relevant series of studies of electron capture by oxidizing agents, using various semiconductor electrodes (TiO_2, ZnO, CdS, GaP, and GaAs) in acetonitrile and dimethylformamide solvents. The results illustrate the transition from (1) quasimetallic behavior, where E^o_{redox} is well above E_{cs}, through (2) "normal" semiconductor behavior, where E^o_{redox} is near E_{cs}, to (3) nonideal behavior, where E^o_{redox} is far below E_{cs}. We will refer to all three regimes here, although we will emphasize case (3). Case (1) will be discussed further in Section 6.6.3. In these studies, the use of nonaqueous solutions permitted high voltage without interference by electron capture by the solvent (by protons). Thus they could sweep the voltage (cyclic voltammetry) very negative to make the semiconductor degenerate, and induce electrons to flow from the *n*-type semiconductors to oxidizing agents with very high E^o_{redox}.

The electrode potential (vs. SCE) where the voltammetry peaks are observed is shown in Figure 6.6, an illustration from the paper by Kohl and Bard.[21] This figure will be used here, but some of its features will not be discussed until later sections. It is clear that the current peaks for the reduction and reoxidation of various species on the metal electrode are usually at a different potential from the same process on the semiconductor electrode. The electron transfer from the semiconductor to an oxidizing agent depends on the relative value of V_{fb} (essentially E_{cs}) and the E^o_{redox} of the ion. If the E^o_{redox} of the oxidizing agent is much higher than the V_{fb} of the semiconductor (indicated as F.B. in the figure), then a very high negative voltage is required for electron flow. The high negative voltage causes an accumulation layer to form, leading to the Fermi energy being in the conduction band region—the semiconductor is degenerate at the surface. Then much of the applied voltage appears across the Helmholtz double layer, the semiconductor acts as a metal, and E^o_{redox} moves down toward E_F. With such metallic behavior the current flow is essentially of the same magnitude, and occurs at the same potential, as that of the platinum electrode used for comparison. This quasimetallic behavior is described in more detail in Section 6.6.3.

If the E^o_{redox} of the ion in solution is near the E_{cs} of the semiconductor, particularly when it is lower than E_{cs} by a few tenths of an eV, the re-

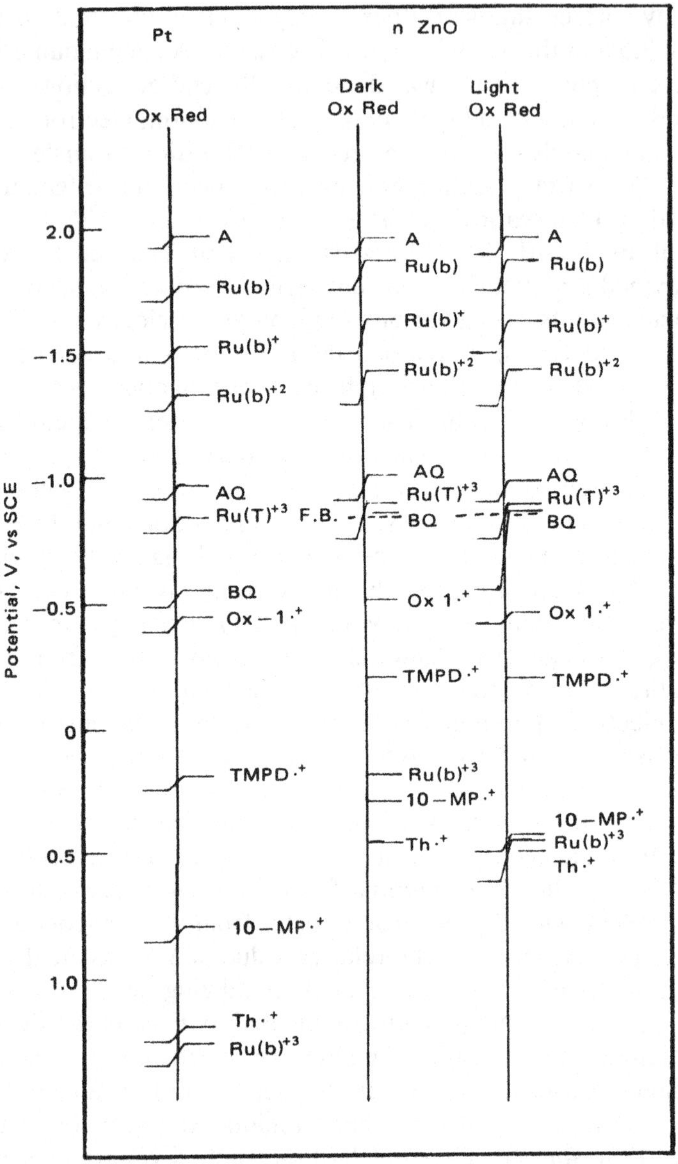

Figure 6.6. The reduction and reoxidation voltammetric peak potentials for some organic compounds on Pt and on ZnO. The reader is referred to the original article (see text) for the actual compounds; here we are interested primarily in the difference in electrode potential on the semiconductor versus metal electrodes for the redox reaction and the lack of reversibility for some reactions on the semiconductor.

duction of the ion becomes irreversible. The result is "normal" for semiconductors, and Kohl and Bard's results on ZnO can more or less be compared to the reduction curves of Figure 6.1, except that in the work of Kohl and Bard, the voltage span was generally much larger; the voltage was swept so negative (the barrier so low) that the realm of diffusion-limited currents was reached (described as concentration polarization in Section 3.1.2).

Of particular interest in this subsection is the case when the energy E^o_{redox} becomes very much lower than the energy E_{cs}. Kohl and Bard and Frank and Bard concluded that the behavior in this region is different from "normal" semiconductor behavior, and invoke surface state effects. Specifically, in the work of Bard and co-workers (e.g., Figure 6.6) it was found that for all species with E^o_{redox} below a certain value, reduction seems to occur at the same value of V_s in the semiconductor. The current density is high, the order of 0.2 A/m^2. They conclude that this pattern becomes evident when E^o_{redox} is 1.6 eV or more below E_{cs} of TiO_2 or ZnO. The critical surface barrier at which the electron transfer begins is high, about 1.3 eV for ZnO and 1.1 eV for TiO_2. There is no evidence of such behavior for the more narrow bandgap materials CdS and GaP; the cathodic current varies with E^o_{redox} on these electrodes.

The question becomes, how can one obtain such a very high cathodic current on the ZnO or TiO_2 electrodes with such a large surface barrier, even when surface state effects are invoked. It makes a valuable exercise to consider how these systems appear on the band diagram.

To illustrate the band diagram for the semiconductors and ions involved in the above paradox we can consider Figure 6.7. The figure shows the band conditions under which high cathodic currents flow according to Bard and co-workers. In the diagram we assume an oxidizing agent with $\lambda = 0.5$ eV, whose E^o_{redox} is located 1.6 eV below the conduction band of the semiconductors. Only for illustration are the conduction band edges of ZnO, TiO_2, and GaP placed at the same energy, each with $V_s \sim 1$ eV. It is easily seen from such a diagram that direct isoenergetic electron transfer from the conduction band edge to the oxidizing agent-in solution is not likely because there is no overlap of the fluctuating energy level with E_{cs}. With GaP there is a possibility of valence band processes (hole injection), but obviously with ZnO and TiO_2 there is no such possibility.

The question is why should current flow, with the bands bent as indicated in the figure? Clearly conditions would be favorable for bulk conduction band electrons to transfer isoenergetically from E_c to the energy level E_{ox} of the oxidizing agent, but this in principle cannot help, because without tunneling the transfer must be from E_{cs}, not E_c (Section

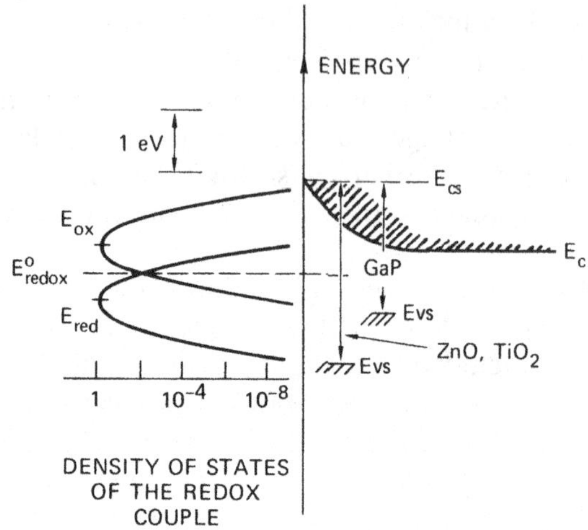

Figure 6.7. Illustrative band model for a redox couple with E^{o}_{redox} about 1.6 eV below the band edge, using the value of λ for quinone in water (~0.5 eV). The conduction bands of GaP, ZnO, and TiO_2 are indicated at a common value for illustrative purposes only. A band bending of about 1 eV is indicated to relate to the discussion in the text.

6.2.1). The electrons have no way to reach the surface except by being thermally excited to E_{cs}.

Bard and co-workers[21,22] suggest that a surface state at about 0.5 eV below the conduction band could explain the results, with the surface state acting as an intermediary between the conduction band and the oxidizing agent in solution. This explananation, however, does not seem to help because, as discussed in Section 3.3.3, electron flow to surface states is also proportional to n_s, and in Figure 6.7 the surface barrier ($E_{cs} - E_c$) is so high that n_s is much to small. For example, let us assume that $\bar{c}$ is 10^5 m/sec, σ has the exceptionally high value of 10^{-17} m^2, and consider the most extreme case where the density of surface states is a complete monolayer, 10^{19}/m^2. From Equation (2.10) with the bulk doping, $n_b = 10^{24}$ m^{-3} and with V_s, as discussed above, 1 eV, we have that n_s is 4×10^6 m^{-3}. Even with these favorable values the current by Equation (3.27) is only about 6×10^{-6} A/m^2 compared to the observed currents of 0.2 A/m^2 or so. Thus a simple surface state does not appear to permit such current flow.

Frank and Bard[22] hinted that the effect may be due to energy levels in the bandgap in the bulk. This may be a better possibility. The TiO_2 and the ZnO samples were mechanically polished, so that a very high dislocation density was produced. They were reported as being etched, but with etches that remove almost no material. And conducting TiO_2 is the classic example of a material with disorder, Magneli phases.[24] Thus, undoubtedly, the samples were heavily disordered in the surface region.

Thus conductance through the dislocation levels (Section 5.2.3.1) could indeed be the mechanism for the observed electron transfer to the surface for all oxidizing agents whose E^o_{redox} is below the energy band of the dislocations.

Clearly more work needs to be done on such systems to analyze the mechanism more clearly. However, the studies of Figure 6.6, indicating the transition from quasimetallic through semiconducting behavior to this odd behavior when E^o_{redox} is deep in the bandgap region, are of great interest.

Vandermolen *et al.*[25] attributed another form of anomalous behavior during cathodic reduction, to surface states. They reduced $Fe(CN)_6^{-3}$, Fe^{3+}(aquo), and $IrCl_6^{2-}$ on TiO_2 and $SrTiO_3$. They etched the samples thoroughly in molten NaOH, so there should have been no dislocation effects. They found the current is proportional to ion concentration but only at low concentration. With a negative bias, at moderate V_s, the current anomalously saturates at high oxidizing agent concentration, saturating at a concentration too low to consider that the saturation is due to concentration polarization. This result clearly does not follow Equation (6.2), and the authors suggest that surface states are an intermediary in the reaction, and that at high cathodic voltage the rate-limiting step is electron transfer to the surface states. In this case, with $V_s = 0.4$ eV at saturation, a value of $N_{ss} = 10^{15}$ m^{-2} would explain the results, and even with a well etched sample, 10^{15} m^{-2} surface states can be expected. However, it is not clear (a) why direct electron capture by these ions does not dominate over surface state capture, and (b) why electron transfer to surface states should be rate limiting at low V_s, but not be rate limiting at high V_s. So again the surface state model does not seem to explain the results well.

To summarize the findings for electron capture by oxidizing agents, where electrons are the majority carrier, the dominant controversial feature is the behavior with highly exothermic reactions. Usually the reduction of an oxidizing agent follows the fluctuating energy level model if E_{ox} is at, or just above, E_{cs}. If, however, E_{ox} is below E_{cs} (so E^o_{redox} is far below E_{cs}) the theory states that the capture cross section σ should decrease exponentially, but it does not always do so.

There is some evidence that σ decreases as expected for perfect, lightly doped crystals. Reviewing the cases where σ appears to decrease as E^o_{redox} becomes lower in the band model, as would be predicted by the fluctuating energy level model, we note several cases already discussed, and more to be discussed later. Memming[26] has shown that Eu^{3+}, with $E^o_{redox} = 0.3$ eV, has a higher capture cross section for electrons from *p*-GaAs than $Fe(CN)_6^{3-}$ with $E^o_{redox} = -0.4$ eV. Vanden Berghe *et al.*[3] in their analysis of electron capture by ferricyanide on ZnO, and Morrison[12] in his studies of electron capture by a series of oxidizing agents on ZnO, both obtained results consistent with a decrease in σ with very low E^o_{redox} relative to E_{cs}. These studies were discussed in Section 6.2.1 above. The studies of Pettinger *et al.*,[19] described in the present section, showed that as the Gaussian for E_{ox} moved below the conduction band, the tunneling probability became lower. And the studies of rubrene by Grabner and of ruthenium by Gleria and Memming[18] on ZnO, also reviewed above in the present section, also suggest the same conclusion, for here capture on the excited state is observed although direct capture on the ground state of the dye could occur. This case was also discussed in Section 6.2.1 above.

On the other hand, there are definitely cases where the electron capture rate does not decrease as E_{ox} moves below the conduction band edge, as would be expected by the fluctuating energy level model. Several such cases were discussed in the present subsection. As has been indicated in this subsection, it is the opinion of the author that these cases can be explained by one of the four mechanisms described in Section 3.4.2: assistance by surface states, by excited states of the ion, by complexes of the ion (as discussed in more detail in Section 7.2), or by dislocations (as discussed in more detail in Section 5.2). However, the conclusion that the fluctuating energy level model does describe electron capture on *n*-type semiconductors if assistance by defects and by the complications described above is accounted for, must still be considered an opinion rather than an established fact.

6.3. Minority Carrier Capture

In this section we discuss the capture of minority carriers, the capture of holes from *n*-type semiconductors, or electrons from *p*-type semiconductors. Minority carrier capture by intrinsic surface states, by one-equivalent ions in solution, and by two-equivalent ions in solution will be reviewed, but not in that order. First, we discuss minority carrier capture on two-equivalent ions in solution, than go on to discuss the other two cases together. The reason for the order chosen is very simple: although minority carrier capture on two-equivalent species should be, and probably is, much more complex than minority carrier capture on one-equivalent species, the "current-doubling" technique provides such a powerful tool for the study of two-equivalent species that much more is known about two-equivalent minority carrier reactions. Additionally, the current-doubling technique has also provided a great deal of our knowledge about the capture of minority carriers on one-equivalent ions. And, as the theory of the two-equivalent process must be examined to understand this technique, we have a second reason to examine the two-equivalent case first. It will turn out that when capture by one-equivalent ions is reviewed in Section 6.3.2, much of the review will be a simple reference to the results obtained while studying two-equivalent ions.

Most of the work to be described will be on n-type material, although a few experiments have been done on p-type, and these will be mentioned.

There are two major differences between majority carrier reactions as described in the preceding section, and most minority carrier reactions. One is that when a depletion layer is present on the surface, photoproduced minority carriers are forced to the surface by the electric field in the space charge region. The second is that usually the minority carrier capture is highly exothermic (especially with oxides, as is clear from Figure 5.5). Thus the carriers are forced (by the field) to remain at the surface until one of the possible exothermic reactions available to them occurs. Normally the question (especially for holes as the minority carrier) is not whether the photoproduced carrier will react, but which reaction will dominate.

One question that will be in the background in all these discussions, especially when discussing hole capture from n-type material, is whether minority carrier processes are always indirect. Is the hole always captured first by an intrinsic surface state, or by a defect level located well above the valence band, before transferring to the ion in solution? There is probably not a single answer to this question, but the experimental evidence, as it is discussed, will be found to suggest that the hole is always captured before being passed on to the solution. The question is analogous to the same question for exothermic electron capture, reviewed at the end of the preceding section, but the answer seems different. The evidence, and the means by which the evidence on this question is obtained, should be noted particularly, while reading the section, as it bears strongly on the corrosion analysis developed in Section 7.3.

6.3.1. Minority Carrier Capture on Two-Equivalent Species: Radical Formation and Current Doubling

6.3.1.1. Model. In Section 2.5 we discussed the energy levels associated with two-equivalent species, species where normally a two-electron reaction occurs. The energy level diagram is indicated in Figure 1.8. In the present section we examine more closely the behavior of electron (or hole) transfer to the energy levels associated with such a two-equivalent species in solution.

First, let us compare metals vs. semiconductors as electrodes for dark two-equivalent reactions, where minority carriers are not present. With a metal electrode, as was pointed out in Section 1.2.5, if the Fermi energy is positioned such that the first electron transfer can occur, then the second electron transfer can immediately occur because it is even more exothermic. With a semiconductor electrode in the dark, however, as discussed in Chapter 3, a second carrier of the same kind may not have a favorable energy for the second step. It is the position of the bands rather than the position of the Fermi energy that determines whether electron (hole) transfer will be possible. In the case of the configuration of Figure 1.8, electron transfer to the As^{5+} level may be possible if the semiconductor is n-type; but reduction of As^{5+} to As^{3+} by electrons from the conduction band would not be favored by the simple fluctuating energy level model because the As^{4+}/As^{3+} level is in the midgap region. Thus oxidation or reduction of two-equivalent species in the dark on wide band-gap semiconductor electrodes (only one band involved) should not, in many cases, be favored according to the theory.

Oxidation or reduction of two-equivalent species on an illuminated semiconductor electrode is common, is of great interest both practically and theoretically, and has been analyzed in many studies. In this case minority carriers are involved. Consider Figure 1.8, again specifying that the semiconductor is n-type. Let holes (minority carriers) be produced by illumination, and let a positive potential be applied, forcing the holes to the surface where they must react. It is found[(7)] that if the holes are captured by two-equivalent reducing agents, majority carrier injection often follows immediately. When the minority carrier is captured, an unstable radical is formed. In the As^{3+}/As^{4+} example of Figure 1.8, As^{4+} is formed. The As^{4+} will now be able to inject electrons into the conduction band of the semiconductor, as is obvious by inspection of the figure. In principle, the As^{4+} could also capture another hole, but the availability of these photoproduced minority carriers is limited, and experimentally the electron injection route is found to be favored.

The two-equivalent reactions are thus considered to have the form

$$h^+ + R \rightarrow R^{\dot{+}} \tag{6.4}$$

$$R^{\dot{+}} \rightarrow R^{2+} + e^- \tag{6.5}$$

or, for p-type materials with photoproduced electrons,

$$e^- + X \rightarrow X^{\dot{-}} \tag{6.6}$$

$$X^{\dot{-}} \rightarrow X^{2-} + h^+ \tag{6.7}$$

where R is the two-equivalent reducing agent (often an organic molecule), X is the two-equivalent oxidizing agent, and $R^{\dot{+}}$ and $X^{\dot{-}}$ are the corresponding free radicals. Reactions such as Equations (6.4) and (6.5) or (6.6) and (6.7) are commonly termed "current-doubling" reactions in the literature. This is because the reaction leads to twice the current flow expected from the illumination of the semiconductor with no current doubling. In principle, for each minority carrier reacting at the surface, a majority carrier is injected. Reactions (6.4)–(6.7) can be compared to a non-current-doubling reaction where the minority carriers are captured by a simple one-equivalent reactant, such as hole capture by Fe^{+2}:

$$h^+ + M \rightarrow M^+ \tag{6.8}$$

Here no free radical is formed, no electron injection step will occur, so the current will be simply that corresponding to the rate at which photoproduced holes arrive and react at the surface.

Although, in principle, the rate of hole capture by a reducing agent M or R is given by the first term of equation (3.29) with n_t the density of reducing agent ions within tunneling distance of the surface, and p_s the density of holes at the surface, the equation is not necessarily realistic. If the rate-limiting step is the rate of hole generation by light rather than the rate of hole capture at the surface, we have that the hole current is simply proportional to the light intensity:

$$J_p = \alpha I \tag{6.9}$$

where J_p is the current of holes reaching the surface to react, I is the light intensity, and α is a proportionality constant, depending on electron–hole recombination processes. Although an "effective hole density at the surface" can at times be defined, it cannot normally be measured, and forcing the mathematical description into the form of Equation (3.29) has little meaning.

If two species are present at the surface, each of which can be oxidized by (can capture) holes, the rate of hole capture on each will be proportional to their capture cross section and to their effective surface density. Thus for the two species, M and R, with capture cross sections σ_M and σ_R, we have

$$J_{pM}/J_{pR} = \sigma_M[M]/\sigma_R[R] \tag{6.10}$$

where $[R]$ and $[M]$ are the effective surface densities, which may or may not imply adsorption of the species. With the total minority carrier current J_p given by

$$J_p = J_{pM} + J_{pR}$$

a simple manipulation yields

$$J_p/J_{pR} = 1 + (\sigma_M/\sigma_R)\,([M]/[R]) \tag{6.11a}$$

The system becomes particularly interesting when, as described in Equations (6.4), (6.5), and (6.8), the species R is a two-equivalent species and species M is a one-equivalent species. Then the minority carrier current J_p can be measured by introducing only the one equivalent ion in solution. J_{pR} is the current of electrons injected into the conduction band, and is measured as the difference between the total current and the minority carrier current. Equation (6.11) is followed experimentally and permits the comparison of the capture cross sections of various species in solution.

Equation (6.11a) is used to determine relative capture cross sections by a "titration" technique. We note that the electron injection current J_e is equal to J_{pR} because each hole captured by R means an electron is injected, so the total measured current is given by $J_p + J_{pR}$. Then Equation (6.11a) can be rearranged as

$$\frac{\sigma_M}{\sigma_R} = \frac{[R]}{[M]} \frac{2J_p - J}{J - J_p} \tag{6.11b}$$

where J is the measured current. In the titration one can begin with a given concentration of the one-equivalent species, M, present with no R, and thereby determine J_p. Then one adds the current-doubling species R and, as holes are captured by R, the current increases. By Equation (6.11b), when $J = 1.5\, J_p$, the holes are captured equally by R and M, and the ratio of the cross sections of R and M is given by the inverse of the concentration ratio at that point of the titration.

The highly reactive radicals formed by minority carrier capture not only can react by injecting a carrier into the majority carrier band, but also can react with another species, Y, in solution. In lieu of the carrier injection process described by Equation (6.5) or (6.7), a simple reduction or oxidation of the ion Y added to the solution is possible, e.g.,

$$R^{\overset{+}{\cdot}} + Y \rightarrow R^{2+} + Y^- \tag{6.12}$$

A very simple analysis can be made, that permits evaluation of a parameter describing the radical reactivity of Y, by observing the decrease in carrier injection as the concentration of Y is increased. We assume the probability of a radical reacting with the species Y is $\gamma[Y]$, proportional to the concentration of Y and to a factor γ proportional to the reactivity. If the radical does not react with the species in solution, carrier injection according to Equation (6.5) or (6.7) is observed. Thus with J_{pR} the hole current to the current-doubling agent, the carrier injection current J_e is given by

$$J_{pR} = J_e = J_p(1 - \gamma[Y]) \tag{6.13}$$

As will be reviewed in the experimental discussion to follow, for the one case quantitatively studied to date, Equation (6.13) is obeyed. However, a comparison of radical reactivity for a series of species has not been made to date. The difficulty in such studies is the selection of current-doubling agents R and agents Y, such that Y can react with $R^{\overset{+}{\cdot}}$ but cannot spontaneously react with R.

Recently it has been demonstrated[27] that the mechanism of current doubling can be chemically much more complex than that shown in Equations (6.4) and (6.5). In rotating ring-disk electrode measurements on ZnO, Fujishima[27] found that the cation of the electrode was dissolved into the solution (the ZnO was corroding) at a rate equivalent to the hole current, even though efficient current doubling was occurring. By Equations (6.4) and (6.5) such corrosion of the ZnO should not be observed. This suggests a mechanism such as

$$h^+ + \mathrm{ZnO{-}H^+} \rightarrow \mathrm{OH\cdot} + \mathrm{Zn^{2+}}\ \text{(in solution)} \tag{6.14}$$

$$\mathrm{OH\cdot} + R \rightarrow \overset{+}{R\cdot} + \mathrm{OH^-} \tag{6.15}$$

$$\overset{+}{R\cdot} \rightarrow R^{2+} + e^- \tag{6.16}$$

where $\mathrm{ZnO{-}H^+}$ is intended to represent an H^+ ion from solution adsorbed on a lattice oxygen ion (Section 2.1.1.1). In the model it effectively behaves as an intrinsic surface state. Now two radicals are sequentially produced. More complexities can be visualized, but Equations (6.14) to (6.16) suggest a minimum (but acceptable) series of steps that will yield the observed decomposition of the solid during current doubling. The critical step in this model is the dissolution of the electrode to yield an oxidizing agent that can still oxidize the two-equivalent molecule or ion. The conclusion is that the hole must be first captured by an intrinsic surface state, and it is the charged surface state that reacts with the ion in solution.

In the subsections below, we will emphasize the three main applications, using the models described above, of the current-doubling observation. They are (i) to observe whether minority carriers react with a given species, (ii) to determine the relative capture cross sections of various one- and two-equivalent ions in solution, and (iii) to observe reactions of free radicals with ions in solution.

6.3.1.2. Observations of Current Doubling on ZnO. Current doubling was first noted during studies of the electrode ZnO,[7] and because the behavior is simpler on this semiconductor than on most others, most experiments and most reviews of current doubling[R33,R10,R34,R31] have emphasized this material.

The current-doubling observation has been useful in determining individual steps in many reactions. In the original work, the objective was to study the oxidation of formate ions by holes from ZnO:

$$\mathrm{HCOO^-} + h^+ \rightarrow \text{products} \tag{6.17}$$

The details of the experiment will be discussed further in Section 6.6, where the use of current-doubling observations in combination with other observations to analyze the photocatalytic oxidation of formic acid will be reviewed. For the moment it is of interest to point out that the observation of current doubling, with the interpretation as given by Equations (6.4) and (6.5), made it clear that the photocatalysis in Equation (6.17) involves the capture of photoproduced holes by the formate. It was found that "current doubling" is not actually an accurate term; for example, in this case the current increases by a factor closer to 1.7 under the most favorable conditions used. Thus the first, and a common usage, of the current-doubling observation is simply to identify the species oxidized.

The second usage of current doubling—the determination of the relative capture cross sections of the species being oxidized, monitoring the hole capture competition between a one-equivalent and a two-equivalent species, and analyzing the results with Equation (6.11b)—has also been done in many experiments, using ZnO. For example, Gomes *et al.*[28] carried out an extensive investigation of the relative capture cross sections for hole capture by this technique. Freund[29] and Vanden Berghe *et al.*[30] used the method, in these cases emphasizing the capture on one-equivalent species. Sancier and Morrison[31] studied the oxidation of organic molecules on TiO_2 using the method, with the object to identify stable solvents (solvents with a low cross section for hole capture) for use in paints. Table 6.2 summarizes the observed hole capture cross sections on ZnO assembled from these various studies, where the results are expressed as σ/σ_I with σ_I the capture cross section for iodide ions. Figure 6.8 shows results obtained by Freund,[29] showing the adequate, but not satisfying, agreement between measurements where various competitive reactions are used to obtain the same series of capture cross sections. By the above direct theory, of course, the three lines drawn, giving the relative capture cross section for one-equivalent species as obtained [from Equation (6.11b)] by measuring them against three different two-equivalent species, should all be parallel. The observed relation between σ and E^o of the ions will be discussed further in Section 6.3.2 below.

Table 6.2. Relative Hole Capture Cross Sections[a] of Various Ions at the ZnO Electrode

One-equivalent ions		Two-equivalent ions	
I^-	1.0 (by definition)	CH_3OH	0.09
SO_3^{2-}	0.2	C_2H_5OH	0.21
Br^-	0.015	$(CH_3)_2CHOH$	0.08
Co(II) (amino)[b]	7×10^3	$(CHOH\ COO^-)_2$	0.04
FeII (cyano)[b]	80	BH_4^-	2.5
Cl^- [b]	0		

[a] The values are in general those obtained by Gomes *et al.*[28] The values from Vanden Berghe *et al.*[30] are essentially in agreement.
[b] From Freund,[29] as derived from Figure 6.8.

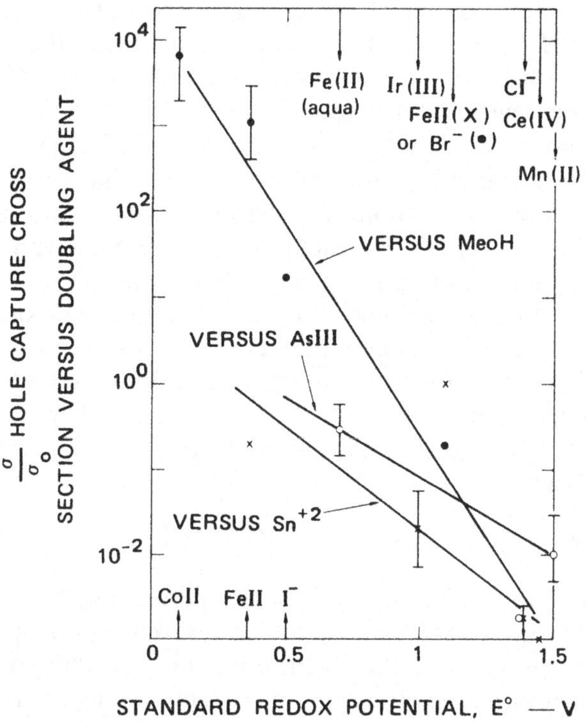

Figure 6.8. The reactivity toward holes from ZnO for various one-equivalent reducing agents as measured by the current-doubling technique, titrating against a two-equivalent species. The curves for three titrants, methanol, arsenic, and tin, are shown. The higher hole reactivity of the stronger reducing agents is unexpected by the simple models (Section 6.3.2).

The third use of the current-doubling technique, discussed in relation to Equation (6.12), is to detect the reaction of the generated free radicals with other ions in solution. If the free radical reacts with an ion in solution, it will not inject an electron, so the reaction will result in a lower "current-doubling" current as the species in solution increases in concentration, in accordance with Equation (6.13). The observation of such an effect was made in the work[7] on the photocatalytic oxidation of formate [Equation (6.17)], where it was noted that the presence of oxygen in solution lowered the current due to electron injection from the radical, and it was therefore concluded that one of the steps in the formate oxidation was the reaction of the radical with oxygen.

Gomes *et al.*[28] explored radical reactions quantitatively, studying the interaction of radicals formed on ZnO with $Cu(NH_3)_4^{+2}$ ions. They found that independent of the doubling agent, $HCOO^-$, CH_3OH, C_2H_5OH, or $(CH_3)_2CHOH$, the interaction was identical, viz., the constant γ in Equation (6.13) was the same. The result was interpreted to mean that during the oxidation of any of these species the radical was the same, most probably H·. However, it is possible, considering the observations of Fujishima[27] that suggest a reaction route of Equations (6.14)–(6.16), that the common feature of the reaction is the radical OH· or some equivalent radical associated with the intrinsic surface state on the ZnO rather than a radical associated with the two-equivalent reducing agent from the solution. Then the competition of Equation (6.13) occurs between the "intrinsic surface state radical" reacting on the one hand with the current-doubling agent, and on the other hand with the $Cu(NH_3)_4^{2+}$ ion.

In addition to the competitive reactions with redox species in the solution, anions are found to affect the efficiency of the current-doubling process. Micka and Gerischer[32] found at the ZnO (0001) face (the Zn face) with no one-equivalent ion added that injection current was variable—the amount of injection current being highest for perchlorate and lowest for halides:

$$ClO_4^- > SO_4^{2-} > NO_3^- > H_2PO_4^- > Cl^- > I^- > Br^-$$

In fact, if the anion in the solution is the perchlorate ion, Micka and Gerischer found they could obtain perfect current doubling. For some reason the other ions lower the electron injection current. At the $(000\bar{1})$ face (the oxygen face) the efficiency of current doubling varied with the anion in a similar order:

$$ClO_4^- > SO_4^{2-} > NO_3^- > H_2PO_4^- > Cl^- > Br^- > I^-$$

The Br^- and I^- results are complicated by the fact that they are oxidized themselves (Table 6.2). The order above more or less reflects the expected strength of anion adsorption, with Br^- and I^- usually the most strongly adsorbed species. If the reaction route of Equations (6.14)–(6.16) is valid, then the observed results could be consistent with ion exchange between OH^- groups at the surface and the anion in solution. The ion exchange could lower the density of reactive hydroxyl ions. However, such a naive explanation does not adequately explain the anion effect, especially considering Fujishima's observation[27] that the oxide decomposes independent of whether current doubling occurs. A satisfactory model must describe why decomposition with no electron injection is favored over decomposition with electron injection, when anions such as Cl^- are in solution.

A further observation of Micka and Gerischer,[32] that may help to understand the anion effect, is that the efficiency of current doubling varies with pH. Specifically, as the pH increases over pH7, current doubling becomes less efficient, especially on the (0001) face. But also on the $(000\bar{1})$ face, the oxygen face, they find only 40% current doubling in reasonably alkaline solution. These facts are not inconsistent with the dominance of OH^- groups in accordance with Equations (6.14) to (6.16). Micka and Gerischer suggest that a strongly adsorbed anion tends to stabilize the holes at the surface. As the corrosion reaction is a higher-order reaction in holes than the current-doubling reaction of Equation (6.4), such hole storage should tend to favor corrosion. However, this interpretation is hard to reconcile with Fujishima's observation that corrosion occurs anyway, even during current doubling. A speculative explanation of the anion effect, consistent with Fujishima's observation, is that the activation energy for removal of the Zn^{2+} ions is lower if suitable ligands such as Cl^- are present, and rapid removal of Zn^{2+} lowers the Madelung potential, facilitating the complete oxidation of the remaining adsorbed OH· groups by holes. But clearly more clarification is required.

As a final topic on observations of current doubling at the ZnO anode, the work of Gomes and Cardon[33,34] should be mentioned, where the lifetime of the free radical was investigated. They used noise measurements, as discussed in Section 4.3.3., to estimate the lifetime of the free radical produced in the current-doubling measurements. They concluded that on ZnO the radical lifetime is less than 10^{-4} sec. By another technique, using short light pulses and measuring the decay of the injected current, they were able to obtain a better estimate, and concluded that the radical lifetime is probably the order of 10^{-5} sec for various organics on ZnO.

On powders, the generation of radicals by the interaction of photoproduced minority carriers with two-equivalent species in solution has important practical implications. This process is involved closely with photocatalysis on TiO_2 (described in the next subsection) but work has also been done on ZnO. Cunningham and Zainal[35] provided an interesting study of ZnO powders. They measured the electron injection from the radicals by measuring the presence of the injected electrons using ESR (electron spin resonance). They determined that when ZnO powder was illuminated in aqueous solution, extra electrons appeared in the conduction band if current-doubling agents such as formate or ethanol were present and if one of the better anions to minimize the anion effect, such as NO_3^-, was present.

Clearly the production and reactions of radicals on ZnO (and on other semiconductors as discussed below) is of great interest, and it is also clear that considerable further work must be done to resolve the details of these interesting reactions.

6.3.1.3. Observations of Current Doubling on Other Electrodes. Current doubling is, of course, not restricted to ZnO. Gerischer and Rösler[R35] review the work of their group on current doubling on CdS, where many species are current-doubling agents. Here again, a *p*H dependence is observed, with a peak of electron injection at *p*H 9. The HS^- ion concentration was not controlled, and it should be noted that more recent results (Section 5.1.1.3) indicate such control may be important for these crystals. Vanden Berghe *et al.*[30] found current doubling on CdSe with formate ($CHOO^-$) and formaldehyde (CH_2O), but not with the alcohols and other compounds so effective with ZnO. From Section 5.3 it is clear that the valence band of CdSe is very high in energy, so it is understandable that holes from CdSe are not very strong oxidizing agents, and may therefore not oxidize alcohols at all. In agreement with this interpretation it was found that the holes from CdSe could not oxidize Br^-, although the ratio of cross sections for $Fe(CN)_6^{-4}$: I^- is about 100, the same ratio as found above with ZnO. The paper by Vanden Berghe *et al.* provides an interesting discussion of hole reactions as affected by the energy of the hole, the valence band energy level. Vanden Kerchove *et al.*[36] find current doubling on $SrTiO_3$ of acids, alcohols, and aldehydes.

Current doubling is not restricted to *n*-type semiconductors. Photoproduced electrons from *p*-GaAs have been shown to reduce bromine and result in hole injection into the valence band.[37] Photoproduced electrons from *p*-GaP have been shown to reduce $S_2O_8^{2-}$ and H_2O_2 and result in hole injection into the valence band.[38] Both these reactions are symmetrical with Equations (6.4) and (6.5), with the radical still injecting the majority carrier, which of course is now holes rather than electrons.

The ions $S_2O_8^{2-}$ and H_2O_2 are exceptional in that upon capture of an electron, they provide such an active oxidizing intermediate that cases are observed where majority carriers seem to cause "current doubling." The reduction of $S_2O_8^{2-}$ or H_2O_2 on TiO_2 [39] or ZnO [40] have been shown to result in hole injection into the very deep valence band of these *n*-type materials. In these cases the hole injection is detected by luminescence. However, Noufi *et al.*[39] find most of the luminescence on TiO_2 is infrared, and they suggest, reasonably, that perhaps hole injection into defects, rather than hole injection into the valence band, may dominate.

Current doubling has, at times, been used to analyze reaction steps, to reject a proposed system, and to help in the analysis of energy levels in solution. Chai *et al.*[41] were able to reject the use of acetaldehyde as an electrolyte in an electrochemical solar cell (Section 9.1) with CdSe as the electrode, because they found current doubling. As emphasized by Freund and Gomes[R31] and Morrison,[2] the observation of current doubling on a semiconductor often implies an irreversible reaction [e.g., the irreversible generation of CO_2 during the oxidation of formate in Equation (6.17)]. In a solar cell, where the electrolyte must be restored continuously, such irreversible current-doubling reactions cannot be allowed.

Kunugi *et al.*[42] were able to analyze the behavior on illuminated Ag_2O by noting that current doubling occurred when ethylene glycol was present as the electrolyte. Peaks in this case were observed during the anodic sweep, one peak being ascribed to the Ag^0 being oxidized to Ag^+ at the silver/silver oxide interface, and the next peak being ascribed to the Ag^+ being oxidized to Ag^{2+} at the silver oxide/solution interface. The oxidized surface state Ag^{2+} was able to oxidize glycol (if present), this oxidation being detected by the resulting current doubling. Actually, the authors suggest that the Ag^{2+} is in the chemical form of AgO, being hydrolized in a manner equivalent to Equation (1.13). This again suggests that "intrinsic surface states" on the semiconductor/electrolyte interface can exhibit chemical complications [as in Equations (6.14)–(6.16)] that must be considered in a quantitative analysis of current doubling.

Memming and Möllers[4] were able to analyze the energy levels of the quinone–hydroquinone two-equivalent couple and determine the four levels necessary to describe the oxidation states (Section 1.2.5) with the analysis based in part on current-doubling observations. The semiconductors studied were GaP, Si, Ge, and SnO_2. This study will be described in depth in Section 6.7.2.

As a final and not the least important topic for this section, we will discuss the formation of radicals on TiO_2. This material has not been studied in the depth that ZnO has, but the radical generation is of greater practical importance because TiO_2 is used for coatings (because of its high index of refraction), and TiO_2 has been of great interest in solar energy conversion (Section 9.1). Photoproduced radicals on its surface can be very undesirable in both these applications. In the coating field, photoproduced free radicals initiate chain reactions causing oxidation of the paint binder (chalking), and in the solar energy conversion area, as discussed above, such irreversible reactions as current doubling lead to permanent changes in the composition of the solution.

Current doubling has been observed by Dutoit *et al.*[43] on TiO_2 when oxalic acid, methanol, formaldehyde, formic acid, citric acid, lactic acid, malic acid, tartaric acid, As^{3+}, Sn^{2+}, or $(Hg_2)^{2+}$ are present in solution. Workers have also observed[44] current doubling with CN^- and with ethyl formate, and it has been reported[45] that isopropanol and ethanol lead to current doubling. The efficiency of current doubling (using alcohols) is highest at low *p*H. Others[31] find that methanol and ethanol are current doublers, but acetronitrile, dioxane, and ethyl acetate are not. The latter study was for the purpose of finding solvents that would help to stabilize paint. It is clear that holes from TiO_2 can oxidize many organic groups.

In their study, Frank and Bard[44] find, as predicted by Freund and Gomes[R31] that interesting compounds can be synthesized during current doubling. Such has been known in the paint industry for many years,[46] but in the paint industry they find the new compounds quite undesirable. The new compounds produced in the paint vehicle cause chalking of the paint during exposure to sunlight.

6.3.2. Minority Carrier Capture by One-Equivalent Ions

The discussion in the present section will center primarily on capture of holes as the minority carrier, because studies on *p*-type semiconductors are not so common.

There are two general ways in which capture of minority carriers by one-equivalent species in solution can be monitored: chemical and electronic. The first is direct observation of the reaction product, and the second is monitoring electrical or optical effects resulting from the minority carrier capture. We will discuss chemical measurement first, but the quantitative data with which to base a theoretical discussion come from electrical measurements.

Direct observation of the chemical products has not been common, because usually the rate of product generation is very low. The slow chalking of paint on TiO_2 can be considered an indirect observation of a one-equivalent oxidation by photoproduced holes, the oxidation of OH^- to OH· radicals. Actually, chalking is fairly good evidence for such capture, although (a) it is not an electrochemical process, and (b) it is quite indirect. Direct observation that anodic hole capture from ZnO yields H_2O_2 was made by Morrison and Freund,[7] using iodimetric titration, but this was considered corrosion, not oxidation of OH^- ions from the solution. Degradation of solar cell characteristics, discussed in Sections 7.3 and 9.1, is another observation of hole capture involving corrosion, and the effect of one-equivalent species in preventing such corrosive degradation is an important indirect method of monitoring the capture efficiency of one-equivalent additives. The problem is currently of considerable interest, and substantial research is in progress.

New techniques will improve such "chemical" methods greatly. Most promising is the recent use of the rotating ring-disk electrode (RRDE), where very sensitive measurements of the concentrations of products are possible. RRDE features the direct observation of the products (either of corrosion or of minority carrier capture for the ion) and thus, in the future, will make an important contribution to our knowledge of carrier capture. Memming[26] has compared the capture of electrons by various oxidizing agents on *p*-GaAs, using the RRDE—analyzing the effectiveness of each in competing with hydrogen reduction—and has shown that Eu^{+3} is much more effective than $Fe(CN)_6^{-3}$. Using the RRDE he monitored the electron capture by observing the generation of the reducing agent. It should be noted that the E^o_{redox} for Eu^{+3} is much higher than that for $Fe(CN)_6^{-3}$, so his observations provide yet another example where the fluctuating energy level model appears to be valid even for exothermic reactions (Section 6.2.1). The SRI group[47] have compared various complexes of iron in hole capture from GaAs using RRDE, and found that FeII(citric acid) > FeII(EDTA) > FeII(cyano) > FeII(aquo) is the order of hole capture cross sections.

The most common electronic technique for monitoring minority carrier capture by one-equivalent species has been current doubling. Most of our quantitative knowledge about minority carrier capture by one-equivalent species has been obtained by this method. Such measurements are reported in Section 6.3.1.2 above, where we note particularly Table 6.2 and Figure 6.8. These results probably represent the most complete studies of the cross section for hole capture by one-equivalent reducing agents. It is noted that the listings include species such as halides, where hole capture may well be two-equivalent (involve two energy levels such as described in Section 1.2.5). There are at least two reasons why a species that can form a radical may behave as a one-equivalent species in current-doubling experiments on an inert electrode: the radical energy level may be below the conduction band edge, or the radical may disproportionate. Koybayashi *et al.*[48] measured halides as stabilizing agents for photocorrosion of ZnO, and showed qualitatively the same order ($I^- > Br^- > Cl^-$) as shown (Table 6.2) by current doubling.

In principle, the theoretical expression for hole capture on the one-equivalent species M^{n+} is given by

$$h^+ + M^{n+} \rightarrow M^{(n+1)+} \tag{6.18}$$

where the rate should be:

$$\begin{aligned} \text{rate} &= k_1[h_s^+][M^{n+}] \\ &= k_2\{\exp[-(E_{\text{red}} - E_{\text{vs}})^2/4\lambda kT]\}[M^{n+}]p_b \exp(eV_s/kT) \end{aligned} \tag{6.19}$$

with the k's rate constants, and where the first factor stems from analogy with Equation (3.46) and the fluctuating energy level model; and where the second factor, with p_b the bulk density of holes, stems from Equation (2.11) and the Boltzmann equation. Unfortunately, such a model does not describe the observations.

There are two problems met in applying this model to hole capture on one-equivalent species (and, actually on two-equivalent species too, although the problems are not so evident there, because the energy levels are not known as well). The first problem is that it is not clear that $p_s = [h_s^+]$ can be defined, and the second problem is that the rates measured to date do not follow the prediction that they should decrease exponentially with increasing $(E_{\text{red}} - E_{\text{vs}})^2$.

If the minority carrier availability is low for any of several reasons (e.g., a low light intensity), the carriers may be consumed as soon as they reach the surface and then a free-hole density may be difficult to define. This is particularly obvious when one considers that a hole is essentially a vacant bond on an anion, and a vacant bond on a *surface* anion is actually an intrinsic surface state, not a level in the valence band. Thus, if the hole has time to become localized on a surface anion, it suddenly is no longer a free hole. Similarly, an electron on a cation can be an electron on a cationic intrinsic surface state when it reaches the surface, rather than an electron in the conduction band. The point is that $[h_s^+]$ in Equation (6.19) cannot be clearly defined. Gomes *et al.*[28] commented that although they used a formal density of minority carriers at the surface when calculating the expressions for minority carrier transfer to ions in solution, it was only a formality. Williams and Nozik[49] extend this thought further, suggesting that the concept of the quasi-Fermi-level (Section 1.1.1.3) that is so popular in discussion of corrosion, may be entirely inapplicable; the charge transfer may be thought of better as individual events.

However, the most important error in Equation (6.19) is not the error of using a formal hole density, but the manifest error in the expression used for the hole capture cross section—the expression that stems from the fluctuating energy level model. The use of a "formal hole density" in Equation (6.19) is reasonably acceptable, and may usually give the right dependences on the parameters of interest. But, as is clear from the curves of Figure 6.8 and from the results of Table 6.2, the observed hole capture cross section varies with E^o in exactly the wrong order to be described as in Equation (6.19). We note (Figure 5.5) that the energy level of the valence band of ZnO is about 3 eV below the conduction band, and we note the result discussed in Section 5.3 that the conduction band for ZnO is near, or just above, the zero of energy (on the hydrogen scale). Then the value of E^o_{redox} for Co(II) (amino), which has an E^o of 0.1 V, should be near the conduction band, not the valence band. The value of E^o for Ce(IV) should be at +1.6 V (NHE), much closer to the valence band than that for Co(II). Equation (6.19) says the current should decrease exponentially as the square of the difference between the band of interest and the energy level of interest (in our case $E_{\text{red}} - E_{\text{vs}}$), but we find that the current increases exponentially with such a term.

The influence of intrinsic surface states is probably the cause of the complete reversal of the behavior from that expected. Such surface state dominance can be rationalized in the case of hole capture, for anionic surface states can represent fairly stable species such as O^- or As^0 or S^0 for oxides, arsenides, and sulfides, respectively. The concept is discussed further in Section 6.4.1.

6.3.3. Photocatalysis

When a semiconductor is illuminated and the electrons and holes move to the surface, the electrons will reduce species at the surface; the holes will oxidize species at the surface. The two processes can be the same reaction (the forward and reverse direction, respectively, of the same reaction). If not, however, permanent chemical changes will occur in the medium in which the semiconductor is immersed. Such photoinduced chemical changes are termed photocatalytic effects. Reviews of such processes in liquid or resinous media have been prepared by Hauffe,[R33,R36] Morrison,[R4,R10] Bard,[R37] Gerischer,[37] Volkenshtein,[R39] and Freund and Gomes.[R31] A special case has recently become very important in solar energy conversion, the case of the photocatalytic generation of fuels from solar radiation, viz., the direct conversion of optical energy to chemical energy. This energy conversion will be discussed in more detail in Section 9.1.

We will discuss briefly the important concepts in photocatalysis. First, the flow of electrons and holes to the surface must be equal at steady state. If they are not, either (a) the surface becomes electrostatically charged and the process stops, or (b) another electrode must be present to complete the electrical circuit and the process is termed photoelectrocatalysis, not photocatalysis. A practical difference between photocatalysis and photoelectrocatalysis is that the former can occur on powders, as no counterelectrode is required.

The requirement that the rate of majority carrier flow equal the rate of minority carrier flow can be expressed quantitatively by the various equations developed for such carrier capture. For example, when electrons are the majority carrier, we can use Equation (4.31) or (4.32) to describe the rate of hole capture, and Equation (3.34) to describe electron capture. Equivalent expressions must be used if holes are the majority carrier. In other words, with electrons the majority carrier, and selecting Equation (4.31), we must have

$$J_{n,p} = J_1 \exp(-qV_s/kT) = bI_0(L_p + BV_s^{1/2}) \tag{6.20}$$

where $J_{n,p}$ represents the (equal) electron and hole currents, where the first expression comes from the first term of Equation (3.34), with J_1 given, for example by Equation (3.52). The second expression is obtained from Equation (4.31), in which C_{sc} is evaluated using Equation (4.19) to give the $V_s^{1/2}$ dependence. The constant B in Equation (6.20) represents a collection of remaining constants. We note that for a high current flow to the surface, L_p should be high: the bulk minority carrier lifetime should be high. We also note that V_s depends on I_0 from the last equality of Equation (6.20), and when this dependence is inserted into the expression for $J_{n,p}$, $J_{n,p}$ is not linear in I_0.

From the above argument, it is observed that a photocatalytic reaction will not be proportional to the light intensity but will be a more complex function. This complication does not necessarily appear with a photoelectrocatalytic reaction, for in the latter case a constant voltage may be applied, so V_s in Equation (6.20) is constant.

Equation (6.20) gives the total rate of carrier flow to the surface without regard for whether the carriers, once there, react by the desired photocatalytic reaction, react via an undesirable reaction (e.g., corrosion, as in Section 7.3), or recombine on recombination centers (Section 6.4.2). The higher the current flow to the surface, the higher the majority carrier density at the surface, and the higher the relative recombination velocity. Miyake *et al.*(45) suggest that photocatalysis may never be as practical as photoelectrocatalysis because in the latter the applied voltage can be used to keep majority carriers away from the surface and avoid surface recombination [see Equation (6.33) below and the accompanying discussion]. Wilson(50) has made a calculation of the competition between surface recombination and a desired reaction of the minority carrier.

A feature of photocatalysis or photoelectrocatalysis is that the catalytic rate is essentially independent of the sample area, except insofar as the area controls the absorption of the light. The area in photocatalysis is never the dominating feature that it is with dark catalysis. Thus, for example, photocatalytic reactions can be studied as well on single crystals as on powders, because a cubic single crystal, with 6 cm^2 area, absorbs the same light as the same volume of powder, with, say, 10 m^2 total area, for in both cases the light will fall on 1 cm^2.

From our knowledge of electrochemical reactions, we can usually anticipate whether a photocatalytic reaction will occur. Freund *et al.*(104,R31) pointed out that the observation of current doubling, while studying a photocatalytic system, was almost a guarantee that a photocatalytic process would occur if that same system was at open circuit. The reason is the irreversibility of the minority carrier capture by a current-doubling species. With reference to Figure 1.8b, it is clear that if (following hole capture) the occupied level associated with the radical is high enough on the band diagram to inject an electron, then the unoccupied level on the radical is usually too high to capture an electron. The product cannot be rereduced, so the oxidation represents a permanent chemical change.

Another approach, from a theoretical point of view, to predict whether a photocatalytic reaction or a photoelectrocatalytic reaction will occur, is to compare the "thermodynamic" energy levels of the redox couples in solution with the energy of the band edges. The thermodynamic energy levels are essentially the negative of the redox potential [Equation (1.68)] as discussed in Section 1.2.1.2. If the E^o_{redox} for the one-equivalent or two-equivalent couple, the thermodynamic energy level for the couple, is above E_{vs}, then holes will, in principle, be able to oxidize the reducing agent of the couple to a first approximation, independent of whether the holes are a majority or a minority carrier. Similarly, if E_{cs} is above

E^{o}_{redox}, it is normally exothermic for electrons to reduce the oxidizing agent of the couple. The reaction, of course, may not occur for any of several reasons. The real energy levels (as opposed to the thermodynamic levels) may be so far from the band of interest that the activation energy is too high. Capture of the carrier by an intrinsic surface state may make the energy of the carrier unfavorable such that the reaction cannot occur. Or, a competitive reaction may dominate over the reaction desired. But, at least the thermodynamic energy levels offer a basis for suggesting which reactions might be favorable.

Both desirable and undesirable reactions can occur, either in photocatalysis of photoelectrocatalysis. The techniques of semiconductor electrochemistry have proved very powerful in analyzing undesirable as well as desirable reactions. Recently, much of the interest has been in desirable reactions, specifically photocatalytic or photoelectrocatalytic conversion of solar to chemical energy. However, as will be discussed in more detail in Section 9.1, to date electrochemical solar energy conversion has not been found attractive from a practical point of view because of the low efficiency and/or corrosion problems. On the other hand, there has been a substantial interest in undesirable photocatalysis, in particular on TiO_2, specifically the cases of undesirable photocatalytic oxidation of resins and polymers in paints and plastics. In these cases we are interested in how to prevent the photocatalysis. With TiO_2 it is generally accepted[46] that OH^- ions at the surface are oxidized by the photoproduced holes to the radicals $OH\cdot$, and these latter can oxidize a whole spectrum of organics or other reducing agents. For example, Irick[51] and Cundall *et al.*[52] have studied the oxidation of isopropyl alcohol to acetone, and Irick provides a long list of references to other photocatalytic oxidation processes. Others are described below. As discussed in Section 6.3.1.3, other studies using current doubling indicate further the oxidizing power of the holes from TiO_2 in photocatalytic processes. At present the most significant of these is the oxidation of water, the observation of which has led to a great many studies in the area of solar energy conversion.

We will here list example studies of photocatalytic reactions, many of which will be described in more detail in other sections. The first study of photocatalysis, where electrochemical techniques were used to analyze a reaction in detail, was done by Morrison and Freund[7]:

$$O_2 + HCOOH \xrightarrow[\text{ZnO}]{h\nu} H_2O_2 + CO_2 \qquad (6.21)$$

This work will be described in detail in Section 6.7.1, where it is used as an example to show how electrochemical measurements can be used to analyze relatively complex chemical and electronic behavior. The various steps can be individually identified as hole or electron processes or radical reactions. Hauffe and Volz[53] studied a similar photocatalytic process on ZnO, including the oxidation of organic molecules and the reduction of oxygen, monitoring the zeta potential (electrophoresis); this will be described in detail in Section 7.3.3. Bard[R37] reviews photocatalytic reactions on TiO_2, referring to studies of the oxidation of acetate to ethane and of other organic molecules. Clechet *et al.*[54] describe hole capture by OH^- and electron capture by O_2 on TiO_2 leading to H_2O_2 generation. Miyake *et al.*[55] describe oxidation of methanol to aldehydes on TiO_2, with simultaneous reduction of oxygen. It is of interest that on TiO_2 the oxygen cannot be reduced without the photoproduced holes present. The authors suggest either that methanol blocks the adsorption sites for oxygen until the methanol is removed by oxidation, or that oxygen is only reducible if it forms a complex with an intermediate of the methanol oxidation reaction. Krautler and Bard[56] describe the photocatalytic decomposition of carboxylic acids to alkanes and carbon dioxide on TiO_2.

Many cases of photocatalysis can be interpreted with the simplest model: the holes simply oxidize one species in solution; the electrons simply reduce another. Studies of solar energy conversion, where most often hydrogen generation is sought—involving the reduction of protons and the oxidation of OH^-—will be described in Section 9.1. Another common photocatalytic process, the deposition of noble metals on ZnO, TiO_2, or III/V compounds, will be discussed in Section 8.3.

A popular class of photocatalytic reaction—because it is so easily observable optically—is the reaction of dyes at the surface. Dye reactions in general are discussed in Section 6.5.3; here we emphasize results showing photocatalytic behavior. For example, Yoneyama *et al.*[57] and Miyake *et al.*[58] studied the reduction of methylene blue dye in methanolic and aqueous solutions, with TiO_2 as the photocatalyst. The reaction results in the leuco form of the dye (a one-equivalent reduction), the holes simultaneously oxidizing water to oxygen, or methanol to formaldehyde. Using a powdered ZnO catalyst, Cunningham and Zainal[35] found that the electron flow to the surface reduced a dye PNDA (*p*-nitrosodimethylanaline). They also examined reduction of NO_3^- and MnO_4^-. They checked the influence of hole-accepting ions in solution on the rate of the reduction process. The influence of the reducing agents seemed to be more or less in line with expectations: good current-doubling agents increased the rate of dye reduction; ions that showed less current doubling also showed lower reduction; and finally, one-equivalent ions that could act as recombination centers (Section 6.4 below) decreased the photocatalytic reaction. They monitored the electron density with ESR, and indeed there was a direct correlation of dye reduction with the ESR signal (the density of conduction band electrons). A particularly interesting observation is that the efficiency of photocatalysis (the number of dye molecules reduced per photon) decreased with increasing light intensity. This supports the concept that surface recombination may become more dominant as the carrier concentration increases, as indicated in the introductory paragraphs of this subsection. Along the same line is the observation of Williams and Nozik,[49] that if the wavelength of illumination is very short, so the light does not penetrate, the quantum yield goes up (because there is no bulk recombination). In fact, they report unit quantum yield for the photocatalytic decomposition of water on TiO_2.

6.4. Intrinsic Surface States and Recombination Centers

Intrinsic surface states have been introduced and their behavior reported in earlier sections. In Section 1.1.2 intrinsic surface states were defined, showing that, for ionic crystals, states associated with the cation are acceptor states, those associated with the anions are donor states, and that they are usually in the hydroxylated form when in contact with an electrolyte [Equations (1.6a) and (1.6b)]. The fact that normally the heterogeneity of the surface leads to a spectrum of levels was also discussed. In Section 2.1.2 the possible effect of the polar liquid medium in causing the surface states to fluctuate was considered. In Section 3.3.3.4 a simple numerical example was provided to illustrate that surface states could, in theory, accelerate charge transfer to ions in solution. In Section 5.1.1 we discussed the experimental evidence that V_H is controlled by intrinsic surface states in their role as adsorption sites for H^+ and OH^- and also discussed experimental evidence that intrinsic surface states do *not* affect V_H by exchanging electrons with redox couples in solution. And finally, in Sections 6.1 and 6.2, we offered a few examples to elucidate the role of surface states in electron transfer from the conduction band to ions in solution, and the role of surface states in hole transfer from the valence band to ions in solution.

In the present section, we will discuss the expected energy levels associated with intrinsic surface states, attempting to reconcile experimental observations of hole and electron capture. We will especially consider hydroxylation and the chemistry of the surface cations and anions, and the resulting surface state levels.

Hydroxylation affects both the chemical activity and the surface state energy level of the intrinsic surface state. With ionic solids, the chemical form of the intrinsic Tamm surface state, and of the resulting free radical, will depend on the *p*H because hydroxylation depends on pH. Thus, for example, on a hydroxylated surface (Figure 2.1) the capture of a hole on a lattice oxygen ion, $O_L^{2+} — H^+$, effectively results in the hydroxyl radical[59,R4]:

$$O_L^{2-} — H^+ + h^+ = O_L^- — H^+ \qquad (6.22)$$

is basic so that the lattice oxygen ions are in the nonprotonated form of O_L^{2-}, we may have

$$O_L^{2-} + h^+ = O_L^- \tag{6.23}$$

The capture of an electron by a surface lattice cation of valence 2 in a solution much more acid than the pzc will also result in a radical. For example, with ZnO, we may obtain lattice zinc with the formal valence +1:

$$Zn_L^{2+} + e^- = Zn_L^+ \tag{6.24}$$

or if the solution is much more basic than the pzc, so the zinc is hydroxylated:

$$Zn_L^{2+} - OH^- + e^- = Zn_L^+ - OH^- \tag{6.25}$$

The chemical form of the intrinsic surface state will clearly affect the energy level. However, little is known about the effect. The problem is complicated because ions (hydroxylated or otherwise) on surface steps, or even on kinks, will have different energy levels from ions on planes, and there will be enough step sites to provide a significant surface state density. When ions on the planar surface are hydroxylated, a simple calculation[60] suggests the levels will be in the same range as crystal bands, but, as discussed below, there appear many cases when the levels are in the bandgap region. These may be surface states associated with steps, or the simple calculation may be too naive. A further problem is that, although from the pzc we have some idea whether the surface states should be dominantly in the hydroxylated form, in most cases we have no knowledge whether the species will remain hydroxylated following carrier transfer. We know the various species in the above equations will have different pH ranges where they are the most stable form. If local hydroxylation or dehydroxylation occurs during carrier transfer, the description of the energy levels of the intrinsic surface states must be modified as in Section 1.2.4.5, and the energy levels cannot be easily described on the band diagram. The observation, that recombination of minority carriers from ZnO in an inert electrolyte[61] apparently depends on pH, may be related to differences in hydroxylation with pH.

Because of the above chemical complications, theoretical description of intrinsic surface states in carrier transfer, in recombination, or in corrosion is very difficult. An analysis of electron transfer vs. recombination has been attempted by Wilson.[50] He first analyzes the probability that a minority carrier reaches the edge of the space charge region, and then calculates the competititon between the two above-mentioned electronic routes of removal of the hole. Unfortunately, to keep the calculation simple, all chemical transformations are neglected, a "hole density in the valence band" is assumed (see discussion of Section 6.3.2), and the presence of hydroxyl groups on the surface atoms is neglected. The difficulty of including all the known chemical/physical factors quantitatively is almost overwhelming without an experimental justification for any simplification, and the results of the calculation reflect the problem. To fit the experimental curves for the oxidation rate of species on TiO_2 as a function of voltage, a complex recombination center distribution was assumed.

In this section we will first discuss Tamm surface states (Section 1.1.2.1) as carrier transfer centers, then discuss Shockley surface states as carrier transfer centers, then discuss recombination. Surface states in corrosion will be discussed in Chapter 7.

6.4.1. Intrinsic Surface States as Carrier Transfer Centers

Experimentally the role of donor intrinsic Tamm surface states, either their role as hole transfer centers or their role in corrosion, is better clarified than is the role of acceptor intrinsic surface states. Thus most of this section will describe donor intrinsic surface states. We can not normally distinguish between bare and hydroxylated intrinsic surface states.

The most direct evidence for the activity of anionic (donor) surface states, as was discussed in Section 6.3, is the observation of their influence in hole capture. The experimental results on ZnO (Figure 6.8) show that the capture cross section, for oxidation of ions in solution by holes from ZnO, was higher the higher the energy level of the ion. The observation seems general, so far as has been determined, for all hole capture processes on semiconductors. This order is exactly the reverse of that expected from the fluctuating energy level model [Equation (6.19)]. When this discrepancy was first observed[28] it was concluded that the behavior is consistent with "chemical expectation," not the expectation of the fluctuating energy level model. The carrier is localized on an intrinsic surface state before being transferred to the ion in solution. In most chemical molecule/molecule interactions, the most exothermic reaction goes fastest. And the hole capture reaction becomes a classical molecule/molecule interaction if, for example, the capture of a hole at surface lattice $Fe(CN)_6^{-4}$ ion occurs in two steps: first, Equation (6.22) above, then

$$O_L^- - H^+ + Fe(CN)_6^{-4} \rightarrow O_L^{2-} - H^+ + Fe(CN)_6^{-3} \qquad (6.26)$$

where the $HO_L^{\cdot}$ on the surface is simply acting as a particularly active hydroxyl radical in a "simple" chemical reaction.

Another indication that hole reactions occur via intrinsic surface states is the observation of crystal orientation effects. Capture of carriers by intrinsic surface states of either the Tamm or Shockley type can, of course, be expected to diminish the carriers' ability to oxidize or reduce species in solution. The holes will be at a higher energy level and so will be weaker oxidizing agents after being captured by an intrinsic surface state. Similarly, in a reduction process, electrons on surface states will be weaker reducing agents. This may be the reason for the orientation dependence in the water oxidation reaction by ZnO holes, as reported by Pettinger *et al.*[19] When the exposed surface is the $(000\bar{1})$ face (the oxygen face), where surface states in accordance with Equation (6.34) should be found, the rate is lower than on the (0001) face, the zinc face, where anionic intrinsic surface states, capable of capturing holes, may not be found at a convenient level. With no intrinsic surface states [the (0001) face], the holes should be essentially as energetic as the holes in TiO_2 (the two valence bands are at essentially the same energy, as shown in Figure 5.5) and holes in TiO_2 are known to oxidize water easily.

A third interesting observation of intrinsic surface state behavior at the semiconductor/electrolyte interface, specifically the silver oxide/electrolyte interface, is reported by Kunugi *et al.*[42] A hole, reaching the surface of a silver oxide layer on silver, oxidizes the Ag^+ to Ag^{2+}, according to these authors, as observed by voltammetric measurements. Then the Ag^{2+}, hydrolyzed to AgO, in a way similar to Equation (1.13), is able to transfer the hole to ethylene glycol. The latter reaction is observed by the resulting current doubling.

Naturally there is no rule that hole capture by surface states must precede oxidation, but anionic surface states capable of capturing holes may be difficult to avoid. Most anions in semiconductors have multiple valence states, and, for example, localization of the hole on a surface oxygen ion to form the peroxide ion, O_L^- or $O_LH\cdot$ may well be an exothermic reaction at the surface, especially if the anion is at a step or a kink site. The intrinsic surface state, in other words, may be well into the bandgap in these cases.

The above discussion, together with corrosion observation discussed in Section 7.4, is convincing that anionic surface states are active on ionic semiconductors. The transfer of electrons either way between ions and the conduction band, via surface states, is more poorly documented. Some observations indicate the possibility, based on somewhat indirect evidence. In most cases, although the results can be explained based on surface state models, it is hard to be sure that the explanation is the only possible explanation. Examples of such studies are Vanden Berghe *et al.*[62] or Boddy.[63] Vanden Berghe *et al.* studied electron injection into CdS and concluded by analysis that the electron injection must be aided by an intrinsic surface state level 0.12 eV below E_{cs}. However, the injection was independent of the concentration of the reducing agent, and if the surface states are transfer agents, not the origin of the electrons, electron exchange between the solution and the surface state is required. Thus the lack of dependence on electrolyte composition does not appear to fit the model. Boddy[63] suggested tunneling into TiO_2 from a level 3.2 eV below the conduction band. But 3.2 eV below E_{cs} is the neighborhood of the valence band, and it is difficult to reject tunneling from the valence band rather than from surface states.

Several authors have suggested that there are important cationic surface states on TiO_2. The observations of reduction on TiO_2 that was attributed to surface states by Frank and Bard[22] was discussed in Section 6.2.2 and did not seem to fit surface state models. Morisaki *et al.*[64] found photoeffects they attributed to these same levels, but Decker *et al.*[65] have concluded that the effects were due to temperature, not optical excitation. Vandermolen *et al.*[25] suggest that a low density of surface states is present (about 10^{14} m^{-2}) to explain their cathodic reduction process, but as discussed in Section 6.2.2, the evidence does not seem conclusive. Other results of the Ghent group,[66] on the other hand, suggest that many of the observed effects of electron exchange on CdS could be due to dislocations—energy levels associated, not with surface states on the perfect surface, but with energy levels at the end of the dislocation pipes (Section 1.1.1.4). Specifically, they found photoeffects on CdS at long wavelengths (up to 800 nm) where the photocurrent induced dissociation of the crystal, suggestive that photoproduced "holes" trapped in dislocations reach the surface, break bonds, and permit dissolution of the crystal at these sites. Presumably, intrinsic surface states associated with anions on a dislocation would be much higher in energy than intrinsic surface states associated with anions on the perfect crystal face. However, normally such dislocation states are not present in sufficiently high density for such surface states to be dominant; presumably the surface must have a high density of dislocations or the experiment (such as the use of long wavelength illumination in the cited case) must be designed to emphasize defect states.

A different reason for suspecting cationic states are present arises in the work of Tyagai and Kolbasov.[16] They conclude surface states are present on CdS and CdSe, because the transfer coefficient in the current–voltage characteristics is far different from zero. As discussed above, a high density of active surface states should have such an effect because they can permit accumulation of a large charge density at the surface, and lead to changes in V_H with applied voltage. However, as indicated in Section 3.4.3, another phase at the surface may have the same effect, and Tyagi and Kolbasov do observe sulfur or selenium on their surfaces.

Thus there is little evidence that cationic surface states have been observed to affect cathodic reactions. The cathodic reactions that do not follow the fluctuating energy level model in its simplest form can be explained generally by assuming that dislocation levels (Section 5.2.3.1) assist the electron transfer. However, there is no fundamental reason to suggest that cationic surface states cannot be active. The only reason to suspect that it will continue to be rarely observed is that most cations in *n*-type semiconductors do not have stable valences with one electron added to the valence state associated with the compound. Thus, chemically speaking, it is less likely that localization of a single conduction band electron on most hydroxylated cations will be an exothermic process, even at a surface step or kink site.

On covalent semiconductors, intrinsic surface states seem to be involved in injection and the capture of both majority and minority carriers, according to Equations (5.5) and (5.6). However, the reactions described by those equations describe the effect of the states on adsorption, rather than describing a continuous transfer to charge to ions in solution. The three species, Ge·, and the complexes Ge:H or Ge:OH, represent three different forms of intrinsic surface states that capture the electrons or holes according to the equations cited. As will be discussed in Section 6.4.2 below, the Ge· form is the only one that freely exchanges carriers both with the conduction and the valence band. But, during a continuing dc current (e.g., during hydrogen evolution) there is no evidence suggesting surface state participation. The behavior of germanium, however, provides a good example of the importance of the groups bonding to intrinsic surface states in controlling their carrier capture characteristics. Further evidence of the effect of adsorption on intrinsic surface state characteristics is provided by Toshima and Uchida,[67] who show that adsorption of halide ions (even Cl^-) on these Ge· intrinsic surface states changes their energy so that the capacity versus frequency measurements are affected (see Section 7.3.2).

In the above, we have emphasized the role of intrinsic surface states in one-equivalent reactions, but multiequivalent reactions associated with intrinsic surface states can occur. Majority carrier injection can follow minority carrier capture. This fairly complicated process was one of the first studied on semiconductor electrodes appearing in the classic paper by Brattain and Garrett[68] on germanium. The process is actually corrosion, but the role of surface states is particularly clear, so we shall discuss the experiment in the present section. Brattain and Garrett studied *n*-type germanium, using the transistor method (Section 4.3.3) to distinguish hole reactions from electron reactions, particularly electron injection. With a *pn* junction near the surface, they could cause holes to flow to the germanium surface, and observe the electron injection. They found that about one electron is injected into the surface for every two holes reaching the surface. The effect is termed "current multiplication." The overall chemical reaction occurring is presumably

$$Ge + (4x)h^{+} + 6OH^{-} \rightarrow GeO_3^{2-} + 4(1 - x)e^{-} + 3H_2O \quad (6.27)$$

Here x is the fraction of total interface current carried by the holes. The value of x has been found to range from 0.6 with very low hole currents, to 0.8 for higher hole currents.[(69)]

The process occurring in current multiplication may be visualized as follows. As the holes, injected from the p region into the n region, reach the surface, the germanium molecule must go through four stages of oxidation, going from Ge^0 to Ge^{4+} before it is dissolved as GeO_3^{2-}. Intrinsic surface states are, by definition, the energy levels capturing these holes. Some of the several intrinsic donor surface states associated with the various oxidation states of germanium (perhaps the states associated with the less stable oxidation states, Ge^{+3} or Ge^{+1}) are so high in the bandgap that they are more likely to inject electrons than wait to capture one of the holes. Section 1.2.5 discusses the energy levels associated with such unstable intermediates. Pleskov's result[(69)] simply indicates that the higher the hole concentration, the greater the probability that each oxidation step occurs by hole capture. Unfortunately, the complexity of the reaction obviously precludes detailed analysis in terms of the energy levels of the intrinsic surface states involved.

Results on silicon suggesting similar behavior (to the germanium case above) have been observed by Wolkenberg,[(70)] who observed a decrease in the current multiplication when Fe(II) was introduced to compete with the intrinsic surface states in capturing the holes.

6.4.2. Intrinsic Surface States and Ions in Solution as Recombination Centers

Simple surface recombination is a possible fate for a minority carrier that will compete with oxidation of ions in solution, or with corrosion. Minority carriers approaching the surface, especially those captured by intrinsic surface states, can recombine with majority carriers, with no net chemical change in the system. The energy used to produce the electron–hole pair is dissipated as heat. In systems where it is desired that the energy be stored or converted to electrical energy, as in solar conversion, recombination is a very undesirable competitive reaction. In systems where it is desired that there be no chemical changes following the inevitable generation of the hole–electron pairs by light—as for example, in paints and polymers—surface recombination is highly desirable.

Surface recombination can occur at intrinsic surface states, at interface states between a semiconductor and its passivating layer, or by alternating oxidation and reduction of one-equivalent ions in solution. Chemically speaking, the reaction is simply

$$h^+ + M^{n+} = M^{n+1} \tag{6.28}$$

$$e^- + M^{n+1} = M^{n+} \tag{6.29}$$

and it matters not which reaction occurs first. The net result is the removal of the hole–electron pair with no chemical change. Here M can be a surface state, an interface state, or an ion in solution.

The kinetics of carrier capture, either by surface states or by ions in solution, should in principle, involve the fluctuating energy level model. The mathematics comes directly from the concepts of Chapter 3. However, such recombination has not yet been analyzed in the literature in terms of the fluctuating energy level model. The lack of theoretical background may be due to the fact that recombination by surface states located physically at the solid/liquid interface has not been studied. Also, any theory of such recombination will be characterized by a very ill-defined reorganization energy λ. The difficulty with the reorganization energy, arising because the states are at an interface with an abrupt change in dielectric constant, is discussed in Section 2.1.2. Another difficulty is distinguishing between surface states (high λ) and interface states (low λ). For example, in the discussion of surface states on germanium in Section 5.1.1.2, it was suggested that the evidence indicates that the Ge· (to be discussed as a recombination center below) is actually under an oxide layer, and so is really an interface state, rather than a surface state. Many of the other recombination centers on germanium (Cu, Ag) are not surface states, but separate phases. Of course, with some materials, such as oxides, many surface states can be expected to be at the solid/liquid interface, but a study devoted to their recombination properties has not been made.

Although carrier recombination based on the fluctuating energy level model, in particular recombination involving ions in solution, has not been quantitatively analyzed, it is of interest to discuss such recombination qualitatively before going on to the solid state model.

An examination of Figure 1.6 will suggest the possibility of highly efficient recombination by redox couples in solution if $E^o_{\text{redox}} = \frac{1}{2}(E_{cs} + E_{vs})$ especially if $2\lambda \approx E_q$. For in this case the Gaussian for the reducing agent can overlap with the valence band, and in principle, provide particularly efficient hole capture, and simultaneously the Gaussian for the oxidizing agent can overlap the conduction band and provide particularly efficient electron capture. Then the dissipation of the energy of the electron–hole pair should occur very naturally via the reorganization of the solvent molecules. Only the individual steps in the process have been examined experimentally—the study of hole capture on ZnO for example, as a function of the energy level of the reducing agent (Figure 6.6), and the study of electron capture as a function of the energy level of the oxidizing agent (Figure 6.2). These results suggest that contrary to the simple model, the optimum value of E^o_{redox} will be just below E_{cs}, for then electron capture is high, and hole capture is experimentally (Figure 6.8) higher the higher E^o_{redox}.

Some qualitative results have been obtained experimentally on the behavior of such one-equivalent ions, either in solution or in a resin, where $E^o_{\text{redox}} \lesssim E_{cs}$ was found best for recombination centers in preventing photocatalytic changes.[31,71,72] However, these studies were phenomenological and did not delve deeply enough into the behavior to distinguish between the normal recombination theory and a recombination theory based on the fluctuating energy level model. Other indirect observations have been made. For example, there is a decrease in the density of photoproduced electrons in a suspended ZnO powder when species are present that could be recombination centers.[35] However, a definitive study directed at the role of redox couples is yet to come.

It is not expected that recombination will be favored on two-equivalent species, especially current-doubling species.[2] If the levels are such that the radical intermediate will inject a carrier into the majority carrier band (Figure 1.7b), the recapture of a majority carrier on such a level will be highly activated.

In this book we will only review the theory of recombination developed in solid state studies that describes recombination when the states are not in a polar medium. And, actually we will offer a very brief review—sufficient, hopefully, to provide an understanding of when surface recombination will be strongly competing with electrode reactions for the minority carriers, but far from sufficient to be an exhaustive survey of the topic. Therefore we refer the reader to other references.[R4,R15,R16]

The theory of surface recombination is based on the requirement, as discussed earlier for photocatalysis, that the rate of hole and electron flow to the surface is equal. That is, with standard Shockley–Read notation:

$$U_p = U_n = U \tag{6.30}$$

where the U's are the net rates of carrier flow to the surface, given by Equations (3.27) and (3.29) for electrons and holes, respectively. At equilibrium, U_p and U_n are zero, and, as discussed with respect to the Equations (3.27) and (3.29), the parameters p_1 and n_1 can be determined. When electrons and holes are being generated by light, or some other nonequilibrium stimulation, the densities n_s, n_t, p_s, and p_t (the densities of electrons and holes in the bands at the surface and in the surface state that is involved in the recombination) are no longer at equilibrium and are indicated by an asterisk. It is assumed, however, that the surface densities p_s^* and n_s^* follow from the bulk photoinduced concentrations n_b^* and p_b^* in accordance with Equations (2.10) and (2.11), viz.,

$$n_s^* = n_b^* \exp(-qV_s/kT) \tag{6.31}$$

$$p_s^* = p_b^* \exp(qV_s/kT) \tag{6.32}$$

Qualitatively then, the model is that the illumination produces a nonequilibrium bulk density of excess electrons and holes. This bulk density causes a nonequilibrium excess surface density in accordance with Equations (6.31) and (6.32), and in turn this surface excess causes a net flow into the surface states, controlled by Equations (3.27) and (3.29) with the boundary condition of Equation (6.30). Combining all these equations with substantial manipulation, it is found that the surface recombination velocity, s, is given by

$$\begin{aligned} s &= U/\delta n_b = U/\delta p_b \\ &= [K_p K_n N_t (n_b + p_b)/2n_i]\,\{K_n n_b \exp(-qV_s/kT) + K_p p_b \exp(qV_s/kT) \\ &\quad + K_n n_b \exp[(E_t - E_F)_0/kT] + K_p p_b \exp[(E_F - E_t)_0/kT]\}^{-1} \end{aligned} \tag{6.33}$$

where the first two equalities define s, with $\delta n_b = \delta p_b$ because the excess density of carriers of the two types in the bulk is necessarily of equal concentration for electrical neutrality. The parameter n_i is the density of electrons in the intrinsic semiconductor, given by Equation (1.4) with $n = p$. The factor $(E_t - E_F)_0$ is the difference between the Fermi energy and the surface state energy level with no band bending (flat bands). The above equation appears complex, and indeed a more elegant form is available,[R15] but Equation (6.33) permits easy visualization of how the recombination velocity varies as V_s and E_t are varied.

It is clear from Equation (6.33) that there is a maximum in the surface recombination velocity, s_{max}, for an intermediate surface barrier. If V_s is large and negative, the first term in the denominator is large, so s is small. If V_s is large and positive, the second term in the denominator is large, so s is small. An intermediate V_s leads to a maxiumum in s. A similar observation holds for the relative position of the Fermi energy and the surface state through which recombination is occurring. If the Fermi energy (at zero band bending) is much above the surface state, the last term of the denominator is large, so s is small. If the Fermi energy is much below the surface state energy, the remaining term is large, so s is small. Allowing for a significant variation because the preexponentials in the denominator are not equal and depend on whether the sample is p or n type, etc., we can say in general that for an effective hole–electron recombination center, we should look for (a) a modest surface barrier V_s, and (b) a surface state in the middle of the energy gap.

More qualitatively, Equation (6.33) illustrates that for hole–electron recombination at the surface, both holes and electrons must be able to flow into the surface states. If V_s is such as to repel either holes or electrons very strongly, s will be low. And the surface state should be positioned in the gap so that it has a reasonable cross section for both electron and hole capture. If it is too near a band edge, s will be low.

We note that in electrochemical systems, V_s is a parameter at the control of the experimenter. In nonelectrochemical systems, on the other hand, where no net current flow to the surface is allowed, V_s approaches zero as the minority carrier density at the surface increases. V_s becomes lower because the density of minority carriers builds up to make the flow of majority carriers equal the flow of minority carriers to the surface. Thus s in Equation (6.33) is not independent of light intensity (the rate of generation of minority carriers). The rate of recombination can increase more rapidly than first order in light intensity. This higher order can be important when surface recombination is competing with other reactions, such as are found in photocatalysis (Section 6.3.3).

Measurements indicating the validity of Equation (6.33) are plentiful in the old solid state literature,[R15] and a review of them is beyond the scope of this book. However, this omission should not be interpreted as a suggestion that an understanding of surface recombination is not important, for the competition in semiconductor electrochemistry between recombination and electrode reactions will be an important feature of many of the discussions to follow, and to understand the electrochemical results, an understanding of the qualitative features of surface recombination velocity is needed.

One of the more interesting electrochemical investigations of surface recombination via surface states is that of Memming and Neumann.[73,R4] They studied recombination at the Ge· surface state on germanium, the origin of which is described in Section 5.1.1.2. As noted there, it is not clear whether Ge· should be considered a surface state or an interface state. As discussed [Equations (5.5) and (5.6)], the Ge· radical at the surface is a transient intermediate between the bonding of H· and the bonding of OH· to the Ge· covalent dangling bonds. While the surface adsorbates, under the influence of an anodic or cathodic current, are changing from H· to OH· or from OH· to H·, some unbonded Ge· is present, and the dangling bonds on Ge· can alternately capture electrons and holes, and thus act as a recombination center. Memming and Neumann observed recombination that fitted the theory of the transient appearance of the Ge· radicals. They observed the variation of s with N_t, the density of Ge· recombination centers as the adsorbed species switched from H· to OH·.

A variation on the same theme (Ge· states) was the measurement of the surface recombination velocity with H_2O_2 in the solution. Gerischer and Mindt[74] concluded that a steady state number of Ge· radicals should be present during the etching of germanium with H_2O_2, the steady state number being determined by the concentration of peroxide. Memming and Neumann[73] showed that for this case also the surface recombination velocity was described quantitatively by the theoretical density of Ge· as described by Equation (6.33).

It is of interest that Schulte and co-workers[75,76] found that a surface state capacity was also associated with the Ge· surface states. This seems the only proven case where a surface state capacity is found associated with intrinsic surface states (as opposed to the more general case where any observed surface state "capacity" is actually charge storage in a separate phase). The behavior in germanium may be possible because the bandgap is so small. As described in Section 4.1.3.3, for a surface state capacity to be observed, charge stored from the solid on the one-half of the ac cycle must be returned to the solid on the other half. With a small bandgap, where electronic transfer rates across the barrier can be high compared to the rate of interaction of the charged surface state with the solution (corrosion, etc.), the capacity should be measured. The chemical rates may be particularly slow in the germanium case if there is a separate phase, a thin oxide, at the surface, so the Ge· states are actually interface states. And certainly, interface state capacity is commonly observed (Chapter 8).

In summary, the case of Ge· centers on germanium seems to be the only electrochemical case where the system has been so well defined as to permit quantitative comparison of recombination theory and experiment.

6.5. Carrier Injection

6.5.1. Direct Electron and Hole Injection

Direct injection of electrons or holes into the semiconductor means that the species in solution are highly reactive, relative to the semiconductor. They must be reactive enough so that the E^o_{redox} of the reducing agent is close to, or above, the conduction band edge, permitting electron injection; or the E^o_{redox} of the oxidizing agent is close to, or below, the valence band edge, permitting capture of valence electrons (hole injection). Naturally, the requirements on the chemical activity of the species in solution depend on the semiconductor. Electron injection into a semiconductor with a low (in energy) conduction band edge will be possible from a much weaker reducing agent than required for injection into a semiconductor with a high conduction band edge. The conduction band edges and the valence band edges of the various semiconductors commonly studied to date are reviewed in Section 5.3; here we simply want to review whether electron injection and hole injection are indeed observed with ions in solution that have the necessary oxidative or reductive power.

The value of comparing the carrier injection versus the energy levels of the band edges is not only to confirm or reject the simple theories described above, but also the appearance of injection can be used in the estimation of unknown energy levels. For a first example, carrier injection can be used to determine the energy of a band in the solid that is "completely" occupied or unoccupied. In such a case the Fermi energy relative to the band edge, μ, cannot be determined, so the determination of E_{cs} or E_{vs} by V_{fb} measurement (capacity/voltage measurement) is not possible [see Equation (4.29)]. The hole or electron injection process can be used to provide a rough estimate of the band edge because it does not depend on μ. Table 6.1 illustrates the thesis: the band edge is expected to lie between the values of E^o_{redox} where injecting agents can inject, and the values E^o_{redox} where, with a nondegenerate material, accepting agents can accept. Frese and Morrison[77] have used this technique to estimate the energy level of a conducting band of levels in the anodic oxide covering GaAs. As a second example, carrier injection can be used to provide an estimate of unknown energy levels of ions in solution when the bands in the solid are of known energy. For example, we will discuss, both in this section, and elsewhere in Section 6.5, the injection of carriers from excited states of dye molecules into the conduction or valence bands of semiconductors whose band edges are known. The observation of injection by the excited state only indicates the ground state is below, the excited state above, E_{cs}. More quantitatively, results on electron injection by tunneling have been used (Section 6.6.2) to determine the energy levels of quinone. As another example, this time associated with lack of injection: Vanden Berghe, Cardon, and Gomes[3] noted the absence of electron injection from ferrocyanide into ZnO and, therefore, questioned their assignment of energy levels for iron cyanide.

We will first discuss observations of electron injection, then observations of hole injection.

Qualitatively the observations of electron injection into semiconductor conduction bands are as expected, with high values of E^o_{redox} giving injection, low values not. Table 6.1 provides an example. Many other examples have been noted. Dutoit *et al.*[78] observed electron injection into TiO_2 by Ti^{+3} and V^{+2}. Gomes[79] studied V_2O_5 and found an abundance of reducing agents that could inject electrons, as could be expected from the low conduction band edge as reported in Table 5.1. For example, ferroine, iodide, ferrocyanide, and the ferrous ion could all inject. The bromide ion ($E^o = 1.07$ V) and Co^{2+} ($E^o = 1.8$ V) could not inject. He concluded that, in general, species with E^o_{redox} less than 1.1 eV below the hydrogen zero could inject, although Mn^{2+} was found to inject despite its low E^o_{redox}, corresponding to an E^o of 1.5 V. If we compare the 1.1 eV value to the value of E_{cs}, which is 0.9 eV or so below the hydrogen zero, as tabulated for V_2O_5 in Section 5.3, this means $E_{cs} - E^o_{redox}$ should be more positive than about -0.2 eV for detectable injection. Such a value is not unreasonable (in comparison with theory) because this semiconductor exhibits high background currents due to corrosion, and so a high injection current is needed to be detectable.

Electron injection from excited states of ions in solution is also observed. Dyes that can be excited by visible light are commonly used. One is limited to less-than-bandgap photon energy, otherwise the measurement is confused by the appearance of minority carriers. Many cases of electron and hole injection from dyes are observed, and will be described (together with the difficulties in deciding whether direct injection is being observed) in the next section.

Quantitatively, the observations of electron injection into semiconductors are somewhat inconsistent. Table 6.1, reflecting the results of Freund and co-workers,[12,11,83] indicates a set of measurements where the injection current is very low compared to the predictions of Equation (3.55) [evaluated for $E_{cs} = E^o_{redox}$ as shown in Equation (3.63), to be 5×10^8 A/m^2 M]. A current so high should be easily detectable, but Table 6.1 suggests that at $E^o_{redox} \simeq E_{cs}$ no injection current is observed. The observations of Kohl and Bard[21] indicate reasonable electron injection into ZnO when the E^o_{redox} is about equal to E_{cs}. However, when the E^o_{redox} of the couple is below V_{fb} by only about 0.3 eV, injection is no longer observed. These results are shown in Figure 6.6, and again represent much lower injection than the Marcus–Hale model predicts. At the other extreme are the results of Vanden Berghe and Gomes,[80] who obtain high injection currents into CdSe and CdS. They conclude that injection can be measured with E^o_{redox} as much as 0.8 eV below the conduction band edge. By Equation (3.55) the current in this case should be 5×10^{-3} A/m^2 M, a measurable current. Thus their conclusion is consistent with Equation (3.55). However, they use an anomalously low value of E_{cs} to reach this conclusion. If we accept the band edge values of Figure 5.5, rather than their lower values for CdSe and CdS, there are inconsistencies in Vanden Berghe's results when compared to Equation (3.55). Consider, for example, the observed injection by Fe(II) (EDTA) into ZnO, CdS, and CdSe. By Equation (3.55) and Figure 5.5 (assuming $\lambda \sim 0.7$ eV), the currents should be 2×10^{-16} A/m^2 M for CdSe, 2×10^{-9} A/m^2 M for CdS, and 3×10^5 A/m^2 M for ZnO, assuming $pH = 7$. The observed values were 2.3, 0.5, and 25 A/m^2 M, respectively. The values for CdSe and CdTe are clearly much higher than predicted by the Marcus–Hale model of Equation (3.55), and the ZnO value is much lower, probably diffusion limited. In summary, the theory does not quantitatively agree with the experimental results, and the experimental results do not agree with each other very well. Such inconsistencies will be clarifed, undoubtedly, in the near future.

Other experimenters observe similar, sometimes somewhat higher, current densities for electron injection. For example, Beck and Gerischer[81] observe 10^3 A/m^2 M with V^{2+} ions injecting electrons into germanium, a value probably limited by the diffusion of V^{+2} to the surface. Gomes[79] observes 50 A/m^2 M from ferrocyanide ions into the V_2O_5 conduction band (much lower than theory).

Hole injection by strong oxidizing agents has been a common observation, particularly on germanium. Data on germanium are plentiful because the germanium valence band is not very low, so hole injection is easy, and because much of the early work of semiconductor electrochemistry was done on germanium. Hole injection is, more often than not, accompanied by corrosion of the solid. The rate of hole injection does not seem usually to be limited, as in the case of electron injection, but can be high, in the diffusion-limited regime (concentration polarization, as described in Section 3.1.2), with currents upwards to 10^3 A/m^2 M. It should be noted as a caution, that in most experiments to be described, a negative bias is put on the electrode to induce hole injection, and such a bias can induce proton diffusion into surfaces which complicates the interpretation. This factor is not considered.

Three measurement techniques have dominated the study of hole injection. They are the transistor technique, corrosion, and luminescence. The transistor technique, as described in Section 4.3.3, differentiates cathodic current arising from electron capture from that arising from hole injection. As the technique is only (almost) applicable to germanium, and in the late 1950s and early 1960s germanium was of great technical importance, it is understandable that much of our early knowledge was generated using this technique. Corrosion, associated with the decomposition or oxidation of the lattice by the injected holes, is another excellent method of determining whether holes are being injected. And, finally, luminescence of n-type semiconductors is often observed when holes are injected, arising from radiative recombination processes during the deexcitation of the crystal.

Gerischer and Beck,[82,R10] studied by current/voltage measurements the capture of electrons from the germanium valence band (hole injection) and concluded they were actually measuring corrosion. They concluded this because they observed electron injection. As described in Equation (6.27), the capture of holes by the intrinsic surface states on the germanium surface results in electron injection. Thus the overall process is as follows: first, hole injection, followed by corrosion due to the hole, followed by electron injection due to the corrosion. Thus the hole injection by an oxidizing agent may be monitored by the resulting electron injection. Gerischer and Beck observed electron injection when they added the electron acceptor ferricyanide to the solution, and thus were able to conclude that the ferricyanide was injecting holes at a rate limited by the diffusion rate of ferricyanide to the surface.

Pleskov[69,R10] used the transistor method to monitor the rate of hole injection and compare it to the rate of conduction band electron capture by various oxidizing agents. The results are given in Table 6.3, where y is the fraction of electrons captured from the valence band. Now from the more recent estimate of Memming and Möllers[4] the valence band of germanium is near the E^o_{redox} of the hydroquinone/quinone couple. Thus, by the simple energy level model, y should be close to unity for E^o at the value of quinone or higher, and should decrease for lower E^o reaching zero for E^o about zero. The observations, therefore, seem about 100% wrong. Unfortunately, these measurements were done in the early years of semiconductor electrochemistry, and several factors were poorly defined.[R10] In these measurements the influence of electron injection (associated with corrosion by the injected holes) is hard to determine. The change in the Helmholtz potential of germanium, which changes with current as described in Equations (5.5) and (5.6), was not known at that time. The importance of the two-equivalent nature of the adsorbate leading to multiple energy levels was not recognized, and the possibility of adsorption was not analyzed. Thus we can derive little information from these data. However, they point to a powerful tool in basic semiconductor electrochemistry that should be used again, and interpreted according to our present models.

Observation of the luminescence resulting from hole injection is a good way to confirm that hole injection actually is occurring. Pettinger *et al.*[40,R38] studied SnO_2, ZnO, CdS, and GaP and GaAs and discussed the resulting luminescence. The holes were injected essentially by a current-doubling process, with a cathodic current reducing peroxide or persulfate, and the resultant active intermediate injecting holes. They found that the voltage must be more cathodic than flatband (there must be an accumulation layer on the n-type material) for luminescence. The holes must be able to move into the bulk of the sample without a repulsive electric field to excite the characteristic luminescence of the material. That is, surface recombination of the injected holes in n-type material or corrosion of the material does not lead to the characteristic luminescence. Bernard and Handler earlier had shown[84] that characteristic luminescence was observable on GaAs in 1 M sulfuric acid with a very strong

Table 6.3. Ratio y of Hole Injection to Total Electron Capture with Various Oxidizing Agents on Germanium

Surface species	y	E^o (V)
H_2O_2	0	1.77
$K_2Cr_2O_7$	0.05	1.33
Quinone	0.4	0.7
KI_3 (iodine)	0.42	0.54
$K_3Fe(CN)_6$	0.7	0.36
$KMnO_4$	0.8	0.56

cathodic voltage, more negative than −1.5 to −2 V. They found a high cathodic current, associated with the reduction of protons (see the valence band edge for GaAs in Table 5.1 of Section 5.3) independent of whether the sample was *n*- or *p*-type. They concluded, from all this evidence, that at strongly cathodic voltages, when the surface became degenerate *n*-type, the main source of electrons for reduction was the valence band (hole injection). The need for accumulated *n*-type surfaces to observe hole injection arose, presumably, because some of the voltage must appear across the Helmholtz layer to move the rate-limiting energy level, associated with proton reduction in solution, down near the valence band of the GaAs.

Interesting but complex results on hole injection were observed by Jones *et al.*[85] on *p*-type NiO. They detected hole injection by corrosion of the NiO, and described the process in terms of a complicated chemical interaction with the surface. They invoked complexing of the Ni lattice ions, and invoked kink sites. In later work they also invoked blocking of sites by anionic sorbates.

6.5.2. Injection by Tunneling

The qualitative features of the solid that lead to tunneling, particularly injection of carriers by tunneling through the space charge region, were discussed in Section 5.1.2, where the behavior of the "perfect crystal" was discussed. In particular, Equation (5.10) was used to emphasize the qualitative requirement of a high value of N_{sc} so that the depletion layer is narrow enough (less then 30 Å, say) for tunneling, and Figure 5.1 was used to illustrate qualitatively the band picture needed for tunneling, the primary requirement being a narrow surface barrier.

Memming and Möllers[4] and Schmickler[86] have explored the theory of injection by tunneling in a reasonably quantitative way. The first step of Memming and Möllers' theory was described by Equation (3.56). However, the expression for the current at each energy given in (3.56) must be multiplied by the tunneling probability, so Equation (3.56) becomes

$$J_a = J_0 = qZc_{\text{red}}(\pi kT\lambda)^{-1/2}$$

$$\times \int_{-\infty}^{\infty} \kappa'(E)N(E)\,(1 - f)\exp[-(E - E_{\text{red}})^2/4\lambda kT]T(E)dE \quad (6.34)$$

where $N(E)$ is the density of states available for electron transfer, as in Equation (3.56), and $T(E)$ is the tunneling probability at that energy. Here we are considering electron injection into the conduction band from ions at the surface. The probability $T(E)$ increases exponentially with decreasing thickness of the barrier at the energy E. The point is, of course, that $T(E)$ thus increases rapidly with applied voltage, because greater band bending narrows the potential barrier for tunneling at any given energy. As can be visualized from Figure 5.1, a higher voltage causes a sharper band bending, a higher field at the surface [from Equation (5.10)], and the whole surface barrier becomes sharper, with the accessible conduction band levels closer to the surface.

For a triangular-shaped barrier, the tunneling probability $T(E)$ is shown[4] to vary for a particular energy level in solution as

$$T(E) = T_0(E_0)\exp\{-(8\pi/3h)\,[2m(E_{cs} - E)]^{1/2}x(E)\} \quad (6.35)$$

where $E_{cs} - E$ is the distance between the top of the barrier and the energy E of the tunneling electron (the energy of the electron on the ion in solution). The parameter $x(E)$ is the thickness of the barrier at the energy E (plus perhaps the distance of the ion from the surface), and the other symbols have obvious meanings. Memming and Möllers approximated the surface barrier as triangular because a more accurate description [e.g., Equation (2.22)] makes the analysis more complex than necessary. In summary the tunneling probability increases rapidly with decreasing x, and of course, x at a given E decreases rapidly with increasing voltage as described in the preceding paragraph.

The solution of an equation such as (6.35), inserted into an equation such as (6.34), obviously requires computer work, and there is little gained by further quantitative discussion. Each set of parameters, in other words, requires an individual calculation. Details of the methods are available in the reference[4] given.

Similarly Schmickler[86] obtains an expression for tunneling that is complicated:

$$j = C\int_{E_c}^{\infty} dE[1 - n(E)]\,D(E,\eta)\exp\Big((-8m^*/4\pi^2h^2)^{1/2}$$

$$\times \int_0^{\infty} [(E_c - E)^{1/2}dx]b^{-1}\{1 - \exp[-b(E - E_c)]\}\Big) \quad (6.36)$$

where C is a collection of constants, D the probability of a state fluctuating to the energy E, and b is a constant. Again, clearly, a very complex expression is obtained when one attempts to insert the tunneling probability into the fluctuating energy level model, because the tunneling probability depends exponentially on the barrier thickness x, and the thickness depends on the energy E, leading to an integration in the exponent.

A slightly simpler model arises if one assumes that the tunneling arises at surface states that are not fluctuating. Vanden Berghe *et al.*[62] report the theory of Franz[87] on tunneling from surface states, and use this theory to estimate an energy level at the CdS surface. As discussed in Section 2.1.2, it is a questionable assumption to assume no temporal fluctuations unless one can show that the surface states are interface states, buried under a surface phase with no rotatable polar species nearby.

Schmickler's analysis has bearing on the role of dislocations, or other defects, in semiconductor electrodes. Dislocations have been discussed earlier both from a theoretical (Section 5.2.3) and an experimental point of view (Sections 5.2.3.1 and 6.5.1) as playing an important role when present in carrier transfer processes. Schmickler emphasizes strongly the possibility of "resonance tunneling," wherein localized levels in the bandgap can act as stepping stones to aid the tunneling electron on its way between the bands of the solid and the solution. As he points out, this can be particularly important in the case of tunneling through oxide layers. It may also be important in tunneling into a badly disturbed crystal, where levels associated with dislocations are numerous. The possibility is discussed in Section 5.2.3.1.

Experimentally, several studies have been made of tunneling with an SnO_2 electrode. Memming and Möllers' theoretical work,[4] described above, was used to analyze their results on tunneling from various reducing agents into SnO_2, and from the comparison of theory with experiment they evaluated the reorganization energy λ of various reducing agents. The parameter λ appears prominently in the exponent in Equation (6.34), and because the other parameters were obtainable by theory, or by analysis of the voltage dependence of the current, λ could be evaluated. They found $\lambda = 0.4$ eV for $Fe(CN)_6^{-4}$, 1.2 eV for Fe^{+2}, and 1.7 eV for Ce^{+3}. In other studies of tunneling at the surface of the SnO_2 electrode,[20,88,R32] tunneling was invoked qualitatively to explain the high reversible currents that make SnO_2 a practical electrode, rather than used quantitatively to derive information about the occurrence of tunneling.

Pettinger *et al.*[19,89] studied tunneling at ZnO extensively. They examined tunneling from the conduction band to the ground state of dye molecules; they examined tunneling from the conduction band to $IrCl_6^{2-}$ and from I^- and Fe^{2+} to the conduction band; and they examined what they concluded was Zener tunneling from the valence band to the conduction band yielding holes. The dye studies of Pettinger *et al.*[19] will be reviewed in the next section together with other studies of dye reactions. The observations of Zener tunneling[89] were "confirmed" by the observation of luminescence. However, luminescence should not be observed with Zener tunneling, for the holes and electrons do not occupy the same region of the crystal. Avalanche breakdown may be more probable as an explanation of the observations than tunneling. In any case, the interpretation of the results is not clear.

The electron tunneling that Pettinger *et al.* observed between ions in solution and the conduction band of ZnO is particularly instructive, especially the electron transfer to the $IrCl_6^{-2}$ ion. This observation was reviewed in Section 6.2.2, with particular reference to Figure 6.5.

6.5.3. Injection by Optically Excited Ions: Dye Injection

Carrier exchange with optically excited ions in solution provides an interesting application of the concepts that have been developed. Of particular interest are dye molecules, large organic molecules that often have energy levels for the excited state and the ground state that do not change when an electron is added to or removed from the molecule. Dye molecules have two features that make them useful for study: first, they absorb in the visible, so light of the right wavelength can be passed through a wide bandgap semiconductor electrode with no minority carrier generation; second, they have a high extinction coefficient so the light is absorbed by the dye very near the semiconductor surface. If an electron is raised to the excited state of such a molecule in solution, several reactions can follow: (a) The excited electron can transfer to the conduction band, changing the charge on the molecule by $+1$; (b) the dye molecule can become deexcited to its ground state, (c) an electron from the semiconductor can transfer to the ground state of the dye, changing the charge on the molecule by -1 and leaving an electron in the excited state; or (d) an electron from a reducing agent nearby can transfer to the ground state of the excited dye, again changing the charge on the molecule by -1. Other reactions can occur, but these four are of most interest to us. Additionally, when the dye molecule becomes positively charged, losing its electron say by reaction (a), electrons can be captured either by the ground state or by the excited state. In a sense, the dye molecule acts as a semiconductor, with the ground state equivalent to the valence band of the semiconductor and the excited state equivalent to the conduction band.

There have been many reviews in the literature about dye behavior at semiconductor electrodes. This is because dye behavior on semiconductors, in particular on ZnO, has been of tremendous technical importance in electrophotography, and semiconductor electrochemistry has been a powerful tool in analyzing this system. Dye injection of electrons is also important in normal silver-based photography, and of course, the observations on ZnO and similar semiconductors can be helpful in analyzing the behavior of dyes on silver salts. Because silver salts are not in general semiconductors, the powerful electrochemical tool is not available for their study. Hauffe and co-workers[R33,R40,R41] in particular, have provided us with many studies and reviews of the problem on ZnO. Inoue[R42] presents a good description of the history of the development of ideas on spectral sensitization of ZnO. Supplement #3 of *Applied Optics* (1969) is devoted to electrophotography and provides a substantial background.

The main thrust of research in dye sensitization of semiconductors has been with ZnO, and perhaps the main problem from a fundamental point of view has been to determine whether an excited dye injects its electron directly to the conduction band, or whether the excitation energy from the dye is transmitted to a surface state, and the surface state injects the electron. This problem will be discussed in this section, but first we will look at two examples of research regarding injection or capture of electrons by optically excited ions in solution which are not "dyes" in the commercial sense, but show similar behavior. Then we will discuss dye work on materials other than ZnO. Finally, we will summarize the extensive literature on electrochemical measurements of the ZnO/dye system.

Freund[90] observed that even ferrocyanide ions would cause electron injection into ZnO when excited. The injection is indeed low, because the extinction coefficient of the ferrocyanide is low and because, according to the discussion below, a good injector should be adsorbed, and probably the ferrocyanide ion is not adsorbed on ZnO.

Gleria and Memming[91] studied electron exchange between ruthenium ions in solution and semiconductors. The ruthenium was in the form of a bipyridyl complex. They found that electrons could be transferred to the ruthenium excited state from those semiconductors with a very high conduction band energy (SiC or GaP), and electrons could be transferred from the ruthenium excited state (in this case optically excited) to the conduction band of those semiconductors with a relatively low energy conduction band edge (SnO_2). An energy level picture is derived with the ground state of the ruthenium at about 1.5 eV below, and the excited state at about 1 eV above, the hydrogen reference. Thus (comparing the conduction band edges of the materials studied as presented in Section 5.3) the energy level of the excited ruthenium is at the right energy for the above observations. Also, in agreement with the general philosophy of the expected reactions as described in the preceding chapters, Gleria and Memming found that the introduction of an oxidizing agent, Fe(III), to capture the electrons from the excited Ru(II) prevented the electron injection.

In a similar fashion, Yoneyama and co-workers[57] observed the color change resulting from the reduction of the dye methylene blue by illuminated TiO_2. They studied the reaction under open circuit (no net current) conditions, which implies that the rate of electron transfer from the conduction band of TiO_2 to the unoccupied energy level of methylene blue must be equaled (as with any photocatalytic reaction) by the rate of hole capture by some reducing agent. Specifically, they concluded that the solvent was being oxidized: when methanol was used as the solvent, it was oxidized to formaldehyde; when water was used as the solvent, it was oxidized to oxygen.

Although work on semiconductors other than ZnO has been reported (for example, Watanabe *et al.*[92] have studied dye sensitization of CdS), probably the most interesting study has been that of Arden and Fromherz[93] with In_2O_3. This was a study of the influence of intervening monolayers of paraffins on the dye injection into In_2O_3. The paraffin monolayer separates the adsorbed dye from the semiconductor so one can tell whether the dye can inject over "long" distances. A particularly instructive series of experiments was one where, in some cases, a monolayer of dye molecules (thyacyanine) is adsorbed first on the In_2O_3, and in other cases, no thyacyanine is present. With or without the thyacyanine, a double monolayer of paraffin is used to separate a layer of oxacyanine dye from the surface. With light of a wavelength to excite the oxacyanine dye, electron injection does not occur unless the thyacyanine is present. The conclusion is that the oxacyanine dye absorbs photons, transfers the energy to the thyacyanine layer, and the thyacyanine layer injects an electron. The oxacyanine cannot inject directly, either because it is too far from the surface (a very low transmission coefficient), or because a double layer across the paraffin moves the energy level into the bandgap.

It is concluded in the case of ZnO, as in the In_2O_3 case above, that the dye molecules must be adsorbed for effective injection. Danzmann and Hauffe[94] indeed show that chelated adsorption provides particularly effective sensitization. Supersensitization is observed when, in addition to the dye, a reducing agent is present. The theory is that the reducing agent contributes an electron to the ground state of the dye (or forms a donor–acceptor complex with the dye[95]) after the dye is excited by the photon. Then the dye (now the leuco form) cannot be deexcited by the electron returning to the ground state. This increases greatly the probability that the electron in the excited state will be injected into the conduction band of the ZnO. An example of supersensitization and its analysis is presented by Bode and Hauffe.[95] Fujishima *et al.*[96] used the RRDE to show that the reducing agent is oxidized during the sensitization.

For many years the argument has raged as to whether dye injection of electrons into ZnO occurs by direct electron injection from the excited state of the dye to the ZnO conduction band, or occurs by energy transfer. In the case of "energy transfer" the energy absorbed by the dye transfers to a surface state on the ZnO surface, with the surface state injecting the electron. The electron can, of course, be observed in either mechanism by the current flow in an electrochemical cell.

Most workers believe in direct electron injection. For example, Daltrozzo and Tributsch,[97] confirming the observations of Danzmann,[98] find a *p*H dependence of electron injection analogous to the *p*H dependence of electron capture by ferricyanide observed by Vanden Berghe and coworkers[3] (see Figure 6.3 and the accompanying discussion). Daltrozzo and Tributsch suggest that for the dye, as well as for the ferricyanide, such a *p*H dependence could not be explained by the hydrolysis of the dye molecule and must be due to a shift in E_{cs} [Equation (2.31)]. This interpretation immediately implies electron injection, because a shift in E_{cs} relative to the solution should not affect injection from surface states. Pettinger *et al.*[19] conclude that electron transfer must be the mechanism based on two pieces of evidence: (a) the excited state energy levels of the dyes studied are above the energy E_{cs}; (b) the oxidized dye molecules must be in good contact with the surface, because when the ZnO is heavily doped, electrons can be made to tunnel from the bulk conduction band to the ground state of the excited dye molecules.

Some authors lean toward a mixed mechanism. The results of Arden and Fromherz[93] have been discussed above, and indicate reasonably clearly that energy exchange can occur, but also indicate that the excited state of a dye (thyacyanine in their case) can inject. Other workers using nonelectrochemical systems (e.g., Vodenicharova and Jensen[99]) suggest a mixed mechanism; in the reference given, the authors find a complex photoconductance rise and decay that they find hard to interpret by a simple electron injection model.

In general, the evidence seems to favor an electron transfer mechanism for adsorbed dye molecules, but arguments can be presented that most of the results used as evidence for an electron transfer *could* be explained by an energy transfer mechanism. It could be that, as with most things in nature, if two routes are possible, both will be used, and the dominant one will depend on the existing conditions.

6.6. High-Current, High-Voltage Processes

6.6.1. Introduction

There are various conditions under which semiconductor electrodes can be induced to pass large currents at high voltages. The various possibilities are discussed in this introductory section, and examples are given in the following subsections. High currents are usually associated with quasimetallic behavior of the semiconductor electrodes. Such behavior occurs when, owing to surface states, or, more often owing to degeneracy, a large part of any applied voltage change appears across the Helmholtz potential instead of across the space charge layer.

Various origins of degeneracy can be identified. One, discussed earlier, is an accumulation layer, where high forward voltages lead to the Fermi energy moving into the majority carrier band. However, there are other conditions that allow high currents. And, in fact, most of the "practical electrodes," where semiconductors are used for battery terminals, for fuel cells, or for electrocatalysis electrodes, involve these other conditions, possibly together with an accumulation layer. Specifically, there are two conditions; first, a doping level so high that tunneling occurs easily in both directions; and second, the presence of a dominant, partially filled *d* band, where the Fermi energy is in this band right up to the surface. Both these conditions involve degeneracy, in that the Fermi energy is in, or almost in, an allowed band of levels.

The semiconductor with a degenerate accumulation layer was described in Sections 1.1.1.3 and 3.4.3. It was noted there that Dogonadze *et al.*[100] provide a quantum mechanical analysis. As the semiconductor becomes degenerate, it acts more and more like a metal, with the transfer coefficient (Section 3.2) approaching the metallic value of $\frac{1}{2}$ (as opposed to the semiconductor value of zero).

Tunneling, associated with high doping levels, was reviewed qualitatively in Section 5.1.2, and was reviewed more quantitatively (but not emphasizing high currents) in Section 6.5.2. When an accumulation layer in the forward direction becomes a thin tunneling barrier in the reverse direction (Figure 5.1), the electrode will show quasimetallic behavior insofar as high currents are concerned.

Another mechanism for high currents is breakdown, and this is difficult in many cases to distinguish from tunneling. Avalanche breakdown, as discussed in Section 2.2.4.3, requires that carriers, before being scattered, attain an energy at least as great as the energy required to excite the avalanching carriers. If the excitation is band-to-band (excitation of valence electrons to the conduction band), then the voltage drop in the space charge region must be at least the energy gap E_g (in volts). Only then can a carrier gain sufficient energy to excite a valence electron to the conduction band. However, as discussed in Section 5.2.3.1, it is possible that a quasibreakdown can occur at lower voltages, with electrons excited from trapping levels to the majority carrier band, in particular from dislocation levels or from grain boundary levels. Thus quasibreakdown can occur with qV_s less than E_g. Zener breakdown can also occur. Thus tunneling can occur and lead to high currents, where the carriers may originate at dislocations (Section 5.2.3.1) or at surface states, and tunnel to the majority carrier band.

The third case that should be distinguished, is that of semiconductors with partially filled *d* bands. Because of their low carrier mobility, they must have a high carrier density in order to have a reasonable conductivity. These are nonstoichiometric transition metal oxides in general. The distinction between SnO_2 that is heavily doped because of nonstoichiometry, and RuO_2 that has a partially filled *d* band because of nonstoichiometry, is not that great, however. They differ primarily in carrier mobility. In the discussion of Section 6.6.4, we will not distinguish very much between the two cases, but refer to both as "practical electrodes."

A final mechanism for high currents, which is somewhat unique to semiconductors, is the reduction of the semiconductor, or the plating of a metal that leads to metallic layers on the surface. Thus when materials such as Ag^+ are reduced to form a metal, the semiconductor quickly loses its characteristic properties because it takes on the character of the metal overlayer.[(R43)] As another example, Gerischer[(101)] describes unpublished work, indicating that ZnO becomes reduced if the cathodic polarization is high enough to give a degenerate surface. Then, by his decomposition model (Section 7.4.2), when the Fermi energy becomes higher than the critical energy for decomposition, the zinc is reduced to a metal, and the semiconductor acts in many ways as a metal electrode.

These various conditions leading to high currents, and a few practical examples of cases where semiconductors provide high currents, will be reviewed in this section.

6.6.2. High Currents with Accumulation Layers

Most studies of the high-voltage, high-current region on nontransition metal semiconductors have been made using cyclic voltammetry, as discussed in Section 4.2.1. There it was pointed out that a strong accumulation layer makes the surface degenerate or near-degenerate. For a narrow bandgap semiconductor where there are plenty of minority carriers, a strong inversion layer could behave similarly. Then a great portion of any change in applied voltage appears across the Helmholtz double layer, and the material behaves in a quasimetallic way. As described with reference to Figure 3.4a, when the applied voltage appears across V_H, the energy levels in solution can move to an energy equal to the allowed bands in the semiconductor and substantial currents can flow.

Madou *et al.*,[(102)] for example, have studied both *n*- and *p*-type GaP using cyclic voltammetry in an aqueous indifferent electrolyte (no active ions in the solution). At potentials greater than -1.5 V with *n*-type material, where an accumulation layer forms, they get a high cathodic current, attributable to reduction of protons. The transfer coefficient for the reduction approaches the metallic value of $\frac{1}{2}$. In effect, the n-type GaP reduces protons or H_2O in a way analogous to a metal. Such behavior in voltammetry is typical of n-type semiconductors under high cathodic applied voltage. Qualitatively similar behavior is observed with GaAs.[14,103,105] With p-type GaP, corrosion dominates the accumulation layer region, and the electrode behavior is more complicated. In most pH regions when holes reach the surface, a layer of Ga_2O_3 forms, as will be discussed in Chapters 7 and 8.

An interesting series of measurements on n-type samples with accumulation layers was introduced in Section 6.2.2. These are the studies by Frank and Bard,[22] Landsberg *et al.*,[106] Kohl and Bard,[107] and Yeh and Hackerman,[23] using voltammetric techniques with primarily aromatic hydrocarbons as the species to be reduced in nonaqueous solvents. As an indication of the potential where the hydrocarbons would expect to be reduced (as the voltage is swept negative) and reoxidized (as the voltage returns positive), all the results on semiconductors were compared to results on platinum electrodes in the same solvents. Their results for ZnO (compared to Pt) showing the electrode potential when this reduction and reoxidation occurs, are presented in Figure 6.6. In the present section we want to emphasize that portion of Figure 6.6, when an accumulation layer is required to induce reduction of the ion—that is, the case where E^o_{redox} is well above E_{cs} (the potentials on Pt are above the flatband potential for ZnO). Several characteristics can be noted. First, if the energy level E^o_{redox} of the ion, as determined by the reaction on platinum, is well above V_{fb}, the reduction and oxidation on ZnO (Figure 6.6) and CdS (not shown) are completely reversible.[107,23] This reversibility is understandable because, to reduce the ions, the surface of the n-type semiconductor must have an accumulation layer and probably be degenerate; the Fermi energy must be in the conduction band; and the material therefore must be acting like a metal. Thus, on the return cycle when E_F again crosses E^o_{redox}, E_F is still in the conduction band region, so there are plenty of empty states for electron injection from the reduced ion to occur (just as in the case of a metal). This feature of reversibility of a redox couple, when the sample is strongly accumulated, is not observed with GaP or with GaAs. It is suggested[23] that during reduction on these materials an insulating film develops, and on the return sweep the oxidation is blocked by the insulating film.

Another interesting feature of reduction with an accumulation layer, luminescence, is noted by Landsberg *et al.*[106] They studied the reduction of aromatic hydrocarbons on GaAs and GaP, and found reduction as described above, but they found also that while simple reduction is observed for voltages more negative than -1 V (SCE), if the reduction is induced at −1.9 V (SCE), a luminescence was observed. This they attributed to electron capture by the excited (singlet) state of the hydrocarbon, followed by radiative deexcitation to the ground state. Such a route for reduction is consistent with the fluctuating energy level model as applied to exothermic reactions.

Noufi and co-workers[39] observed the transfer coefficient during reduction of several species on TiO_2: $Fe(CN)_6^{-3}$, Ce^{+4}, Ag^{2+}, $IrCl_6^{2-}$, and Fe^{3+}. They found that the reduction was associated with quasimetallic transfer coefficients with α approaching $\frac{1}{2}$ in many cases. Such transfer coefficients indicate a strongly degenerate surface layer. The observations supported the earlier studies of Dutoit *et al.*[108] It is not clear in the studies of Noufi and co-workers whether the sample surface was accumulated, or whether the observations related to a band of dislocations near the surface, for the samples were not treated in one of the TiO_2 etches that are effective in removing dislocations. Also, with TiO_2, as was mentioned in Section 5.2.2.2 above and will be described further in Section 7.3.4, ions move readily into the lattice while the crystal is cathodically polarized. Thus the metallic transfer coefficients observed may well have been due to the presence of diffused hydrogen causing a strong degeneracy, or due to a badly damaged surface with a pinned Fermi energy.

6.6.3. Tunneling and Breakdown on Non-Transition-Metal Semiconductors

A few cases have been observed where high currents have been reported on semiconductors other than *d*-band semiconductors, and the high currents are not associated with accumulation layers, but with reverse biased materials. This almost inevitably implies either (a) tunneling or (b) breakdown. A third possibility, seldom identified to date, is (c) leakage of current through flaws. For example, there is the work of Pham and Hueller,[109] who found high currents using an ion-implanted silicon electrode. Presumably, this is an extreme case of current due to flaws in the semiconductor, as described in Section 5.2.3.1.

But, in general, the cause of high currents in the reverse direction of the surface diode is assumed to be tunneling or breakdown. For example, Boddy[63] observed high reverse current with TiO_2 and assumed that tunneling occurred from ionic surface states to the conduction band. The resulting "trapped holes" in the ionic surface states then oxidized OH^- to O_2. Then Morisaki and Yazawa[110] studied TiO_2 under 8 V (SCE) reverse bias. (The concept of reverse bias was introduced in Sections 3.3.3.2 and 2.4.) Earlier Pettinger *et al.*[89] had studied heavily doped ZnO under reverse bias. Both these groups concluded that the high current was due to holes arriving at the surface of the *n*-type semiconductors, where the holes were generated by avalanche breakdown of the material or (for the heavily doped material) by Zener breakdown. In the TiO_2 case they all conclude, with Boddy, that the surface reaction was the oxidation of OH^- by the holes. This surface reaction is now well known on TiO_2 following the suggestion by Fujishima and Honda[111] that the reaction could result in solar energy conversion (Section 9.1). The most interesting contribution of Pettinger *et al.* and of Morisaki and Yazawa was the observation that, when the hole generation was forced by the high voltage to be a high-current process, luminescence was observable. Morisaki and Yazawa attribute the luminescence to electrons falling from anionic surface states to the holes in the valence band. The resulting oxidized ionic surface state (essentially O^-) can react with the OH^- from solution as a step in the O_2 production. On the other hand, Pettinger *et al.*, observing the luminescence of ZnO bandgap energy, feel it must arise from electron–hole recombination, but find it difficult to suggest where this recombination could be occurring. Another example, where breakdown is considered to lead to high current flow, is in the work of Transchart *et al.*[112] In this case the breakdown is associated with flaws in the semiconductor GaAs, as evidenced by the observation of pitting at dislocations after the current flow. A further example is the observation by Morrison and Freund,[113] who observed exceptionally high reverse currents when dislocations were introduced at a ZnO surface. Here it was concluded that electrons were injected at dislocation levels, moved along the dislocation pipes, and finally were excited to the bulk conduction band. The process can be considered a form of breakdown.

6.6.4. Practical Electrodes

The electrodes described in the earlier sections of Section 6.6 pass a fair amount of current and are of great theoretical interest, but a few electrodes are almost metallic in the current densities they can pass, and will be discussed together in this section. They are almost metallic because they are effectively degenerate, either made so by doping, or become so because they have a narrow, partially filled band. Specifically we refer to n-SnO_2, p-NiO (lithiated), RuO_2, and n-In_2O_3. Other compounds are mentioned in the reviews by Stonehart[R67] and Tseung.[R68]

Tin oxide and indium oxide electrodes are presumably dependent on tunneling for their high current densities. Armstrong *et al.*[88] indeed suggest that at low doping they are dominated by tunneling, but at high doping "other factors are important including the composition of the surface," presumably meaning that they are so defective from a semiconductor point of view that they can be considered to be entirely flaws. The bands, in other words, become so distorted that the material behaves in a quasimetallic way because there are allowed levels everywhere. What Armstrong *et al.* consider "low doping" is the material studied by Memming and Möllers[4] and others, as was discussed under tunneling in Section 6.5.2. Laitinen[R32] reviews the use of the "highly doped" SnO_2 as a practical electrode.

Undoubtedly the key feature of SnO_2 as a practical electrode is its inertness. For example, Uchida *et al.*[20] studied tin oxide electrodes at 450°C in molten LiCl–KCl eutectic. They found the redox couples studied (europium, chromium, copper, iron, and platinum) were reversible in their oxidation/reduction behavior. Uchida *et al.* suggested that their SnO_2 was acting as a tunneling electrode. Another feature of SnO_2 electrodes (which gives it an advantage over many metal electrodes) is pointed out by Laitinen,[R32] who used SnO_2 in cathodic-stripping coulometry of PbO_2, to study the oxidation of the lead and the subsequent cathodic reduction. Laitinen points out that SnO_2 has an advantage over Pt as a substrate, in that in such experiments, oxidation of the Pt substrate occurs and confuses the interpretation of the measurement of the Pb oxidation. SnO_2 is preferrable because it has no analogous problem. Hansen *et al.*[114] in using chronopotentiometry (where the time necessary to oxidize or reduce all the ions in a vessel is determined) take advantage of the transparency of SnO_2 to perform simultaneous measurements of internal reflection to study intermediate species (Section 4.3.2). Möllers and Memming[115] also used internal reflection spectroscopy, studying the light-absorbing intermediate semiquinone diamine in the reduction of dimethyl-*p*-phenzenediamine at SnO_2.

Both SnO_2 and In_2O_3 have an advantage over Pt as an electrode in aqueous solution, as pointed out by Armstrong *et al.*,[88] in that they have a greater accessible potential range. Presumably this is because they are much poorer catalysts and so reactions such as hydrogen reduction and evolution will not occur as readily.

Certain Group VIII oxides are of particular interest. These oxides, produced by anodization of the metal, have been found to have a higher activity for oxygen evolution than the bare metal by up to 3 orders of magnitude. The metals studied are Ir, Rh, and Ru. Burke *et al.*[116] and Fraser and Woods[117] suggest that the reason for the enhanced activity is the possibility of a partly filled *d* band (say, Ru^{+4}/Ru^{+3}) acting as the electron acceptor from the OH^-. The presence of a thick oxide further accelerates the reaction because it is a highly hydrated oxide, so the reaction can occur anywhere in the oxide, with the neutral oxygen diffusing out in time.

The RuO_2 electrode has large commercial usage as an electrode in the oxidation of chloride ions to chlorine. However, it is not clear whether it should be considered a semiconductor electrode. According to Doblhofer *et al.*,[118] the electrode features a partially filled Ru(IV)/Ru(III) band. These authors conclude that the oxidation state is primarily Ru(IV), which would make it essentially a very heavily doped n-type d-band semiconductor. Discussion of this important electrode will be postponed to Section 9.2.2.

Finally, the lithiated NiO electrode can be mentioned as a practical high-current semiconductor electrode, of particular interest as a possible oxygen electrode[R34] in fuel cells. Like RuO_2, it is a d-band semiconductor,[119] but NiO is definitely a p-type semiconductor, the lithium dopant acting as an acceptor. Like SnO_2 and RuO_2, NiO is a stable electrode relative to most semiconductor electrodes. However, in an inert electrolyte, decomposition can occur[120] under anodic current flow (holes to the surface). Tench and Yeager, in their studies of this electrode, examined the ferrous/ferric and the chromous/chromic couples. They found that in the forward direction (the anodic direction for this p-type material) the transfer coefficient becomes metallic, approaching a value of $\frac{1}{2}$. Cathodically, they found a linear Mott–Schottky plot, but with a slope too high for the known lithium doping. It is possible that some tunneling was being observed—the work was done presumably without knowledge of Memming and Möllers'[4] study of tunneling. The incorrect slope should not have arisen from the neglect of the Helmholtz capacity[121] for, as discussed in Section 4.1.4, when the space charge capacity approaches the Helmholtz capacity, the intercept, not the slope, of the Mott–Schottky plot should change. Tench and Yeager found that in the oxidation of Fe^{2+}, where the transfer coefficient very closely approaches $\frac{1}{2}$, the exchange current is 10^{-2} that of platinum.

6.7. Analysis of Complicated Electrode Reactions using the Tools of Semiconductor Electrochemistry

In this section we wish to illustrate, with a few examples, the power of the tools available in semiconductor electrochemistry to analyze electrode reactions step by step. The many variables available to the experimenter enable the most complicated tangle of reactions to be unraveled. Indeed, in most cases, there are enough additional variables that one can obtain supporting independent evidence. This ability of semiconductor electrochemistry has not been exploited to its potential to date, perhaps because of the forces currently directing research toward quick, practical rewards, but there have been enough studies (on the inert electrode) to provide a few examples. And later there will be further examples where the techniques have been used to determine corrosion mechanisms (Chapter 7) and to determine the behavior of passivating oxide layers on semiconductors (Chapter 8).

6.7.1. The Photocatalytic Oxidation of Formic Acid

One of the first applications of the tools of electrochemistry to analyze a complicated chemical process was its use[7] to separate the steps of the photocatalytic oxidation of formic acid on ZnO:

$$O_2 + HCOOH \xrightarrow[h\nu]{ZnO} H_2O_2 + CO_2 \tag{6.37}$$

The reaction can be studied, and was studied prior to the referenced work, without the use of an electrochemical cell. Indeed, all earlier work was done using powder suspensions of ZnO. But, with the use of an electrochemical cell, the surface reactions could be separated. Specifically the following observations could be made: By biasing the sample negative in the dark, it was determined that the electrons from the ZnO were able to reduce oxygen, simply by noting the increased steady state cathodic current (at constant n_s) when oxygen was present. It was similarly found that electrons from the ZnO conduction band were able to reduce H_2O_2. This showed that reactions (6.38) and (6.39) below would occur. By biasing the sample positive, it was found in the dark that none of the reactants, intermediates, or products, induced a current (no electron injection), but it was found upon illumination, that current doubling was exhibited in the presence of HCOOH. This demonstrated clearly the reaction of holes with HCOOH, and the generation of a free radical by this reaction. If O_2 was present in the solution, together with the HCOOH, no current doubling was observed, implying that the free radical reacted with the O_2 rather than injecting electrons into the ZnO [following Equation (6.12)]. By chemical analysis (iodimetric titration) it was found that in the latter case, (with O_2), H_2O_2 could be detected.

From these observations the following reaction steps could thus be defined—the first two indicated by the cathodic experiments, the second two by the anodic:

$$2e^- + O_2 + H^+ \rightarrow HO_2^- \text{ (peroxide ion)} \tag{6.38}$$

$$2e^- + HO_2^- + 3H^+ \rightarrow 2H_2O \tag{6.39}$$

$$h^+ + HCOOH \rightarrow 2H^+ + CO_2 + e^- \tag{6.40}$$

$$2h^+ + 2HCOOH + O_2 \rightarrow HO_2^- + 3H^+ + 2CO_2 \tag{6.41}$$

Equation (6.40) is the current-doubling equation, and Equation (6.41) is equivalent to Equation (6.12), suggested because the presence of oxygen stops the current doubling, but causes peroxide ion generation. Whether the radical reduces the hydrogen peroxide ion HO_2^- to water is not clear from the experiments done with the ZnO biased to cause current flow.

However, one more trick is available with the electrochemical setup, and it was used to further define the reactions. This is the measurement of the capacity of the space charge region. This measurement can be made even under conditions of no current flow (photocatalytic conditions). It is found with both oxygen and formic acid in the solution [viz., when Reaction (6.37) is proceeding] that the capacity is very high under photocatalytic conditions (the surface barrier is very low), relative to its value under equivalent cathodic conditions, with the same concentration of oxygen and no illumination, but the same flow of electrons to the surface.

Now from Equations (3.34) and (3.52), the rate of electron capture is an expression of the form

$$J_n = C_1 c_{ox} \exp(-qV_s/kT) \tag{6.42}$$

where C_1 is a collection of constants. The conclusion from the experiment is that if V_s is low under catalytic conditions with J_n maintained constant, then c_{ox} must be low. In other words, during photocatalysis the presence of the oxygen provides virtually no energy levels able to accept electrons. With this and similar results using peroxide, it could be concluded that the radical produced during photocatalysis must form a donor/acceptor complex with hydrogen peroxide, adsorbing in a nonreducible form. The complex keeps the peroxide from being reduced to water, and so the reaction Equation (6.41) yields peroxide, not water. The complex dissociates only upon oxidation by a second hole. Then the complete reaction sequence becomes

$$2e^- + O_2 + H^+ \rightarrow HO_2^- \tag{6.43}$$

$$h^+ + HCOOH \rightarrow HCOO\cdot + H^+ \tag{6.44}$$

$$HCOO\cdot + HO_2^- \rightarrow \text{donor/acceptor complex} \tag{6.45}$$

$$h^+ + \text{donor/acceptor complex} \rightarrow H_2O_2 + CO_2 \tag{6.46}$$

This provides an interesting example of the use of the powerful tools of semiconductor electrochemistry in unraveling the steps of a photocatalytic reaction, up until then,[122] and since[123] studied only on powders. Of course, the final conclusion may have to be modified if intrinsic surface states are involved in the reaction, as more recent work may well indicate. But, it is expected that the modification will be in details of the species involved (whether, for example, the radical is a surface state on the ZnO rather than the HCOO· radical), not the general picture.

6.7.2. Analysis of the Energy Levels of Two-Equivalent Species

This short section will be devoted to a description of the analysis of the energy levels of the two-equivalent quinone–hydroquinone couple by Memming and Möllers.[4,R76] This analysis provides a fine example of the use of the techniques of semiconductor electrochemistry for the purpose of analyzing complicated ionic energy levels. The use of semiconductor electrodes in such an analysis is quite unique, and it is unfortunate that more such analyses have not been attempted, although, as will be seen, the research is very complex.

The interpretation of the observed tunneling of electrons into a conduction band, to determine the value of λ for one-equivalent species, was described in Section 6.5.2. The determination of λ for two-equivalent species is much more difficult, because two levels must be used to describe the current flow—first, the energy level of the reducing agent, and second, the energy level of the intermediate. The concentration of the reducing agent is, of course, known, but the density of the intermediate is a second unknown parameter. Now the density of the intermediate is directly related to the energy level difference between the intermediate and the stable reducing agent, as we will show below. Thus, in actual practice, there are fortunately only two unknowns: the value of λ and the energy level difference between the E^o_{redox} for the intermediate/oxidizing agent couple, and the E^o_{redox} for the reducing agent/intermediate couple. That is, referring to Figure 1.7, the difference between the two levels in Figure 1.7a is unknown, and λ, the reorganization energy that gives both the Franck–Condon splitting and the standard deviation for the Gaussian distribution of fluctuating energies is unknown, and that is all. (This, as usual, assumes that λ is the same for all the species of interest—the reducing agent and the intermediate.)

We will use the following notation for the energy levels of interest: E^o_{redox} for the quinone/semiquinone couple will be designated as $E^o_{\text{redox}}(\text{Q,QH})$, and E^o_{redox} for the semiquinone/hydroquinone couple will be designated as $E^o_{\text{redox}}(\text{QH, QH}_2)$. The value of E^o_{redox} for the system, the quinone/hydroquinone couple [$= -E^o$ for the couple, by Equation (1.68)] will be identified as $E^o_{\text{redox}}(\text{Q,QH}_2)$.

Now each couple must, at a metal electrode, yield the same Fermi energy, so from Equation (1.45) or from Equations (1.78)–(1.84) we can immediately derive the relations

$$E^o_{\text{redox}}(\text{Q,QH}) - E^o_{\text{redox}}(\text{QH,QH}_2) = \Delta E^o_{\text{redox}} = kT\ln([\text{Q}]\cdot[\text{QH}_2]/[\text{QH}]^2) \quad (6.47)$$

and

$$E^o_{\text{redox}}(\text{Q,QH}_2) = \tfrac{1}{2}\{E^o_{\text{redox}}(\text{Q,QH}) + E^o_{\text{redox}}(\text{QH,QH}_2)\} \quad (6.48)$$

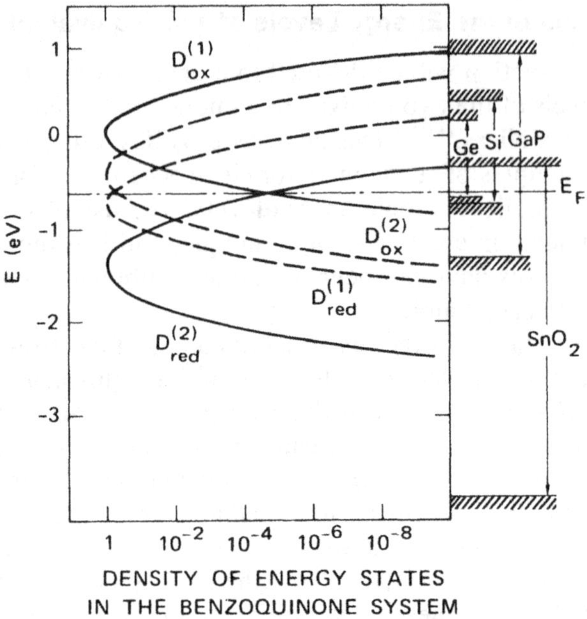

Figure 6.9. Fluctuating energy levels of the quinone/hydroquinone two-equivalent couple. The superscript 2 represents the hydroquinone/semiquinone couple, the superscript 1 represents the semiquinone/quinone couple. Thus $D^{(1)}_{ox}$ represents the unoccupied energy level on Q, $D^{(2)}_{ox}$ the unoccupied energy level on QH; $D^{(1)}_{red}$ represents the occupied energy level on QH, $D^{(2)}_{red}$ the occupied energy level on QH_2. $E^o_{redox}(Q, QH)$ is at the intersection of the $D^{(1)}_{ox}$ and $D^{(1)}_{red}$ curves; $E^o_{redox}(QH, QH_2)$ is at the intersection of the $D^{(2)}_{ox}$ and $D^{(2)}_{red}$ curves. $E^o_{redox}(Q, QH_2)$ is shown as E_F.

where we are defining [Equation (6.47)] the difference in energy levels as ΔE^o_{redox}. From Equation (6.47) the density of the intermediate QH is known, if the separation of the "solution Fermi energies" ΔE^o_{redox} is known. Thus, as described above, ΔE^o_{redox} is the only unknown other than λ.

Memming and Möllers programmed their computer so that it not only optimized the fit of the experimental tunneling data from hydroquinone into SnO_2 to give the best fitting value of λ, as was done for one-equivalent species, but also required of the computer that it simultaneously determine the best fit for ΔE^o_{redox}.

The best fit for the current–voltage characteristics obtainable varying the two independent parameters was found with $\lambda = 0.5$ eV and $\Delta E^o_{redox} = 0.5$ eV. With Equations (6.47) and (6.48) [where $E^o_{redox}(Q,QH_2)$ is known from handbook tabulations of E^o, the standard electrode potential], the individual energies of the levels in solution and the broadening of these energies is then determined. The results are shown in Figure 6.9, where, for comparison, we show the band edges (obtained by the usual methods of Chapters 4 and 5) of silicon, GaP, SnO_2, and Ge, with OH· adsorbed (the germanium E_{cs} depends on H· or OH· adsorption, as discussed in Section 5.1.1.2).

As supporting evidence for the accuracy of the model of Figure 6.9, Memming and Möllers studied the behavior of the quinone–hydroquinone couple at the various electrodes shown in Figure 6.9. The detailed energy level model for the quinones was found to explain many features of various experiments. For example, experimental observation of the two-electron reduction of quinone (using the transistor method outlined in Section 4.3.3) indicated that the reduction occurs entirely by hole injection when the electrode is a Ge–H electrode, but if the electrode is a Ge–OH electrode, one electron must come from the conduction band. The band model clearly indicates why. With E_{cs} and E_{vs} in the Ge–H position, E_c and E_v about 0.4 eV higher than shown in Figure 6.9, so both Q and QH can accept electrons from E_{vs}, but this is not true in the Ge–OH position. There is other supporting evidence. Silicon and GaP also required one electron from the conduction band, followed by hole injection (current doubling) just as in the case of the Ge–OH electrode. Tin oxide, of course, required two conduction band electrons to reduce quinone.

With varying *p*H, the various quinones become hydrolyzed (complexed with H^+ ions) differently. By the *p*H dependence of the current doubling, it was possible to determine the number of protons attached to the species that actually accepted the valence band electron. We will not go into detail, but refer the reader to the original article for the discussion of how it was demonstrated that the species QH_2^+ is the species that actually accepts the valence band electron (injects the hole).

Möllers and Memming[115] continued the studies of two-equivalent species, studying the diamine by reflection spectroscopy at a SnO_2 electrode. They comment that (a) wide bandgap semiconductor electrodes offer a much better vehicle than metal electrodes for such detailed studies of two-equivalent processes, and (b) the E^o_{redox} for the two steps (e.g., As^{3+}/As^{4+} and As^{4+}/As^{5+} in the example of Figure 1.7) can be resolved by conventional techniques if the values are widely separated, but if they are close, conventional methods are unsatisfactory. In the cited paper, to detect the presence of the intermediate species, they use attenuated total reflectance, which helps untangle the steps of the complex electrode process.

6.7.3. The Reduction of Iodine on CdS

Cardon and Gomes[34] have analyzed the reduction of iodine at the CdS electrode, and in this section their analysis will be described. Because the rate is found[124] to be proportional to the concentration of I_3^-, Cardon and Gomes conclude the first electron capture must be a sequence such as

$$I_3^- = I_2 + I^- \tag{6.49}$$

$$I_2 + e^- \rightarrow I_2^- \tag{6.50}$$

This identification of the first reduction is in agreement with an earlier analysis by Morrison,[12] who also studied I_3^- reduction on ZnO. The second step of the reduction could be any of three routes:

$$2I_2^- \rightarrow I_3^- + I^- \tag{6.51}$$

or

$$I_2^- + e^- \rightarrow 2I^- \tag{6.52}$$

or

$$I_2^- \rightarrow 2I^- + h^+ \tag{6.53}$$

By the results of Tyagai and Lukyanchikova,[125] who measured the electrical noise associated with the cathodic reduction current, it could be concluded that in the low-current region the first and second step of the reduction were not correlated, viz., the two processes were independent. In the high-current, diffusion-controlled region, the two steps were correlated: once the first reduction [Equations (6.51) and (6.52)] occurred, the second reduction immediately followed.

Cardon and Gomes suggest by this observation that the second reduction step must be either via Equation (6.51) or Equation (6.52), and cannot be via Equation (6.53). They reason that the improved correlation at high current density must be due to the very high density of electrons in the conduction band of the CdS. Either of the first two mechanisms for the second step could increase in rate if the electron density increases, and thus could show correlation at high current density. The last mechanism involves hole injection and probably should not increase in rate with a high conduction band electron density, and so this possibility is rejected.

We have offered in the present section three examples of how the various techniques of semiconductor electrochemistry can be combined to analyze the details of chemical reactions. Other examples can be found in the body of the chapter. It is to be hoped that this ability with semiconductor electrochemistry to dissect reactions will be utilized much more in future studies.

7

Chemical Transformation in the Electrode Reaction

7.1. Introduction

In the present chapter cases are discussed where substantial chemical changes in the inner coordination sphere accompany electron or hole exchange between the solid and solution. This is in contrast to the last chapter, where to a reasonable degree, the data presented could be interpreted without considering chemical transformations other than simple changes in oxidation state. When a change in oxidation state of the ion is accompanied by the formation of a different inner sphere complex, or when it is accompanied by adsorption or desorption of the ion, or when it is accompanied by electrode dissolution, the influence on the electrode reaction is too strong to be neglected and, in principle, the models must reflect such transformations.

Three classes of chemical transformations will be discussed in this chapter. In Section 7.2 we will describe cases of inner sphere transformations of ions when they exchange carriers with an inert electrode. An example of such a case would be

$$Fe(OH)^{2+} + e^- = Fe^{2+} + OH^- \qquad (7.1)$$

occurring at an inert electrode. In Section 7.3 we will emphasize cases where the ions adsorb or desorb as a result of the redox reaction. An example of such a case would be

$$H^+ = H_{ads} + h^+ \qquad (7.2)$$

as occurs, for example, on germanium [see Equation (5.6) and the associated discussion]. Another example would be precipitation, usually of a metal on the electrode:

$$Ag^+ + e^- = Ag \qquad (7.3)$$

Finally, there are complicated corrosion processes. In Section 7.4 we will describe experiments and theories of corrosion of the solid, emphasizing dissolution of the solid (or at least one component of the solid) during the passage of current.

In considering the role of inner sphere effects for carrier reactions with species in solution or specifically adsorbed, we shall still only discuss reactions where the rate-limiting step is electron transfer. We will not consider concentration polarization, defined in Section 3.1.2, where the rate-limiting step is the formation of a certain complex. The omission of the topic of concentration polarization is reasonable when discussing semiconductors because (unless the semiconductor is degenerate and acting quasimetallic) the low availability of carriers limits the current so electron transfer is the expected rate-limiting step. An additional reason for ignoring concentration polarization is that there seems no case with nondegenerate semiconductor electrodes where this form of concentration polarization has actually been analyzed as a possible rate-limiting step, except, to a limited degree, in corrosion.

The description of inner sphere transformations using an energy level model is not straightforward. In earlier sections, some background was developed. In Section 1.2.1.1 the need was discussed. The distinction was made between outer sphere reorganization effects, the only effects described well by the fluctuating energy level model, and energetic transformations in the inner coordination sphere that interests us in this chapter. In Section 1.2.4.5 a modification of the theory of the fluctuating energy level model to describe such inner sphere transformations was introduced. In Section 2.1.1.2 adsorption was discussed, but only from the point of view of its influence on the Helmholtz layer. In the present section an attempt is made to examine experimental observations where available, and try to provide further details for a band picture.

There are many potential advantages favoring the use of semiconductors and an energy level model to analyze inner sphere changes during redox reactions. A simplification in analysis arises because the enery levels in solution do not shift relative to the band edges when the voltage, or when the concentration of the redox couple, is changed. Then, in principle, it can be determined relatively easily by analysis with a band picture which complex is active in the electron transfer process, by determining which energy level is active and relating this energy level to the right complex. Together with this simplification in theory using the band model, we have significant experimental advantages with a semiconductor electrode over those available with a metal. The details are described in earlier chapters, but as examples we have the ability to alter, easily and reversibly (e.g., by illumination) the density and type of carrier involved in the reaction. We have also the ability, by changing semiconductors, to change the energy levels of interest (relative to the hydrogen zero) in a meaningful way (all inert metals in principle act essentially the same). These parameter variations provide powerful tools to explore any reaction of interest. With these parameter variations we can, in principle, analyze the reaction routes and energy levels involved, and from there characterize the complexes formed or dissociated.

For workers interested in developing applications of semiconductor electrodes, we can identify at least two major reasons for needing a better description of inner sphere reorganization during redox reactions on semiconductors, reasons that are more specific than the ever-desirable "development of basic knowledge." First, reaction rates are expected to be strongly affected by complex formation. With a semiconductor the Helmholtz potential is independent of the energy levels of the redox couple. Hence the shift of energy level due to complex formation can have an extreme effect on the reaction rate. For example, as hinted at in Section 3.4.2, and discussed further below, the formation of complexes in solution may play an important role at times in the failure of the fluctuating energy level model in exothermic processes. Second, corrosion of semiconducting materials is dominated by inner sphere reorganization. During corrosion, the ligands characteristic of the solid are replaced by ligands characteristic of the solution. Actually, as will be discussed in Section 7.4 below, this is the most studied of all semiconductor inner sphere reorganizations, simply because it is of such great technical importance.

7.2. Inner Sphere Changes during Redox Reactions at an Inert Electrode

In this section we discuss cases where the semiconductor electrode is inert but the inner coordination sphere of the ion changes during the reaction. First we develop a simple model to describe the rate of the reaction, then we discuss the few studies that have been done. Semiconductor electrode studies are still in their infancy insofar as models and studies of such complex reactions are concerned. Only a few systems have been examined at all, and the theoretical interpretation is at a very modest level, considering the apparent importance of the process.

In essence, the model we will use is based on the concept that carrier transfer may be accelerated when a particular preferred configuration of ligands (a complex) forms around the ion in solution, shifting its energy level to an energy near the conduction band edge or the valence band edge. Then the current flow is given by the fluctuating energy level model. Consider the example, used earlier [Equation (1.69)], of the hydrolyzed ferric ion and the nonhydrolyzed ferrous ion. For ease of visualization we will consider a carrier injection process, electron injection. The oxidation of the ferrous ion could proceed by any of the reaction routes:

$$m(\mathrm{OH})^- + \mathrm{Fe}^{2+} = \mathrm{Fe(OH)}_m^{(2-m)+} \tag{7.4a}$$

$$\mathrm{Fe(OH)}_m^{(2-m)+} = \mathrm{Fe(OH)}_m^{(3-m)+} + e^- \tag{7.4b}$$

where m can be zero to 2 (or, if precipitation is possible, to 3). The oxidation of the ferrous ion will occur when some preferred configuration of ligands is around the ferrous ion. This preferred configuration will be one with E_{red} near the conduction band edge and a low free energy of formation for the complex (to provide a high concentration of the complex).

Our model will be derived as follows. We will calculate the concentration of the various complexes from known free energy differences, calculate the energy level of the complex from an analysis such as Equation (1.74), and use the theories of electron transfer in Section 3.3.3 and the experimental variables available (e.g., the known band edges of semiconductors) to determine the capture rate for the various complexes.

We will use the Marcus–Hale formulation, as given in Equation (3.55), which describes the rate of charge injection into a band. Then the analysis appears in a particularly simple form. The capture cross section does not enter into the equation, being replaced by a transmission coefficient. In Equation (3.55) the transmission coefficient was set equal to unity, assuming that if the energy level in solution fluctuated into the conduction band region, the energy level density was so high that the probability of transmission would be unity. Application of the same general approach to the other formulas for carrier capture or injection as discussed in Chapter 3 is not difficult for the reader, but the expressions become unwieldy and lead to no new qualitative information.

The anodic current associated with a series of levels i is from Equation (3.55):

$$J_a = qZ \sum_i c_{\mathrm{red}}^i \exp[-(E_c - E_{\mathrm{red}}^i)^2/4\lambda kT] \tag{7.5}$$

where c_{red}^i and E_{red}^i are the concentration and the energy level [as given by Equation (1.74) or its equivalent] for the ith complex. Clearly all these complexes are meant to be variations on the primary reducing agent. The complex can involve a ligand L, or if an appropriate λ can be chosen, the "complex" can involve an adsorption site S. In the present section we emphasize ligands.

For simplicity consider the case where there is only one complex of interest, represented by $L{\cdot}A^{(n+1)-}$, and one aquo (uncomplexed) species $A^{(n+1)-}$, the concentrations of which are determined by the reaction of

Equation (1.70), repeated here:

$$L + A^{(n+1)-} = L\cdot A^{(n+1)-} \tag{7.6}$$

We have for Equation (7.6) at equilibrium

$$[L\cdot A^{(n+1)-}]/[L]\cdot[A^{(n+1)-}] = c_{\text{red}}^{\text{comp}}/[L]\cdot c_{\text{red}}^{\text{aq}} = \exp(-\Delta G_{\text{red}}/kT) \tag{7.7}$$

where the c's, with the appropriate superscripts, represent the concentrations of the two species. From Equations (7.7) and (7.5), using a single λ for simplicity, we have with a small algebraic manipulation

$$J_a = c_{\text{red}} qZ \frac{\exp[-(E_c - E_{\text{red}}^{\text{aq}})^2]/4\lambda kT}{1 + [L]\exp(-\Delta G_{\text{red}}/kT)} + \frac{\exp[-(E_c - E_{\text{red}}^{\text{comp}})^2/4\lambda kT]}{1 + [L]^{-1}\exp(\Delta G_{\text{red}}/kT)} \tag{7.8}$$

where c_{red} is the total concentration of reducing agent. The first term represents the rate of injection for the uncomplexed form, the second term the rate for the complexed form. For the case where the reduced form is the only strongly complexed form, in particular where ΔG_{red} is large and negative and $\Delta G_{\text{ox}} \cong 0$, the use of Equation (1.74) leads to

$$J_a = c_{\text{red}} qZ \frac{\exp[-(E_c - E_{\text{red}}^{\text{aq}})^2/4\lambda kT]}{1 + [L]\exp(-\Delta G_{\text{red}}/kT)} + \frac{\exp[-(E_c - E_{\text{red}}^{\text{aq}} - \Delta G_{\text{red}})^2/4\lambda kT]}{1 + [L]^{-1}\exp(\Delta G_{\text{red}}/kT)} \tag{7.9}$$

Now even if $E_{\text{red}}^{\text{aq}}$ for the uncomplexed form is above the conduction band edge, the addition of a complexing agent L may well prevent injection. The first term, injection by the aquo species, will decrease as $[L]$ increases (because $[L]$ appears in the denominator). If the energy level for the complexed form, $E_{\text{red}}^{\text{comp}}$, given by $E_{\text{red}}^{\text{aq}} + \Delta G_{\text{red}}$, is much below E_c, injection by the complex, the second term, will also be low. On the other hand if, instead of electron injection into the conduction band, we are considering hole capture from the valence band, the formation of the complex may lower the level to the valence band region and permit hole capture.

An analogous shift holds for electron capture by oxidizing agents. An equivalent formalism for an oxidizing agent would indicate that strong complex formation of the oxidized form (together with weak complexing of the reduced form) would provide a higher energy level, $E_{\text{ox}}^{\text{comp}}$, allowing electron capture by species whose aquo form is too far below the conduction band edge.

Figure 7.1 illustrates a complex-induced energy level shift that will allow electron capture by an oxidizing agent that is forbidden for the aquo species. The importance of such a possibility to explain the failure of the fluctuating energy level model was discussed in Section 3.4.2. Route (a) for electron capture is intended to represent capture by the complex, as described mathematically by an equation analogous to the second term of Equation (7.9). Other routes that allow electron capture by the aquo species are possible, such as route (b) of Figure 7.1, involving assistance in the capture by dislocations. The expected behavior of dislocations is discussed in Section 5.2.3.1. Or, as discussed in Section 3.4.2, such exothermic charge transfer to an energy level in the gap region can occur via surface states or via an excited state of the ion.

The potential advantage of semiconductors over metal electrodes in the investigation of complex formation is clear from the simple form of Equation (7.9). In particular we note the simple dependance of the current on the ligand concentration $[L]$. With metals, such a simple equation does not obtain because, as discussed at the end of Section 7.1, E_F of the metal depends on both the concentration of L and the concentration of complexed and aquo ions. With semiconductors, where E_{cs} or E_{vs} is the important parameter, and E_{cs} and E_{vs} are generally independent of $[L]$, the dependence of J on $[L]$ is particularly simple. The band edges of the semiconductor may depend on the concentration of the ligands for the special case when the ligand is OH^-, but in this case E_{cs} varies in a known way with ligand concentration. Naturally, as observed with Equation (1.74), the energy level of any particular complex must be, and is, independent of its concentration (assuming always that the activity coefficient is unity).

Experimentally most of the work to date has been qualitative. Studies have been done of the effect of pH on current doubling and other electrode reactions, but in most cases[1,2] it is not clear whether the change is due to Helmholtz voltage changes of the electrode or due to hydrolysis of the ions in solution. In some studies the influence of complex formation is clear. The interesting work of Memming and Möllers[3] on quinones was described in Section 6.7.2. It was pointed out there, that by varying the pH and thereby changing the concentration of the various forms of the quinones, they were able to show that only one particular complex, $QH_2\cdot$ (a radical), was able to inject holes during a current-doubling process on a p-GaP electrode. The method was simple: because the current doubling stopped at a pH where this particular form of the radical was no longer stable, the identification was obvious.

In studies of electron capture by ions on ZnO,[4] complex formation was considered for the cases of silver reduction and ceric reduction, and the route of the reaction was identified, based on an analysis of the capture

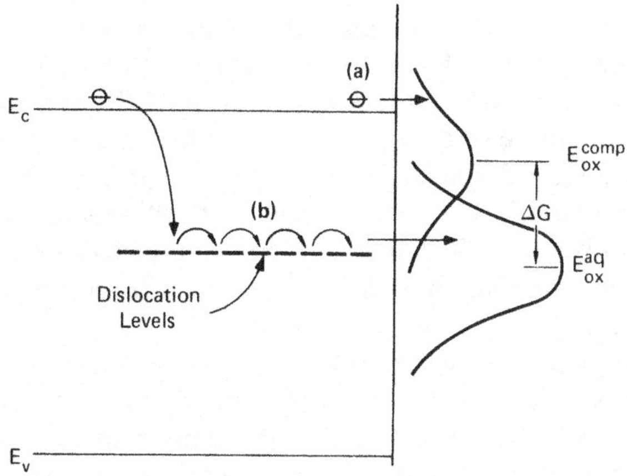

Figure 7.1. Possible electron capture steps for exothermic electron capture by an oxidizing agent with an unsuitable energy level, E_{ox}^{aq}. Route (a) represents complex formation, shifting the energy level up in solution to E_{ox}^{comp} so electron transfer can occur. Route (b) represents electron transfer via dislocation levels. Other routes may involve surface states or excited states as in Figure 3.5.

cross section and expected concentration in accordance with the general concepts described above. Specifically, it was concluded that the reduction of silver must be through the energy level of the $Ag(NH_3)_2^+$ ion, not the Ag^+ ion, because the calculated cross section would be unrealistic if the latter route is assumed. Additionally it was found that the ceric ion in nitric acid, where minimal complex formation could be expected[R72] has a low cross section for electron capture, while the ceric ion in sulfuric acid, where strong complex formation could be expected, has a moderately high capture cross section for electrons. It was concluded that the formation of the complex raises the energy level, as expected by the discussion following Equation (7.9), if the ceric ion is assumed to form more stable complexes than the cerous ion. The experimental result indicates that the level is raised sufficiently that overlap obtains between the conduction band edge and the Gaussian (as in Figure 7.1) for the sulfate complex of ceric ions. This observation was described in Section 6.2.1, in discussions of whether the fluctuating energy level model is valid for highly exothermic electron capture reactions.

Another example of the importance of complexing in electron transfer reactions is the effect of complexing on dye injection reactions, as studied extensively by Hauffe and co-workers.[5–7] Extra sensitization can often be ascribed to complex formation. In some cases[6] the cation of the solid is complexed—in Danzmann and Hauffe's study[6] a chelate complex between dye molecules and zinc ions at the ZnO surface leads to sensitization. They demonstrate the formation of the complex by measuring the absorption spectrum, and showing that it corresponds to the absorption spectrum of the complex that is formed when both the Zn^{2+} and the dye are in solution. In Hauffe and Haeggquist's study[7] a dye complex at the surface leads to increased corrosion. In other cases[5] it is not clear whether the increased sensitivity to corrosion (due to the presence of halide ions, particularly I^-) arises because the halide ions form a complex with the dye, or because the halide ions contribute electrons to the excited dye molecule in accordance with the supersensitization model described in Section 6.5.3.

We can conclude from the semiquantitative formalism of Equation (7.9) and the associated discussion, plus the few experimental observations available, that redox reactions involving inner sphere changes should be susceptible to a particularly simple analysis on semiconductor electrodes, and it is hoped that more basic research in this area will be forthcoming in the near future.

7.3. Adsorption onto and Absorption into the Electrode

The same formalism used for complex formation in the above section is also qualitatively applicable to specific adsorption. Specific adsorption, representing an inner sphere reorganization, is certainly expected to shift the energy levels from their values in solution[8,R4] in a manner similar to complex formation in solution. In addition, other changes will follow adsorption, each of which will affect the electron transfer. First, the Helmholtz potential V_H will change, as discussed in Section 5.1.1. Secondly, the capture cross section will change, because the orbitals of the species overlap the bands of the solid. Third, the effective value of λ can be so low that surface state models appropriate for the gas/solid interface may be required. Fourth, adsorption can mean a substantial change in the density (per unit area) of the reactant. As described in Section 3.3.2, ions in solution within about 15 Å of the surface can be considered available for charge transfer. In an 0.1 M solution this means about 10^{17} m^{-2} ions, whereas a monolayer of adsorbed species will be the order of 10^{19} m^{-2}, a much more concentrated collection of reactant ions. And finally, it should be pointed out that the adsorption of one species, not necessarily a reactant, may have an effect on the rate of redox processes involving another species. Adsorption of the first species may destroy or enhance sites where bridged electron transfer (electron transfer via an energy level of a ligand) can occur, may destroy sites where adsorption of the reactant can occur, and, of course, will undoubtedly affect the electron transfer by changing the Helmholtz potential, and thus the energy level of the reactant relative to the bands in the solid.

7.3.1. Adsorption of Water, Hydrogen, and Oxygen

The discussion of the adsorption of water and its components was presented much earlier because of its importance in most aspects of semiconductor electrochemistry. In particular, the topic was covered in Section 2.1.1, with an introduction to the dominant bonding mechanism (acid/base interaction) indicated in Section 1.1.2.1. These species are so important because their adsorption usually determines the Helmholtz voltage on ionic semiconductors, as discussed in Section 2.2.3. The adsorption on covalent semiconductors is also important. In this case they are adsorbed as radicals, H· and OH·, as reviewed in Section 5.1.1.2.

In this section we will elaborate slightly regarding one feature, the participation of water and its fragments in electrode reactions. And the discussion will be short, because there is little clear evidence whether it is the adsorbed water or its fragments that are actually oxidized or reduced, or whether it is nonadsorbed species. In the case of the H· and the OH· radicals on covalent semiconductors, the evidence is strong that they are the reactive species. The equations for the reaction of adsorbed H· and OH· were described in Section 5.1.1.2, and their validity was confirmed by measurements of the predicted Ge· levels, as described in Section 6.4.

The oxidation of OH^- (or water) to O_2, and the reduction of H^+ (or water) to H_2 on semiconductor electrodes has been of particular interest in recent years because of the interest in solar energy conversion based on the dissociation of water. Here the generation of hydrogen stores solar energy. However, there seems little positive evidence that it is the adsorbed species that are reduced or oxidized, as opposed to solution species. Sakata *et al.*[9] analyzed the position of energy levels associated with H_2O, OH^-, and OH· in solution to see which would be most likely to be oxidized by holes from TiO_2. They conclude the H_2O level is unfavorable, the OH^- level is favorable, and the OH· level is moderately favorable. They suggest that in this case the OH· level is raised by interaction with (adsorption on) the surface; charge transfer from the sea of valence electrons to the OH· level raises its energy. Of course, considerable experimental effort has shown that at the interface between the semiconductor and an organic resin (as in a paint binder) adsorbed OH^- groups are oxidized by holes. In particular we can cite the work of Völz *et al.*,[10] who detected the resulting OH· radicals by ESR. And at the TiO_2/gas interface, also, adsorbed OH^- groups on a hydroxylated surface are oxidized by holes, according to Bickley and Jayantz.[11] But at the TiO_2/water interface there seems no direct evidence to indicate which species are oxidized.

Another action of adsorbed water in redox reactions may be to block surface sites for more reactive species. The work of Micka and Gerischer,[12] on the blocking of current-doubling reactions on ZnO by various anions, will be discussed in Section 7.3.2. Here we simply comment that OH^- adsorption has such an anion effect too, but is more complicated than the other anions to be discussed below. Miyake *et al.*[13] also find a strong *p*H dependance of current doubling on TiO_2, where the current-doubling results from the oxidation of alcohols. However, as discussed above, OH^- ions will capture holes from TiO_2, so here (and possibly also in the ZnO case) any effect of *p*H may represent competing reaction, rather than a blocking reaction. In summary, blocking of reactions by adsorbates on semiconductors is not as well documented as blocking of reactions by adsorbates on metal electrodes.[14,R9]

Other effects of adsorbed water and its fragments common on metal electrodes[R9] do not seem to have been detected yet with semiconductors. For example, there is "cooperative adsorption": in some cases on metals oxygen adsorption causes a dipole that leads to cation adsorption and stabilizes the adsorbed oxygen layer, retarding the oxygen evolution reaction. Analogous effects with OH^- adsorption may occur with semiconductor electrodes, but have not yet been reported.

Finally, it is of value for completeness to mention again another way that water adsorption influences redox reactions. This is by its effect on the Helmholtz potential, two examples of which have been discussed earlier. One is the example of the reduction of ferricyanide ions at ZnO, as measured by Vanden Berghe *et al.*,[15] discussed in Section 6.2 (Figure 6.3 and accompanying discussion), where the object of the investigation was to evaluate the reorganization energy λ. The second example is that of the reduction of quinone on germanium, found by Memming and Möllers[16] to be entirely a valence band process when H· was adsorbed on the germanium, but partly a conduction band process when OH· was adsorbed. This investigation was discussed in Section 6.7.2.

7.3.2. Adsorption of Electrolyte Ions

The adsorption of foreign ions or radicals on the electrode was discussed in Section 2.1.1, and the influence of such adsorption on the Helmholtz double layer was discussed in Section 5.1.1.3. Three major types of adsorption were described: One is the adsorption onto a reasonable covalent semiconductor of radicals and ions, such as iodide ions that tend to form covalent bonds. Examples given were the semiconductors germanium and gallium phosphide. As was discussed in Section 5.1.1.3, it seems probable that a thin oxide must be present on the surface of such materials, judging from the *p*H dependence of V_{fb}. But superimposed on the *p*H dependance, V_{fb} varies due to the adsorption of hydrogen and hydroxyl radicals. A second type of adsorption is the adsorption of ions as a precursor to precipitation, as investigated by James and Healy.[17] In a limited pH range, where a transition metal ion "almost" precipitates as the hydroxide, adsorption will occur. The third type of adsorption is ion exchange, where the ion is so concentrated that it can displace protons or hydroxide ions from the surface. The above discussion is primarily based on knowledge about aqueous solutions, but analogous considerations will be expected for nonaqueous solutions. Of course, the role of protons and hydroxide ions will be somewhat less dominating in nonaqueous solvents, especially in a well-dried aprotic solvent.

In the present section we address the problem of reactions of such adsorbed ions with electrons and holes from the solid. Our interest is primarily in energy level changes, as discussed in Section 7.1, but little quantitative information is available. The energy level information is at best qualitative, and actually we can simply suggest that new energy levels may be formed during adsorption because there are measureable electrochemical effects.

Several authors have noted increased reactivity associated with adsorption of the reactant. Tench and Yeager[18] suggest that the Fe^{2+}/Fe^{3+} couple on NiO has an unusual transfer coefficient because of adsorption. Hauffe[19] relates the increased sensitivity of some dyes on ZnO to adsorption by the formation of chelate complexes.

Other authors suggest that such adsorbed species can act as surface states. Lemasson *et al.*[20] note that anions on WO_3 influence the impedance of the electrode. Yoneyama *et al.*[21] calculate the surface state energy levels expected on ZnO and CdS electrodes for adsorbed halides, and comment that theories of corrosion (next section) must be considered incomplete unless the possibility of adsorption affecting the surface states is considered. Yoneyama and Laitinen[22] describe the slow adsorption (absorption?) of Br^- onto (or into) hydrated SnO_2 that permits the oxidation of the bromide. With the deep valence band of SnO_2, bromide should easily be oxidized by holes, so presumably the major effect of the adsorption in this case is to concentrate the reactant at the surface.

Parkinson *et al.*[23] report the action of Ru^{2+} on GaAs surfaces in increasing PEC solar cell efficiency (Section 9.1.2). Apparently a remarkable improvement is observed. They suggest that the behavior is associated with the ruthenium interacting with surface recombination centers, so that the surface recombination velocity is much lower with the ruthenium present. The additive is also helpful at grain boundaries.

Germanium has been studied by Toshima and Uchida[24] and by Gupta[25] with respect to its adsorption of halide ions from solution. Toshima and Uchida observed the adsorption by frequency dispersion of the capacity (cf. Section 4.1.4). Thus the adsorption of halides apparently represents a confirmed case, where surface states due to adsorbed species apparently store the charge in a reaction such as[24]

$$Ge{:}X^- \leftrightarrows Ge{:}X + e^- \tag{7.10}$$

or

$$Ge\cdot + X^-_{ads} \leftrightarrows Ge^+X^- + e^- \tag{7.11}$$

To observe capacity effects at all, the reaction must be reversible—when an electron is added to the Ge:X complex it must be more favorable in the reverse half-cycle to restore the electron back to the conduction band, than for the X^- to desorb. The fact that both electrons and holes are available in germanium to contribute to the reaction may be the reason for the reversibility. It is also possible that the halides are at the interface between Ge and a thin oxide and because of this adsorption/desorption is very slow. In any case, the observation is interesting enough to review a few related experimental observations. In Section 6.5.1 it was shown that the reaction (7.10) or (7.11) involves both electrons and holes, and that the reduction of iodine on germanium occurs half by conduction band, half by valence band electrons (see Table 6.3). Efimov and Erusalimchik[R24] show that when holes are numerous the oxidation of I^- on n-germanium occurs at least partially by hole capture. Gupta[25] finds that I^- and Br^- are also able to inject electrons into the germanium conduction band when present in concentration over 0.05 M. From such observations it seems quite reasonable that in this case surface states due to such halide/germanium bonds reversibly store charge, and the rate of electron transfer to and from these surface states is such as to give frequency dispersion of the capacity in the 100-Hz to 1-MHz range as observed.[24]

It may be of some interest to compare the covalent interaction of adsorbates on covalent semiconductors such as germanium to the interaction of adsorbates with metals, such as platinum. For example, Conway[R44] discusses in length the dissociation of HCOOH, when adsorbing onto platinum, into H· and HCOO· forming covalent bonds with the platinum. It could be expected that semiconductors with a high covalent character in their bonds, and semiconductors with d levels available for bonding (Section 5.3 below), might have much in common with metals that bond with induced covalent bonding[R4] or crystal field bonding. However, the possibility of an oxide-free semiconductor, to compare to the "oxide-free" noble metal, seems unlikely, so such intimate adsorption of large molecules may not be common.

Above we have emphasized the reactions of the adsorbed species themselves with electrons or holes. A second role of the adsorbed foreign species can be to affect the reactivity of another reactant. A common way in which this effect appears, is in current doubling. For example, Micka and Gerischer[12] have found that anion adsorption affects the current doubling associated with formic acid or methanol on ZnO. As discussed in detail in Section 6.3.1.2, the competitive reaction in this case (competitive to current doubling) is usually the photolytic dissolution of the ZnO. Anion adsorption affects this competition. On the other hand, Miyake *et al.*[13] find no difference of chloride ions vs. perchlorate ions on current doubling on TiO_2. In this case the competing reaction is the oxidation of OH^-, not corrosion of the TiO_2. As discussed in Section 6.3.1.2, it is possible that anion adsorption (bonding of ligands) is very important in corrosion.

A third way in which adsorption affects redox reactions is when one of the products of the redox reaction is adsorbed. The work of Hauffe and Volz[26,R33] shows an example of this type of reaction. By measurement of electrophoretic mobility they were able to detect the adsorption of the product and the sign of the charge on the adsorbed species. They studied a photocatalytic reaction where isopropyl alcohol is oxidized by holes from ZnO, TiO_2, or CdS. The matching electron flow to the surface (required for photocatalysis as discussed in Section 6.3.3) is taken up by oxygen that is bubbled through the solution when desired. By the resulting sign of the surface charge, they find that the oxygen remains absorbed on the surface, perhaps as O_2^-.

7.3.3. Action of Deposited Species

By the term "deposited species," we mean cases where "adsorption" is induced to the extent of building up many layers, or when small amounts of foreign material are sputtered, precipitated, or reacted with the surface. Such deposited species change the properties of the surface, hopefully in a favorable way. Such deposited layers (or crystallites) will be discussed in the present section.

Germanium is unique with respect to deposition as well as being unique (as discussed in Section 7.3.2) with respect to adsorption. In particular, a significant amount of work was done on the adsorption of Cu and Au on germanium[27,28,R10] where the "adsorption" turned out to be involuntary deposition, in the sense that the metal was formed on the surface (out of solution) in the form of crystallites. The crystallites formed with a density of about 10^{15} crystallites/m^2. Simple immersion of a germanium crystal, in cupric solutions of greater than about $10^{-6}\,M$, suffices to cause these crystallites to deposit on the surface. The crystallites strongly affect the surface properties. They substantially increase the surface recombination velocity (Section 6.4). This is another of the few cases observed where "surface states," if we can call these crystallites surface states, affect the capacity (Section 4.1). Being a separate phase, such crystallites would be more likely than real surface states to affect the capacity, for the arguments presented in Section 4.1 do not apply to this case. Memming[28] suggests that the crystallites form at dislocations, but their density appears much too high to be associated with dislocations on a reasonable germanium crystal, and they may more likely form at surface steps, or even kink sites.[R4]

The process leading to the copper deposit on germanium may be analogous to the photodeposition of noble metals on, say, TiO_2, as discussed in Section 8.3. Of course with the small bandgap of germanium, minority carriers are plentiful without photoexcitation, so the deposition of metals on germanium can occur in the dark, if the levels of the metal are located favorably relative to the germanium bandgap.

The intentional deposition of metals on semiconductors has been used for various purposes. As described in Section 5.2.2.2, Yoneyama *et al.*[29] deposited metals on GaP and then biased the metallized GaP cathodically to reduce protons to hydrogen atoms. The metal provides adsorption sites for the hydrogen atoms and thus provides a concentrated source of atoms for diffusion into the GaP with the cathodic bias still applied. Thus the presence of the metal enhances hydrogen doping (see Section 5.2.2.2). The approach is similar to one used in catalysis termed "hydrogen spillover," where metal deposits, on an oxide, are used to dissociate hydrogen molecules, and the atoms can then diffuse over the oxide. Sancier[30] describes the use of spillover from metals onto oxide substrates in reduction of the oxides; Batley *et al.*[31] review hydrogen spillover in general.

Another use of intentionally deposited metals has been made by Nakato *et al.*,[32,33] who were interested in solar energy conversion. The concept was to use the metal film to resist corrosion of the semiconductor, but have the film thin enough so that "hot" holes (holes maintaining the valence band energy as they pass through the metal) could reach the surface without "cooling." Unfortunately, with reasonably thick layers (200 Å or so) the holes cool and the electrode becomes an ordinary metal electrode with a semiconductor contact. With thin layers, (30 Å or so) the metals are in the form of cystallites.

As a third example of the use of deposited metals, the work of Schrauzer and Guth[34] should be cited, who used iron in, or on, TiO_2 to enhance the photocatalytic activity of the TiO_2 in fixing nitrogen to NH_3.

As a final topic in this section, and probably an important area for future electrochemistry studies, the concept of "chemically modified electrodes" will be described. Most of the work on such chemically modified electrodes has been done on metals to date, but some has been done on SnO_2,[35–40] on TiO_2,[41–43] on SiO_2, and on In_2O_3.[37] The overall concept in these experiments is that a reaction between an organosilane and the semiconductor surface leaves the organosilane securely bonded to the surface, and some special group attached to the silane can be chosen to be electronically or chemically active. Specifically, for example, we can consider the reaction between the hydroxylated TiO_2 surface and a chlorosilane:

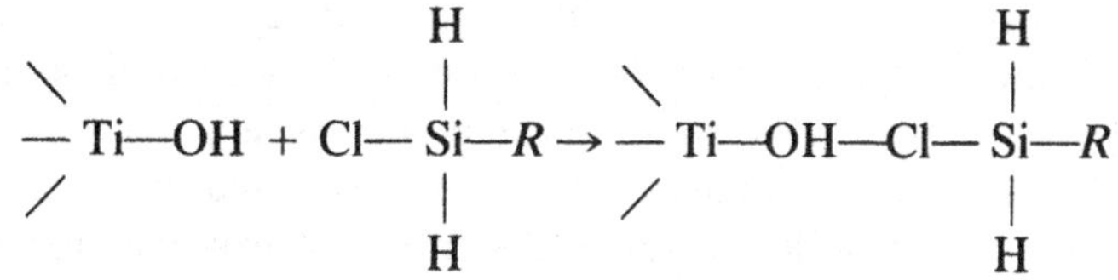

(7.12)

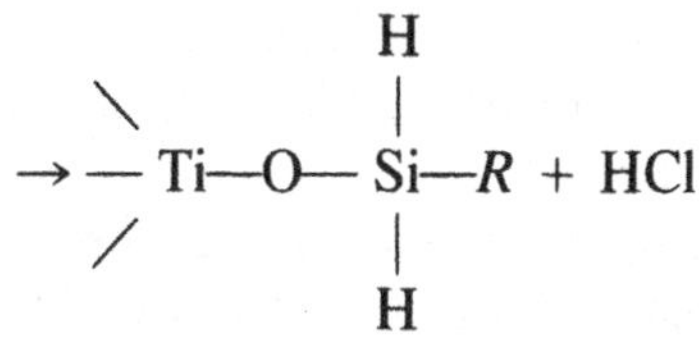

(7.13)

where —Ti represents a surface titanium atom on the TiO_2, and the *R* represents some organic group on the silane that can be chosen, in principle, arbitrarily. After the reactions (7.12) and (7.13) proceed, this organic group *R* is bonded strongly to the surface through the oxygen linkage shown in Equation (7.13). The group *R* can be chosen to provide catalytic character; it can be chosen to act as a recombination center; it can be chosen to protect the surface from corrosion; or, it can be chosen to induce adsorption of some other reactant of interest for redox reactions (enhancement of the electrocatalytic behavior of the surface).

To date the use of chemically modified electrodes is still in its infancy, as is the use of semiconductor electrodes for most of the above practical applications. Bocarsely *et al.*[(44)] and Wrighton *et al.*[(45,R45)] have found that ferrocene, so coupled to silicon, acts as a corrosion inhibitor preventing the anodic formation of SiO_2, but this seems the only positive result to date. Finklea and Murray[(43)] find no effect on the electrode properties of TiO_2 from deposited silanes. But, in agreement with Moses *et al.*[(40)] we can anticipate that organosilanes will provide a versatile route to many varieties of modified electrodes.

7.3.4. Movement of Impurities and Defects into the Electrode

Changes in the bulk semiconductor (near the surface, of course) following or preceding redox processes, represent yet another possible class of transformation associated with redox processes. The in-diffusion of ions or defects, or the formation of a new phase by absorption of reactants into the crystal, provide examples of such transformations.

The in-diffusion of ions associated with field-aided processes was discussed in detail in Section 5.2.2.2. However, in that discussion no distinction was made as to whether the ions diffused in as a result of a redox process at the surface, or whether the ions diffused in directly from the solution. In Section 7.3.3 we mentioned the one case where it seems obvious that the in-diffusion results from a surface reaction, namely, the in-diffusion of hydrogen into GaP where the surface concentration of hydrogen was made large by (a) reducing the protons to hydrogen atoms, and (b) providing metal atoms for surface sites to adsorb the hydrogen atoms. It may well be that many of the cases discussed in Section 5.2.2.2 were analogous, in that the surface concentration was magnified by oxidizing or reducing the species in solution and forming an insoluble product which would then reside on the surface. However, it is not known which, of the many cases discussed, depend on reduction to provide the diffusing species.

A second form of change of the bulk properties due to redox reactions is an increase in defect structure. This is almost analogous to the introduction of impurities, so need not be discussed at length. An example of the process is found with Cu_2O, where copper ion vacancies are very soluble in the solid—copper oxide can have major changes in stoichiometry without changing phase. Anodically, holes oxidize lattice oxygen to molecules and decrease the copper vacancy concentration, and cathodically the reverse can occur. O'Keefe *et al.*[46] find this to occur at temperatures slightly above room temperature at the gas/solid interface. With the help of an electric field, such stoichiometry changes corresponding to reactions, such as the oxidation of lattice oxygen to O_2, should be observable at room temperature under normal electrochemical conditions. The author of this book, in unpublished work, attempted to characterize Cu_2O electrochemically, but was forced to abandon the attempt because of continually changing electrode characteristics that were attributable to such stoichiometry changes. Whenever the voltage was changed for a few seconds, the electrode characteristics changed in an irreversible manner. Other materials, such as V_2O_5, can be expected to show similar characteristics (with V^{+5} easily converting to V^{+4}) and indeed difficulties have been found[47] in any but the simplest attempts to characterize V_2O_5 electrochemically. Another case where defects are apparently affected by redox reactions is the result of Sayers and Armstrong.[48] They reduced TiO_2 by hydrogen treatment, but found that after using the TiO_2 to photolyze water for a day or so, the TiO_2 was again stoichiometric. It should be mentioned, however, that it is not clear whether the effect was a change in stoichiometry of the TiO_2, or whether it was the removal of absorbed hydrogen donors. Sayers and Armstrong reported this oxidation of the TiO_2 as an undesirable feature because they were interested in energy conversion and the reduced TiO_2 is more efficient.

Finally, the behavior of CdSe and CdTe should be discussed as an example of a major phase change due to redox reactions. Chai *et al.*[49] report that these electrodes change character under illumination: the capacity becomes insensitive to the voltage and the photocurrents become very low. Cahen *et al.*[50] and Gerischer and Gobrecht[51] studied CdSe in a highly reducing sulfide/polysulfide electrolyte, and found that sulfur substitutes for the selenium under illumination and an amorphous material is formed at the surface. In fact, Gerischer and Gobrecht find that illuminated CdS in a sulfide/polysulfide electrolyte, forms an amorphous CdS phase at its surface too. The amorphous layer extends several tens of angstroms into the lattice in the dark, and much further under illumination. Perhaps the transformation results from oxidation of the lattice and immediate rereduction to form the amorphous deposit. Cahen *et al.* conclude that such a layer increases the photovoltage obtainable and could be beneficial on CdSe. Gerischer and Gobrecht find from Mott–Schottky plots that the surface layers lead to a large defect density. The CdSe turns yellow, and these authors conclude the layer in general is undesirable for photovoltaic conversion.

7.4. Corrosion

7.4.1. Introduction

In redox reactions at a semiconductor electrode, the possibility of corrosion of the electrode—either the development of a new phase or the dissolution of the electrode—is always present. This possibility has been referred to many times in the earlier sections and chapters, often as a competing reaction to the reaction under discussion. Such corrosion is in general undesirable, but at times is used to advantage. A specific case where dissolution of the semiconductor is desirable is electropolishing of the semiconductor. This application will be discussed in Section 9.4.

The emphasis of the present section will be the discussion of electrochemical corrosion during current flow, with little discussion of strictly chemical etching of the semiconductor. However, electrochemical dissolution of the electrode and chemical etching may in many cases have much in common. As will be described, electrochemical dissolution is usually related to a carrier from the solid being localized at the surface and breaking a bond; and chemical etching of the semiconductor seems often to be related to a carrier injected from the solution being localized at the surface and breaking a bond. The carrier is injected by one component of the etch. In these cases electrochemical etching and chemical etching differ only in the source of the bond-breaking carrier. All other functions, such as complex formation and solvation of the departing ions, are common to both processes. Actually, electrochemical techniques have proven valuable in the analysis of chemical etching processes. For example, the analysis of the etching of germanium by ferricyanide ions was discussed in Section 6.5.1. Here the ferricyanide injects a hole, which in turn can oxidize the germanium to a soluble valence state. Another example is the reaction of hydrogen peroxide on germanium. The mechanism of this complex reaction was traced out in the studies of Gerischer, Maurer, and Mindt.(52) Their mechanism requires the dissolution of germanium in the presence of H_2O_2 and, therefore,(R10) makes unattractive the later suggestion of Brouwer(53) that the germanium/H_2O_2 system makes a useful *p*H sensor. Other examples of carrier injection during chemical etching appear in the studies of Jones *et al.*,(54) who observe that NiO decomposes in the presence of Co^{3+}, and of Hauffe and Reinhold,(55) who observe that Cu_2O decomposes in the presence of O_2. In both cases the ion injects holes. Hauffe and Reinhold's experiments indicated clearly the similarity in corrosion, whether holes are produced by light or by oxidizing agents. Tributsch and Gerischer(56) studied the chemical etching of a series of sulfides in water and observed hole injection by a peroxide intermediate of the reaction.

Electrochemical corrosion occurs by various routes(R46) but basically the routes entail either the movement of the electrons to the surface, to reside in cationic surface states and interfere with the bond between the cation of a compound semiconductor and the neighboring anion, or the movement of holes to the surface to reside in anionic or covalent surface states, and (a hole being the absence of a valence electron) to result in the bonding electron being no longer there.

The effect of the electron reaching the surface is analogous to the effect with metal electrodes, where cathodic currents are frequently used to reduce an oxide layer. The use of cathodic currents to remove oxide layers from metal electrodes is a time-honored procedure,(R9) although the detailed steps of the mechanism are not always clear. For example, MacDougall and Cohen(57) in investigations of the cathodic breakdown of oxide films on nickel find the mechanism depends on whether the film was prepared by anodizing or electropolishing. The mechanism also depends on the *p*H in an important way. The mechanism changes as the film is removed—there are several steps in the electrode potential/time curve that indicate changes in the mechanism. Finally, they conclude that undermining of the film plays an important role. So reduction of an oxide layer by a cathodic current can be complicated.

Cathodic decomposition of semiconductors (as opposed to the reduction of thin semiconducting oxide layers on metals) has been studied very little. We can cite a few example studies. We have the work of Dyrek *et al.*,(58) who photolyzed V_2O_5 and found (by ESR measurements) that V^{+4}, in the form of VO^{2+}, was dissolved into the solution. The exact steps through which the VO^{2+} is formed by electron capture are not clear. The photoproduced holes in this case (photolysis requires an equal rate of hole and electron flow) are assumed to oxidize water or OH^-, and indeed, as indicated in Section 5.3, holes in V_2O_5 should be powerful oxidizing agents, able to oxidize water. Markham *et al.*(59) have shown, and Gerischer(60) comments, that ZnO can be reduced by electron flow to the surface. Eger *et al.*(61) note that a pulsating or continuous cathodic dc bias applied to a ZnO electrode leads to surface damage. But, they conclude that such damage requires 10^{26} to 10^{27} m^{-3} electrons, a strong accumulation layer, as calculated by Stein's analysis.(62) And finally, Madou *et al.*(63,64) observe that oxide films on GaP and GaAs can be removed by cathodic current in a manner similar to stripping oxide films from noble metals, although the exact mechanism is not clear. However, results showing cathodic corrosion of semiconductors are rare, and these few examples are somewhat unique.

Anodic processes in metals (the bias that brings holes to the surface for semiconductors) have been well studied. Either the metal can dissolve or a surface oxide (anodic oxide) can form. In an indifferent electrolyte, in which the metal ion is soluble, the anodic corrosion reaction will depend on the applied voltage. At low applied potential the Fermi energy of the metal is often not low enough to form an oxide, but may be low enough that the metal can go into solution as a cation. However, under certain conditions, above a potential known as the Flade potential,(R13,R7) the formation of the oxide (in an aqueous medium) occurs faster than the dissolution of the metal, and a passive layer forms—the metal becomes anodically oxidized. Interesting reviews are available by Hoar(R47) or Erdey Gruz.(R9)

In anodic processes on semiconductors, where the semiconductor is biased so that holes will reach the surface, the applied voltage (as usual) does not normally make much difference to the route of the chemical reaction. The preferred reaction will depend on the valence band energy, the energy levels associated with possible reactions, and the solubility of the cation and the anion of the semiconductor. In general, in the discussions of hole reactions to follow we must distinguish three cases: first the case where the ions from the semiconductor are soluble when oxidized by the holes to a higher valence state, so the semiconductor can dissolve; second, when the ions from the semiconductor are insoluble, and a new phase (probably an oxide in aqueous solution) can form; and third, when the holes find it thermodynamically more favorable to react with a species in solution than to react with the semiconductor itself, so the semiconductor is unchanged. Normally any of the three reactions can occur if the holes are very energetic, but one reaction will usually dominate. GaP is an example of a case where a semiconductor can either dissolve, as in the first case, or form an oxide, as in the second case, depending on the nature of the solution. Ga^{3+} is insoluble in the *p*H range 3.5–9 and Ga_2O_3 forms,[63] but in more acid or basic solutions the Ga^{3+} dissolves. Silicon almost always provides an example of the second case, forming an oxide rather than going into solution. In the case of silicon a high voltage can induce a change in the corrosion mechanism—an example of a voltage-dependent mechanism as is found for metals. In the case of *p*-type silicon, but not in the case of *n*-type, a very high voltage can induce anodic dissolution.[65,66] The surface is presumably degenerate (accumulated), and thus quasimetallic, before this voltage-induced electrochemical etching occurs. And even then, the presence of HF is required to make the silicon soluble in the +4 valence state. Titania offers an example of the third case, where it is essentially inert in aqueous solution, it being energetically more favorable for the holes to react with OH^- ions than to decompose the titania. Yet when sulfate ions are present so the titania can form a soluble complex, decomposition is observed.[67,68]

A fourth case, discussed further in Section 9.1, is the case where the holes (or electrons, if appropriate) are in nonbonding orbitals, so that no bonds are broken if the carrier reaches the surface. An example might be MoS_2, where a *d* band of Mo is the highest energy occupied band (the valence band), and the *d* electrons are not involved in bonding. Then, when a hole reaches the surface, there is little possibility that the suflide ions of the lattice will be oxidized and go into solution.

In open circuit photocorrosion processes, it is required, as indicated in Section 6.3.3, that the rate of hole and electron flow to the surface be equal at steady state. With open circuit conditions, in order to have corrosion by holes, we must provide an oxidizing agent to capture the electrons; to have corrosion by electrons, we must provide a reducing agent to capture the holes. Thus Dyrek *et al.*[58] in their studies of the corrosion of V_2O_5 to VO^{2+} by electron capture on the vanadium assumed the holes were trapped by OH^- ions in the solution. Markham *et al.*[59] in their studies of the photochemical decomposition of ZnS and CdS, where the holes oxidize the sulfide to sulfate ions (and the Sn^{2+} and Cd^{2+} ions from the solid go into solution with no valence state change), found, in general, that the electrons reduced dissolved gaseous oxygen to peroxide. Only in one special preparation of ZnS was the cation zinc itself reduced by the electrons, forming a layer of zinc metal, while the holes oxidized the sulfide ion.

An important feature of photocorrosion is the fact that one of the photoproduced carriers, usually the minority carrier, is attracted to the surface by the band bending. Such band bending can also occur and attract majority carriers to the surface with a biased electrode, but the electrode must be biased to an accumulation layer. With photoproduced carriers the feature is inevitable; one carrier is always attracted to the surface and thus is available to cause corrosion. Much of the experimental work on corrosion has been done on photocorrosion because it is technologically so important—in so many semiconductor/liquid systems photocorrosion is a major problem.

7.4.2. Theory and Observations of Semiconductor Corrosion

In this section we discuss first the free energy change in corrosion, then the activation energy (rate) of corrosion, and finally, the influence of defects on the corrosion rate. But for each of these topics in turn, a more detailed model of the corrosion process must be assumed. The problem in developing the models arises because of the complexity inherent in the process.

Consider corrosion by holes. As discussed in Section 1.1, a hole is a missing valence electron. Thus when a hole arrives at the surface and becomes associated with the surface atom, this means that one of the valence bonds between the surface atom and the rest of the crystal is broken. The hole has been captured by an intrinsic surface state. Because the donor surface states are just perturbed valence band orbitals (Section 1.1.2.1.), they may be expected to have a high hole capture cross section. Capture of holes in such surface states is equivalent to localization of the empty valence bond (the hole) on a surface atom. Capture of holes by such "intrinsic" donor surface states, or capture of electrons by "intrinsic" acceptor surface states, can lead to species that are soluble in the solvent and that are so weakly bonded to the surface that the gain in solvation energy is greater than the loss in adsorption energy. Then dissolution will follow. For example, on an oxide, MO, a reaction such as

$$h^+ + O_L^{2-} \rightarrow O_L^- \tag{7.14}$$

$$M_L^{2+} \rightarrow M^{2+} \tag{7.15}$$

$$H^+ + O_L^- \rightarrow (1/2)H_2O_2 \tag{7.16}$$

might occur where we have used the subscript L as usual to denote a lattice ion.

Of course Equations (7.15) and (7.16) will each represent several chemical steps. Indeed all three equations, (7.14)–(7.16), represent substantial oversimplification of the exact reaction sequence. Normally the surface is solvated. In the case of the oxide, the surface is hydroxylated, as in Figure 2.1. Then the reaction of a hole is to oxidize a surface complex, rather than a simple ion as in Equation (7.14). Many complexes may form, depending on the ions in solution, the location of the ion relative to crystal defects, such as steps, and the oxidation number of the ion.

Each of the complexes is characterized by an energetically different intrinsic surface state for the electron or hole capture. The point is that we can calculate the overall energy change for the complete dissolution, but cannot calculate energy changes for the individual steps of the reaction, because we cannot visualize the reaction route that is kinetically most active.

In our theoretical development we will be able to indicate, reasonably quantitatively, whether the reaction is thermodynamically favorable, but we will be able to discuss reaction rates only qualitatively. When comparing corrosion with competitive reactions, such as reaction of the carrier with ions in solution, we will be able to determine whether corrosion or the competitive reaction is energetically the most favorable. However, we will not be able to identify the activated complex, and so will not be able to calculate the relative rates of the two reaction routes. On our energy level picture we will be able to indicate "thermodynamic energy levels" for corrosion in the sense of Section 1.2.1.2, but will not be able to determine the real energy levels—the real intrinsic surface states involved in the corrosion reaction that determine reaction rates and probabilities.

7.4.2.1. Overall Energy Changes in Corrosion. First, the overall energies of corrosion involving hole or electron capture will be evaluated in such a way that they can be illustrated on a band diagram. For this purpose we must determine the free energy change during the oxidation or reduction of the atoms of the semiconductor to a solvated, soluble form. If the free energy change is negative, the reaction will be thermodynamically favorable, although the rate will not be defined. If the free energy change is positive, the reaction will not be favored—at least until enough voltage is applied to make the free energy change negative. In order to describe the reaction with an energy level model, we calculate and show on the band model the thermodynamic energy level of the surface species, and compare this energy level to the energy of the electron or hole from the solid.

Gerischer[60,R21] and Bard and Wrighton[69] have formulated the expressions. Following Gerischer, we will specify a compound semiconductor MX and consider two reactions:

$$MX + Ze^- + \text{solv} \rightarrow M + X^{Z-}_{\text{solv}} \tag{7.17}$$

represents corrosion of the host crystal by its conduction band electrons, resulting in reduction of the cation and dissolution of the anion. Here Z conduction band electrons are captured at intrinsic acceptor states associated with the cation. Similarly the reaction

$$MX + Zh^+ + \text{solv} \rightarrow M^{Z+}_{\text{solv}} + X \tag{7.18}$$

represents the corrosion of the host crystal by its valence band holes, resulting in oxidation of the anion and dissolution of the cation. Here Z holes are captured at intrinsic donor surface states associated with the anion. In both Equation (7.17) and (7.18) the influence of the solvent in complexing, hydrolyzing, or compound formation is written in the shorthand form with the term "solv."

Now as usual in electrochemistry, the energies of all levels are referred to the hydrogen scale. This is conveniently done by considering that the electrons or holes in Equations (7.17) and (7.18) come from the proton/hydrogen reaction. From a thermodynamic point of view, as mentioned in Section 1.1.1.3, the Fermi energy is the source and sink of electrons (on n-type material) or holes (on p-type material). That is, if the Fermi energy of the electrode happens to be at zero on the hydrogen scale, there is no difference in energy whether the electron (or hole) comes from the H^+/H_2 couple (with $pH = 0$), or comes from the Fermi energy. Thus for the special case where $E_F = 0$, we can obtain the free energy of charge transfer by considering the source of electrons as

$$(1/2)ZH_2 + \text{solv} = ZH^+_{\text{solv}} + Ze^- \qquad (7.19)$$

and the source of holes as

$$ZH^+_{\text{solv}} + ZY^-_{\text{solv}} = (1/2)ZH_2 + \text{solv} + Zh^+ + ZY^-_{\text{solv}} \qquad (7.20)$$

In Equation (7.20) the anion Y, as yet undefined, is introduced on both sides of the equation because it is convenient to consider the solution as electrically neutral before and after the reaction. With this anion present, the substance MX in the crystal is considered to be transformed to the substance MY in the solution.

First we will consider the free energy for electron capture, reduction of the cation of the crystal, when the conduction band electrons have the same free energy as if they were on the H^+/H_2 couple. Combining Equations (7.18) and (7.20) we have for a reaction equivalent to reduction of the crystal by electrons of zero energy on the hydrogen scale:

$$MX + (1/2)ZH_2 + \text{solv} = M + ZH^+_{\text{solv}} + X^{Z-}_{\text{solv}} \qquad (7.21)$$

The free energy for this reaction is obtainable from the handbooks or from Latimer.[R17]

The free energy of reaction (7.21) at $pH = 0$ gives the thermodynamic energy level for electron-induced corrosion, ${}_nE_{\text{decomp}}$, relative to the hydrogen scale. If reaction (7.21) is exothermic by an energy ΔG, then if the Fermi energy is at zero on the hydrogen scale, the corrosion reaction will be exothermic by an energy ΔG. Thus, on an energy level diagram we place the thermodynamic energy level for corrosion, the "decomposition energy level," ${}_nE_{\text{decomp}}$, at an energy ΔG eV below zero on the hydrogen scale. If the Fermi energy of the solid is above the hydrogen zero, the free energy for corrosion is higher by the corresponding amount. If the Fermi energy of the solid is lower than zero on the hydrogen scale, then the free energy for corrosion will be lower. From Section 5.3 we know (in principle) the E_F's for the various semiconductors MX on the hydrogen scale. Thus we can determine whether corrosion is exothermic or endothermic by noting the relative energy levels of E_F and ${}_nE_{\text{decomp}}$ on the band diagram. It is to be noted that although we presumed $p\text{H} = 0$ in the calculation of the free energy for Equation (7.20), this presumption is not required [if E_{cs} varies in the normal way as per Equation (5.3)], because for a specific reaction $E_{cs} - E_{\text{decomp}}$ will be independent of pH.

For an example of the calculation of ${}_nE_{\text{decomp}}$ using Equation (7.21), consider the free energy change for the reaction:

$$ZnO + H_2 = Zn + H_2O \tag{7.22}$$

The free energy of the reaction as written gives ${}_nE_{\text{decomp}}$ on the hydrogen scale for electron capture on an intrinsic surface state that leads to zinc deposition on the ZnO. If E_F is higher than ${}_nE_{\text{decomp}}$ for reaction (7.22), then the capture of electrons is exothermic, and thermodynamically self-reduction of ZnO will be favorable. The electrode will corrode. If E_F is lower than ${}_nE_{\text{decomp}}$, the reaction will be endothermic and the electrode will not cathodically corrode. It turns out for ZnO that E_{cs} (Section 5.3) is below ${}_nE_{\text{decomp}}$, so it is thermodynamically unfavorable for ZnO to decompose by cathodic reduction.[51]

On the other hand, the free energy associated with hole capture (oxidative corrosion of the crystal) when the holes have exactly the free energy of the H^+/H_2 couple (zero on the hydrogen scale) is given by combining Equations (7.19) and (7.20):

$$MX + ZH^+_{\text{solv}} + ZY^-_{\text{solv}} = M^{Z+}_{\text{solv}} + (1/2)ZH_2 + X + ZY^-_{\text{solv}} \tag{7.23}$$

In analogy to the electron capture case, we define a decomposition energy level ${}_pE_{\text{decomp}}$ equal to the free energy ΔG of this reaction and, therefore, equal to the free energy of hole capture if the Fermi energy is at the zero on the hydrogen scale. If the Fermi energy is lower than zero on the hydrogen scale, corrosion will be more exothermic than ΔG for the reaction (7.23); if the Fermi energy is above zero, corrosion will be less exothermic than the free energy change in reaction (7.23).

For an example of a calculation of ${}_pE_{\text{decomp}}$ corresponding to the corrosion of the ZnO crystals by holes, we can suggest

$$ZnO + 2HCl = ZnCl_2 + \tfrac{1}{2}O_2 + H_2 \tag{7.24}$$

The free energy change gives ${}_pE_{\text{decomp}}$ on the hydrogen scale. The practical importance of the anion Y is now clear. The free energy change for Equation (7.24) where the soluble products are in solution is immediately obtained from handbook values, but the choice of complexing or reactive species in the solution has a direct influence on the decomposition energy level calculated for the corrosion. The energy level of ${}_pE_{\text{decomp}}$ on the hydrogen scale, given by ΔG for Reaction (7.24), is well above the energy of valence band holes in ZnO (Figure 5.5), so ZnO is susceptible to corrosion by its holes.

In the above discussion we have spoken somewhat loosely of the "free energy of electrons" or the "energy of holes" reaching the surface. If the Reactions (7.17) and (7.18) are reversible, then the Fermi energy for the carriers can be used as the "energy" of the carriers and the free energy change will describe whether corrosion will be favored. However, usually corrosion reactions are not reversible, and the Fermi energy only provides an indication (and only for majority carriers) of the carrier energy available to be captured by surface states and initiate the reaction. From a practical point of view the energy levels of greatest interest, the energy levels that determine whether the reaction is effectively favorable, are the conduction band and valence band edges. The relation of the conduction and valence band energies E_{cs} and E_{vs}, to the decomposition energy levels, is a practical comparision that does not depend on equilibrium assumptions.

The case where photoproduced minority carriers are the reactive carriers is a case where obviously nonequilibrium conditions obtain. Here again, to say corrosion will occur if there is a favorable "free energy" for carriers is misleading, because in this nonequilibrium case a free energy cannot be defined. Certainly the quasi-Fermi-energy, defined in Section 1.1.1.3, is not the free energy of the minority carriers, it is simply a convenient way of describing the density of minority carriers in the minority carrier band. So the quasi-Fermi-energy relates to the rate of the reaction, not whether the reaction will proceed.[70] The relative positions of the band edges and the corrosion energy levels give again an indication, and a good indication undoubtedly, of whether the reaction will be energetically favorable or not, but not a rigorous thermodynamic proof. Because a higher density of minority carriers implies the rate of any reaction will be higher, we conclude that a quasi-Fermi-energy near the band edge only suggests the rate will be higher.

In summary, in general a practical indication that holes will be corrosive is if $E_{vs} < {}_pE_{\text{decomp}}$ and a practical indication that electrons will be corrosive is if $E_{cs} > {}_nE_{\text{decomp}}$.

With the above cautions regarding the application of the model, we can discuss the results of calculations based on the model.

Figure 7.2 indicates the model pictorially. To generate such a model for a given semiconductor we first determine the band edges on the hy-

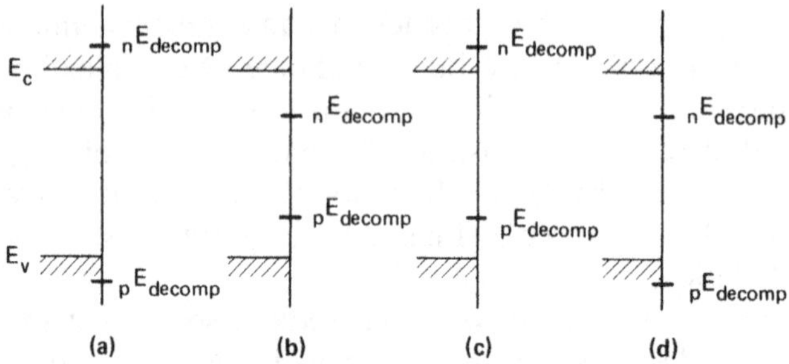

Figure 7.2. Possible positions of the decomposition energy levels relative to the band edges. If condition (a) obtains, the semiconductor is relatively stable; condition (b) implies instability if either electrons or holes reach the surface; conditions (c) and (d) imply instability to holes and to electrons, respectively.

drogen scale, as discussed in Section 2.3 from a theoretical basis, and later in Section 5.3 from an experimental basis. Then the decomposition energy level is determined on the same hydrogen scale using the free energy change of Equation (7.21) for $_nE_{\text{decomp}}$, and using that of Equation (7.23) for $_pE_{\text{decomp}}$. If $_nE_{\text{decomp}}$ lies below the conduction band edge, then one can expect decomposition to be favored during electron flow to the surface; if $_pE_{\text{decomp}}$ lies above the valence band edge, then one can expect decomposition to be favored during hole flow to the surface.

Gerischer[(R21)] has summarized the positions of $_nE_{\text{decomp}}$ and $_pE_{\text{decomp}}$ [assuming specific forms of the reactions (7.21) and (7.23)] for many semiconductors, assuming a particular form of the dissolved ions. He thus predicted on this basis whether they should be particularly susceptible or resistant to electrochemical corrosion. Figure 7.3 and the accompanying table, Table 7.1, indicate the values of the decomposition energy levels for a series of compound semiconductors as calculated by Gerischer,[(R21)] and the particular reactions whose free energy change was used to determine the energy levels. It is observed from this figure and table, for example, that Cu_2O should be reduced by conduction band electrons, that GaAs should be reduced by conduction band electrons, but that ZnO and TiO_2 should not. All species should be decomposed by valence band holes. However, if the solvent is water, the holes should prefer to oxidize the water over oxidizing the solid for ZnO, TiO_2, CdS, and MoS_2. The other semiconductors in the listing should decompose if holes reach the surface.

Park and Barber[(71)] have improved on Gerischer's analysis by considering in more detail the form of the product as it depends on the pH of the solution. Thus the value of $_nE_{\text{decomp}}$ and $_pE_{\text{decomp}}$ depends on pH because the solvation of the product depends on pH.

The model above provides a good approximation to predict whether

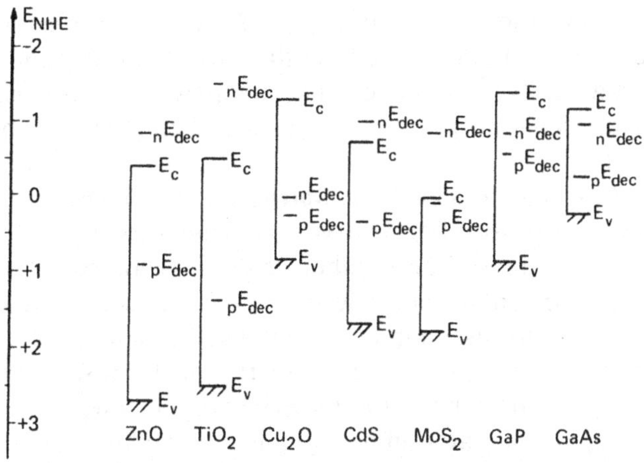

Figure 7.3. Decomposition energy levels for various semiconductors as calculated by Gerischer. (See text for reference.)

a semiconductor is stable against hole or electron flow to the surface. To the same approximation, qualitative modifications can be suggested. For example, the model predicts that ZnO should not be reduced by electrons. However, experimentally if the ZnO becomes so cathodic that an accumulation layer develops, reduction is observed.[60–62,72] We rationalize that if the ZnO becomes so cathodic that V_H is affected by the applied voltage, then there is a gain in energy $(-q\Delta V_H)$ per electron when the O_L^{2-} goes into solution. Then if $(E_{cs} - {}_nE_{decomp} - q\Delta V_H)$ is greater than zero, the corrosion is exothermic. As another example, Madou *et al.*[63] suggest the

Table 7.1. Reactions Assumed in the Calculation of the Decomposition Energy Levels Shown in Figure 7.3

$ZnO + 2HCl \rightarrow ZnCl_2 + (1/2)\,O_2 + H_2$
$ZnO + H_2 \rightarrow Zn + H_2O$
$TiO_2 + 4HCl \rightarrow TiCl_4 + 2H_2$
$TiO_2 + 2H_2 \rightarrow Ti + 2H_2O$
$Cu_2O + H_2O \rightarrow 2\,CuO + H_2$
$Cu_2O + H_2 \rightarrow 2\,Cu + H_2O$
$CdS + 2HCl \rightarrow CdCl_2 + 2S$
$CdS + H_2 \rightarrow Cd + H_2S$
$MoS_2 + 2H_2O \rightarrow MoO_2 + 2S + 2H_2$
$MoS_2 + 2H_2 \rightarrow Mo + 2H_2S$
$GaP + 6H_2O \rightarrow Ga(OH)_3 + H_3PO_3 + 3H_2$
$GaP + (3/2)H_2 \rightarrow Ga + PH_3$
$GaAs + 6H_2O \rightarrow Ga(OH)_3 + H_3AsO_3 + 3H_2$
$GaAs + (3/2)H_2 \rightarrow Ga + AsH_3$

cathodic decomposition potential ${}_nE_{\text{decomp}}$ for GaP is in the bandgap, but only anodic corrosion is observed. In this case the reduction of protons apparently is an effective reaction that competes with corrosion for the available electrons—the protons are a "stabilizing agent" in the sense of Section 7.4.3.

7.4.2.2. The Activated State in Corrosion. The kinetics of corrosion cannot easily be described because we cannot identify the kinetically most rapid sequence of steps. The number of possible reaction routes is too great. This is unfortunate, because in the case of competitive reactions (the usual case, as discussed in more detail in Section 7.4.3), the kinetics can be more important than the free energy of the respective reactions in determining whether or not the material will corrode.

The concept of the rate-limiting step in a competitive reaction, where for example, corrosion is competing with a reducing agent or a recombination center for the photoproduced holes, must be defined more clearly. As discussed in Section 7.4.3 in more detail, when the hole is attracted to the surface by the surface barrier, it is expected always to react with something, as it approaches the surface. The real rate limitation for hole reactions then, must be the rate at which holes approach the surface, that is, the rate of their generation by the light in the bulk. But what interests us more is the fate of the holes after reaching the surface. Thus we wish to compare the slowest step in the sequence leading to corrosion reactions, with the slowest step of the competitive reaction, where in both cases the hole is already available. In other words, in this section when we discuss the rate-limiting step for corrosion, we are referring almost always to the slowest step *after* the initial creation of the minority carrier.

The problem in determining the rate-limiting step for corrosion is that there are so many possible ligand configurations (complexes) for an atom at the semiconductor/liquid interface. As the surface atom becomes dissociated from the crystal, passing through a series of oxidation states, many different inner sphere configurations can be envisaged. Further complications arise, because for a given oxidation state the complex will differ, depending on whether the atom is at a flaw such as a surface step or on the uniform surface. Now each of these many complexes would have a different probability of carrier capture and/or a different activation energy for the next step in the dissolution. To suggest the correct sequence in detail, and then quantum mechanically determine the rate-limiting step, would require a remarkable insight.

Two general classes of possible rate-limiting steps can be expected to dominate the kinetics of dissolution, and it is helpful to suggest which is most probably rate limiting in corrosion reactions. The first class is a rate limitation due to carrier capture, the second is a rate limitation due to inner sphere reorganization. The same two possibilities were pointed out earlier for redox reactions at an inert electrode, as discussed at the beginning of this chapter, but for that case we concluded that because of the low carrier density in semiconductors, we could assume electron transfer is the usual rate-limiting step. In photocorrosion, it seems the other is more likely: reorganization of ligands is the best possibility for the rate-limiting step. This is a reasonable assumption when (a) we stipulate as above that, in the corrosion analysis, the "rate-limiting step" is not to include the creation of carriers, and (b) we consider only cases where the reacting carrier is attracted to the surface by the surface barrier, so there is no activation energy needed as the carrier moves to the surface. Then, in order to rationalize the assumption that some ligand reorganization step is rate limiting, we must only assume that all the intrinsic surface states are in the bandgap. In other words, if all intrinsic surface states can be assumed to be in the bandgap region, independent of solvation, then every carrier capture step (where the carrier is attracted to the surface by the appropriate sign of V_s) will be highly exothermic and may be assumed to have negligible activation energy. This assumption has been used in models of photocorrosion and is helpful. Unfortunately, the assumption cannot be proven. It has been shown at the gas–solid interface, both experimentally and by quantum analysis (cluster models),[(R4)] that the back bonds that must be broken during the corrosion are actually much stronger than the normal bulk valence bonds. Thus the intrinsic surface states that must be occupied for decomposition by holes may be far below the valence band edge, and if so, hole capture will be an activated process. However, the aforementioned experiments and theories do not include the energy gain by the complexing of the surface ion, by species from solution, that follows or precedes the hole capture. Thus, a really viable theory of the energy of the back bonds for a surface atom exposed to a solution has not been developed. At the present time in photocorrosion, in cases where the thermodynamic energy levels are deep in the bandgap, the assumption that ligand reorganization is rate limiting may be the best simplification available.

Even assuming that a ligand reorganization step is the rate-limiting step, two features must be kept in mind—the large variation of ligand configurations available, and the influence of flaws at the surface. We will discuss models and experimental evidence that suggest ligands and ligand configuration are important, and then discuss models and experimental evidence that suggest that flaws at the surface are important.

Partial theories to describe the complexity involved in bonding of ligands from the solution to the corroding surface atom have been suggested by several authors. For example, Yoneyama *et al.*[(21)] note that adsorption will complicate Gerischer's[(60)] and Bard and Wrighton's[(69)] simple theories. Experimentally, Pettinger *et al.*[73] developed a model to explain why on ZnO the ratio (oxidation of water)/(decomposition of ZnO) is higher on the Zn surface [the (0001) surface] than the oxygen surface [the $(000\bar{1})$ surface]. Such an observation is not predicted at all by the thermodynamic theories, although such kinetic differences could be expected from the discussion above. Their model to describe the difference is based on the different hydroxylation to be expected on the two surfaces. On the Zn surface, OH^- groups will be expected to dominate as the adsorbate; on the O surface, H^+ groups will expected to dominate. Finally, Gerischer,[R21] noting that corrosion is expected to occur at defects preferentially (see below), discusses the possibility that holes will preferentially be captured at kink sites. He suggests a series of solvation steps in the removal of surface atoms at kink sites, that he believes will be characterized by donor levels always above the valence band edge, so that some ligand reorganization step will be rate limiting.

Some experimental evidence that the exact configuration around the surface atom is critical arises in observations with ligand substitution. It is well known that anions that form strong ligands, such as the chloride ion, are important in accelerating the corrosion of some metals, such as iron. Erbse *et al.*[74] find a similar effect with ZnO. Similarly, some chelating dye complexes have been shown[7] to accelerate the photocorrosion of ZnO. Others[75] attribute the accelerated decomposition of ZnO in the presence of dyes to the acidic nature of the dyes. Gerischer[8] discusses the influence of dissolved bromine in accelerating the rate of etching of GaAs, but it is not clear whether the bromine acts by hole injection or by providing energetic ligand formation.

7.4.2.3. The Role of Solid State Defects in Corrosion. The importance of steps and kinks in lowering the activation energy of corrosion has been long recognized.[R4,R6,R13] A step is the edge of a layer of atoms partly covering the surface. A kink is a point where a step changes direction. At a step, a surface atom is bonded to fewer atoms of the crystal than on a perfect surface plane, and hence fewer bonds need be broken for the atom to be removed. At a kink, specifically at an outside corner where the step changes direction, the surface atom has one less bond to the crystal than an atom at a straight step. So the kink site, with the outside corner, is particularly susceptible to corrosion. And, of course, it is clear that when an atom is removed from a kink site at least one more kink site is formed (Figure 7.4).

At the end of a dislocation pipe, an atom is found that has one less bond to the crystal than an atom at a perfect planar surface. The situation is sketched in Figure 7.5; the poorly coordinated atom where the extra half-plane in the bulk intersects the surface is indicated as atom *A*. Now when this atom is removed by corrosion, automatically a step is formed

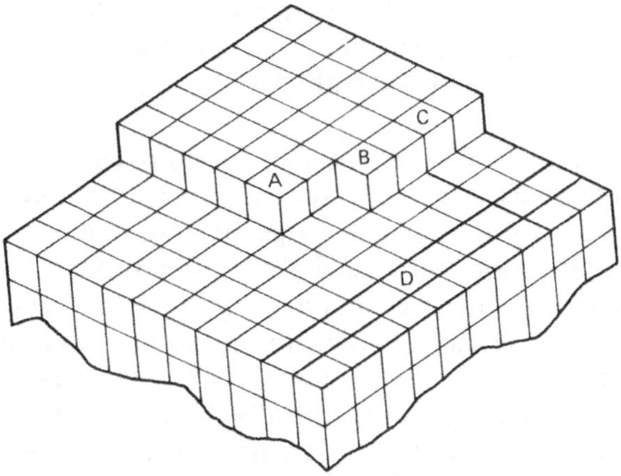

Figure 7.4. Steps and kinks on a surface plane. Atoms A and B are at kink sites. When atom A is removed, two kink sites are generated. When atom B is removed, one kink site is generated. Atom C is on a step, atom D on a terrace.

around the dislocation pipe. As the atoms are removed from the step by corrosion, the step simply expands in diameter. Eventually the step sweeps across the whole surface and the surface plane of atoms is removed. In the meantime another step is generated, centered at the dislocation. Thus the dislocation is expected to be dominant in corrosion

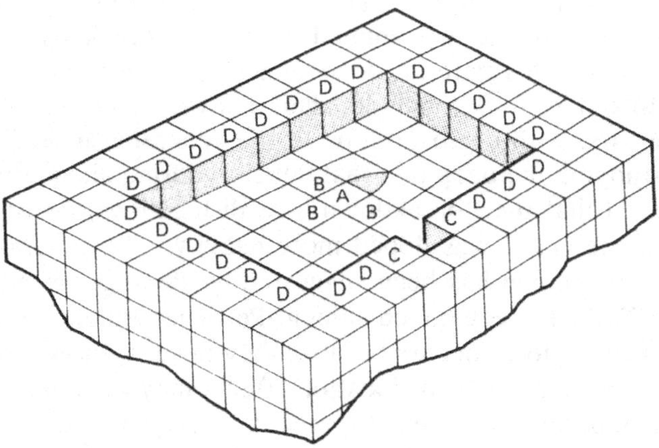

Figure 7.5. Generation of steps by dislocations. Atom A is the poorly coordinated atom at the end of the dislocation pipe that can easily be removed. When it is removed, then atoms B are all located on a step. An expanding step is shown in the top layer of atoms. All the D atoms are on a step, and one atom from the step is removed, forming two kink atoms, the C atoms.

because it provides a source of steps and kinks. In the above discussion we emphasized an edge dislocation; a screw dislocation will also be a source of steps and kinks by a slightly different argument that we will not detail.

In addition to providing active sites for corrosion, dislocations and other flaws in the bulk semiconductor can generate minority carriers, in breakdown processes or in space charge generation processes, that can in turn induce corrosion. Steps and kinks offer surface states that can also lead to minority carrier generation. Space charge generation arises because of Equation (1.5a) or (1.5b). In the space charge region the minority carrier density is below equilibrium if the carriers are consumed at the surface, so, in accordance with the discussion of Equation (1.5), any centers in this region will become generation centers. Of course, bulk flaws such as dislocations will also provide bulk recombination centers, which will remove minority carriers when there is an excess of minority carriers. Thus the influence of bulk defects on minority carrier density is complicated. As a first approximation, bulk defects will increase the minority carrier density in the dark, because they act as minority carrier generation sites, and will decrease the density of photoproduced minority carriers because they act as recombination sites.

The action of dislocations and other flaws (bulk recombination centers) in generating minority carriers for corrosion has been noted experimentally in several cases. For example, Wolkenberg[(76)] found agreement with the theory of space charge generation of carriers on silicon when studying the corrosion of the material with a reverse bias applied. Then there is the work of Faktor *et al.*,[(77)] who studied GaAs and found that when sufficient voltage was applied to induce "breakdown" currents at the dislocations in his material, deep tunnels were formed by the corrosion. Yamamoto and Yano[(78)] conclude, also, that with GaAs space charge generation occurs, with corrosion due to the generated holes. Tranchart *et al.*[(79)] in studying the deep pits that develop during anodic breakdown in GaAs, found that the "breakdown" (at dislocations) occurred at lower voltages if the GaAs was doped more heavily. The generation of holes by Zener breakdown has been suggested by Pettinger *et al.*[(73)] for ZnO, and Meek[(65)] for heavily doped silicon. Pettinger and co-workers suggest that Zener breakdown occurs because they observe current doubling and light emission. However, dislocation effects may be more likely in their studies also, because hole–electron recombination to produce light should not occur with Zener breakdown. Meek[(65)] suggests that the minority carriers leading to the corrosion of silicon must be generated at surface states, because the corrosion occurs with V_s at only 0.25 eV, much too low for any breakdown. However, the reason he rejected dislocation effects is not clear. Finally, we can refer again to the work of Morrison and

Freund,[80] who observed high corrosive minority carrier currents to the ZnO surface under low reverse bias (a few tenths of a volt) when dislocations were intentionally introduced.

As a final topic, we will describe a very clever technique by Rode *et al.*[81] to use breakdown-induced corrosion of GaAs to produce field effect transistors (FETs). The object is to prepare a very thin *n*-type layer over a semi-insulating substrate of GaAs. A relatively thick *n*-type layer is prepared by epitaxial growth. Then a voltage is applied anodically to induce breakdown of the *n*-type GaAs and produce holes. These holes migrating to the surface corrode the *n*-type material away. When the *n*-type layer is very thin, the current must flow through the semi-insulating substrate, and the voltage is dissipated in the substrate and cannot cause hole generation by breakdown in the epitaxial layer. Thus the breakdown-induced electrochemical etching quenches itself at about the right thickness of the surface layer for a good FET. A modest further etch can be made using optically produced holes.

7.4.3. Stabilizing Agents to Prevent Photocorrosion

In this section we emphasize photocorrosion, first qualitatively discussing the general principles and comparing the theory to that described earlier for photocatalysis, and then discussing the use[R45] of stabilizing agents to prevent photocorrosion of the electrode. In our dicussion we will assume *n*-type material, as the concepts are most often applied to this case, but the concepts are symmetrical for *p*-type material.

Photocorrosion is a problem in several areas of technology. The most important currently, and the reason why considerable effort is currently being applied to the problem, is in the area of solar energy conversion (Section 9.1) using electrochemical cells. The electrochemical solar cell can be designed to convert solar to electrical energy, in which case one introduces a redox couple in solution to capture the holes at the semiconductor and the electrons at the counterelectrode. Alternately, the cell can be designed to convert solar to chemical energy, in which case the holes should oxidize one species at the semiconductor and the electrons reduce another species either at the semiconductor or at the counterelectrode. In either case, corrosion of the semiconductor itself is obviously undesirable.

It is clear that in such solar cells the rate of consumption of photoproduced holes, and of photoproduced electrons in solution, must be equal. However, for the above configurations, the photoproduced electrons need not be consumed at the semiconductor surface, but may be consumed at the counterelectrode surface. Thus we are mostly interested in the fate of the corrosive photoproduced hole. In the discussion below we will, therefore, examine competitive hole reactions, but will not restrict the discussion to solar cells, but include simpler cases where a strong reverse bias is applied to the cell, such that the electrons cannot reach the surface at all.

Photoproduced holes that are attracted to the surface (by the surface barrier) must at steady state be consumed by some reaction. The reaction can be simple recombination with electrons at the surface, it can be oxidation of a species in solution, or, it can be decomposition of the crystal to form a soluble species or another phase. The first two cases of minority carrier capture were discussed in Sections 6.3 and 6.4, and the qualitative and semiquantitative models of Sections 6.3 and 6.4 are applicable to photocorrosion with moderate and fairly obvious revision. If there are no processes competitive to the corrosion, there must, of course, be 100% corrosion efficiency of the holes that reach the surface. This is to be expected with a completely inert electrolyte (nonoxidizable), and with sufficient anodic applied voltage to repel all electrons (say, qV_s is greater than 1 eV). However, if the corrosion leads to an insulating film, eventually the applied voltage will appear across the insulating film instead of across V_s, and the holes will be consumed by simple recombination (at interface states). If the corrosion potential ${}_pE_{\text{decomp}}$ is below E_{vs}, and there are no competitive processes, holes will accumulate in the space charge region until ${}_pE_{\text{decomp}}$ is at a higher energy than the surface holes, and then corrosion will start. But very often it is found[82,72] that with a modest anodic voltage, and a soluble corrosion product, corrosion occurs with 100% efficiency. Gerischer[R6] claimed, in 1968, experimental confirmation of 100% corrosion efficiency for all semiconductors used to that date with E_g greater than 1 eV.

For many years the use of additive redox couples to consume photoproduced carriers has been suggested as a means to stabilize a semiconductor against photolytic decomposition. Originally the technique was used to prevent decomposition in vacuum and photocatalytic oxidation of a resin,[83,84] but more recently it has been used to prevent decomposition in solution. The principle is the same in all cases. If photoproduced holes are the active species yielding photolysis, a reducing agent additive is introduced as the stabilizing agent that has a high capture cross section for holes, and has an energy level higher than the decomposition energy level of the semiconductor. Then the holes are consumed by the additive and in principle should not decompose the crystal so rapidly. If the additive is not meant to be sacrificial, then the oxidized form of the additive should be able to capture electrons and restore the reducing agent. Under these conditions the additive acts as a recombination center, according to Equations (6.28) and (6.29).

An additive will have advantages and disadvantages in competing with the corrosion process for holes. Its disadvantage is usually the lack of "intimate contact." The corrosion process, being the breaking of valence bonds, may be considered to be a process of localizing the hole on one atom rather than capturing the hole, and so many have a higher capture cross section than an atom in solution, or even one specifically adsorbed. The additive's advantage is that it is usually chosen to have a one-electron reaction, whereas usually several carriers must be localized on one surface atom for corrosion.

Thus we can suggest the rate of oxidation of the stabilizing agent should be first order in carrier concentration, whereas the corrosion rate should vary as a higher order in carrier concentration. Thus, for example, it has been observed[85,86] that the ratio of photocorrosion to total current flow in the presence of a stabilizing agent increases with light intensity. Kobayashi *et al.*[87] find that with ZnO the ratio increases with light intensity with I^- as the stabilizing agent, but decreases with Br^- as the stabilizing agent. They suggest that the latter can be explained if the quasi-Fermi-energy is considered a thermodynamic free energy, but this explanation does not seem valid, as discussed in Section 7.4.2.1.

The introduction of reducing agents into the solution to compete with, and prevent corrosion, is relatively straightforward. In Section 6.3.2 it was indicated that, contrary to the simplest fluctuating energy level model, hole capture cross sections increased the higher the energy level. In fact, the intrinsic surface states involved in corrosion were suggested as the reason for this behavior. Then, with one-equivalent species the higher the level, or in other words the more negative the value of E^o, the more effective the electrolyte additive in competing for hole capture. Of course, in principle, the energy level of the additive reducing agent should not be so high that electron injection into the conduction band of the semiconductor is a problem. And naturally, one expects no appreciable hole capture from a species with E^o_{redox} below E_{vs}.

Semiconductors with a high E_{vs} require a reducing agent with a high E^o_{redox}. Thus Memming[88] suggests Eu^{2+} ($E^o = -0.43$ V) to protect GaAs ($E_{vs} \cong 0$). Madou *et al.*[63] find ferrocyanide ($E^o_{redox} \approx -0.4$ eV) quite effective for GaP ($E_{vs} = +1$ eV). Madou *et al.*[64] find FeII(EDTA) effective with GaAs. Zinc oxide was shown in early investigations[89] to be protected against decomposition by many species; the most effective found at that time was iodide ions, although later[R4] ferrocyanide was shown to have a higher σ. Even TiO_2 can corrode to some extent, in the presence of sulfate ions, and Harris *et al.*[67] have shown that Co^{+2} will offer protection. Wrighton and co-workers[R45] have introduced a stabilizing agent with the "chemically modified electrode" technique, depositing ferrocene on silicon as discussed in Section 7.3.3.

CdS, CdSe, and CdTe have been studied extensively with solar energy conversion in mind, and it has been found that several ions offer protection. Nakatani *et al.*[90] provide a long list of references to studies of this system in an aqueous solution, and they themselves discuss the protection offered by iodide ions in nonaqueous systems. Memming,[91] studying CdS in an aqueous system, finds that ferrocyanide, aquo iron (II), and Ce (III) all offer some protection. As Memming points out, by the model of Figure 7.3, ${}_pE_{decomp}$ is calculated to be substantially above E^o_{redox} for aquo iron and cerous ions. However, it must be recalled that the model of Figure 7.3 is thermodynamic, not kinetic, and if the reducing agents have a very high capture cross section, and if (as is the case) their energy level is above E_{vs}, there is nothing inconsistent about this result. It only serves as a reminder that the model of Figure 7.3 is only to be considered, at best, as an indication of the probable reactions, and kinetic factors may dominate the selection of the actual reaction that dominates.

It should be noted that the position of E^o_{redox} relative to the band edges can be higher (so the stabilizing agent can be more effective) if a surface film is present. Thus it has been shown,[R74] for example, that a thin oxide layer on n-type silicon leads to stabilization with reducing agents of E^o_{redox} below E_{vs}.

Apparent protection against corrosion does not always mean effective protection. Two cases can be cited. First there is the observation with CdS, CdSe, and CdTe discussed at the end of Section 7.3, where there is apparent corrosion protection, but in actuality an amorphous layer is forming at the surface. Second, there is the example of protection of ZnO by current-doubling species, as discussed in Section 6.3.1.1, where it was clear that the holes from the ZnO were oxidizing the current-doubling agent (because the current increased and the production of H_2O_2 ceased) but, as described at the end of Section 6.3.1.1, it has been found by Fujishima that the ZnO goes into solution even while the current-doubling agent is being oxidized. Thus in this case the additive apparently does not prevent the corrosion: instead of reacting with the holes, it reacts with the corrosion product (H_2O_2).

8

Coated Electrodes

8.1. Introduction

In this chapter we explore the use of the energy level model to describe electron transfer through a coated electrode, emphasizing oxidized electrodes, where the electrode can be a metal or a semiconductor. Oxide layers on a metal electrode have not commonly been described in the literature by the band model. The reason is probably the highly defective nature of the solid, making quantitative analysis of the conduction process difficult. However, there are many cases where the band model has provided a useful description[R34] and these will be discussed. On the other hand oxide layers on semiconductors, silicon in particular, have been described in detail by the band model, and a reasonable band model for oxide layers, in general, can be extrapolated from the silicon analysis. In addition to the input from those electrochemists who describe their passivated electrode with the help of the band model, and the input from physicists interested in the passivating oxide on silicon, there are other sources where pertinent models and concepts have been generated. One important source of background concepts is that of studies of amorphous semiconductors. Many, if not most coatings of interest, particularly coherent oxide coatings on metals and semiconductors, are amorphous. A coherent film on a metal or on a semiconductor will often be amorphous because, in general, only an amorphous film can grow thick enough to passivate the substrate without being so strained that it breaks up. Thus we are interested primarily, but not exclusively, in amorphous films—amorphous passivating oxide layers, as grown or deposited on a metallic or semiconductor substrate. A second source of concepts is a system that has been studied reasonably quantitatively (at least in terms of theoretical analyses), namely, very thin oxide layers, where tunneling of electrons is the dominant electron transfer mechanism. A third important source of information stems from the large background of material that describes trapping and conductance in semiconductor films.[R1] These studies represent an introduction to the complications stemming from the defects usually found in an oxide film on a metal or semiconductor.

It will be concluded in this chapter that the band description, as developed in the preceding chapters, can be applied at least qualitatively to improve our understanding of electron transfer through a covering oxide layer between a conducting electrode and an electrolyte. However, it will also be concluded that usually one must look to models of defects in semiconductors rather than the simple theories for the best description of the electronic behavior of oxide layers. Vijh[R34] distinguishes 11 ways that oxide layers can become conductive, mostly associated with defects in the oxide. Thus in Section 8.1 we begin with a discussion of the defects found in oxide layers. In Section 8.2, where the actual transport of electronic current through such layers is discussed, the behavior of three forms will be distinguished: first, a thin oxide layer, thin enough for electron tunneling; second, a semiconductor oxide, where in principle simple semiconductor theory could apply, but where it will be concluded that generally defects dominate; and third, an insulator oxide, where if there is electronic conductance, it must be through the assistance of defects in the layer, either gross flaws or mobile ions.

In this section the various features of the band model that must be considered in analyzing the conductivity are described. Section 8.2 utilizes the models developed here to describe conductance mechanisms and to compare theory with observation.

8.1.1. The Band Model for Oxide Films

The possible energy levels for amorphous oxide films are shown in Figure 8.1. The figure shows normal band models for a metal or a semiconductor in contact with the oxide. The band model for the amorphous film[R2] and for the interface of the film with the semiconductor are fairly complex, for there are many sources of energy levels.

In the amorphous film there are allowed energy levels corresponding to the orbitals of the valence electrons on the anion and to the orbitals of the next excited state, when the electron is essentially on the cation. Thus there are "bands" more or less equivalent to the valence and conduction bands of a crystalline semiconductor. However, as each lattice site in an amorphous film is not identical with respect to its nearest-neighbor distance and orientation, the energy levels are not necessarily identical in the sense of the Pauli exclusion principle. Most of the orbitals are approximately the same, but there will always be a few that have somewhat unique bonding configurations. Some of these unique orbitals will provide the levels A and E. The energy levels A represent cation levels,

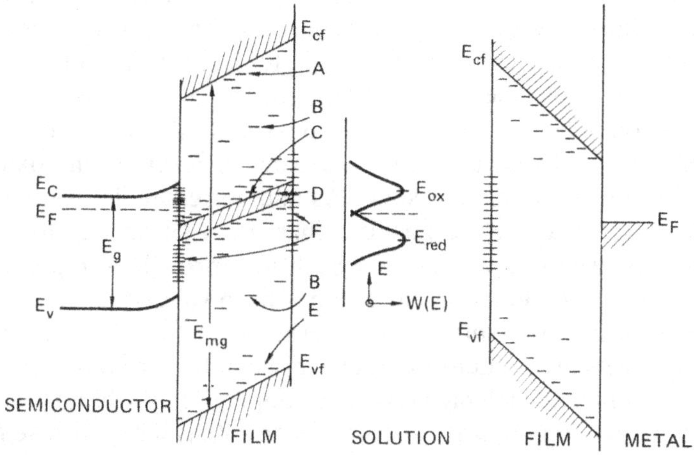

Figure 8.1. The band model for an amorphous or highly defective semiconducting or insulating layer over a semiconductor or a metal. The left side of the figure shows the normal band model for a semiconductor, with a surface film illustrating the various localized levels, A through F, that are discussed in the text. On the right is shown the normal band model for a metal with an amorphous surface film with no bulk defects (although surface states at the solution/oxide interface are shown).

almost, but not quite the same as the "conduction band" levels, that have a bonding configuration resulting in a somewhat higher electron affinity than most of the cations have. The levels E represent those few valence electron orbitals where there is a lower ionization potential than most of the anions have. The energy E_{mg} is effectively the energy gap, but is called a "mobility gap" that represents upper and lower energies beyond which carriers can move. Conduction band electrons at an energy higher than the top edge of the mobility gap, E_{cf}, can move. They are in a region of energy where there are enough "identical" orbitals that there is overlap between levels, and the electron can hop from one end of the film to the other through these identical orbitals. Holes with an energy level below the bottom edge of the mobility gap, E_{vf}, can also conduct: they are in a region of energy where the orbitals overlap sufficiently for transport of the holes. Returning to our discussion of the energy levels A and E, we repeat that they are almost the same as the conduction band and valence band energy levels, respectively, but are rare enough that if an electron becomes trapped in an energy level A, or a hole in an energy level E, they cannot easily hop from one site to the next across the film.

The levels D are included to reflect the possibility that an impurity band, or a narrow band associated with another oxidation state of one of the ions of the lattice, could be present with an energy level density high enough for conductance. If the density of impurities, or of ions in a different oxidation state, is, say, the order of 10^{26} m^{-3}, or 1%, then there could easily be enough overlap of orbitals for carriers to hop from one level to the next across the film. For example, in later discussions of the anodic oxidation of GaAs, we review the possibility that the arsenic could be incompletely oxidized, and lead to a band of levels in the oxide associated with the As^0 oxidation state. As another example, we will later discuss the case of attractive positive charges in silica, due to sodium ions or protons, where in low density such ions provide a trapping level at 2.4 eV below the conduction band. It has been suggested[1] that if such ions are present in sufficiently high density, say, 1 to 10% of the atoms, the trapping levels are so dense that carriers can hop from one center to the next. Of course, at such high levels of "doping" it is arbitrary whether we call the additive an impurity or part of a new phase, but such semantic problems do not change the concept.

Again in the case of such an impurity band, there will be localized levels above and below the band associated with atoms that do not have a typical bonding structure and so have a different energy level from the norm. The case is analogous to that of levels *A* and *E*, and again electrons or holes in these levels do not have wave functions that overlap, and so cannot hop from one level to the next across the film. These levels are indicated as *C* in the diagram.

Then there are impurities or imperfections in the film that are relatively rare (say, less than 10^{24} m^{-3}) that provide traps in the material in the same sense that the word was used to describe normal semiconductors and insulators in Section 1.1.1.4. The possibility (or actually the probability) of such levels is included in Figure 8.1, with the levels designated by *B*.

This lists the various types of levels often present in the "bulk" of the amorphous film. In Section 8.2.2 we will discuss the influence of these bulk levels on electronic conductance through the film. It will be pointed out that two processes are necessary for conduction in such a "perfect" amorphous film. First, carriers must be injected into one of the "bands," and such injection is not necessarily a simple step—for example, a high activation energy is required to inject an electron into the E_{cf} band, or a hole into the E_{vf} band of the film in Figure 8.1. Second, the carriers must be transported across the film, and this also may be difficult. Trapping of charge will block such carrier motion.

We now return to the discussion of Figure 8.1 and consider states at the interface between the semiconductor and the film and between the film and the solution. These states are designated by *F* in Figure 8.1. Surface states between the oxide and the solution are expected to behave similarly to surface states at any semiconductor/solution interface, and are usually expected to be dominated (in terms of ability to develop a charge) by adsorbed ions such as H^+ and OH^-. Interface states between the semiconductor and the oxide are much more important. They may arise because of mismatch between the semiconductor and the film that leaves dangling bonds or other nonbonding orbitals. They may also arise because of nonstoichiometry of the oxide in a transition layer between the conductor and its oxide. They may also arise because of impurities that become localized at the interface. In any case, they are undoubtedly present. In the Si/SiO_2 interface they can be as low in density as 10^{14} m^{-2} eV^{-1}, but that is by far the most perfect interface known, and more commonly densities (e.g., at the GaAs/oxide interface) the order of 10^{16}–10^{17} m^{-2} eV^{-1} can be expected (Section 8.2.3).

Many properties of the semiconductor/oxide/solution system can be dominated by such interface states. First, if they are present in sufficient density, interface states will pin the Fermi energy, with all that that implies. It implies, for example, that a capacity/voltage measurement will show a constant capacity for the semiconductor space charge region (Section 5.2.2.1 and Figure 5.2). It implies that any change in electrode potential will appear across the oxide layer, assuming the oxide is thicker than the Helmholtz double layer, and so the current will be dominated by conductance through the oxide. It implies, finally, that breakdown will also be controlled by the oxide, for the high applied voltages are not able to appear across the semiconductor space charge region. Second, the states, if they are there in sufficient density, provide another important location for charge storage. As in the Helmholtz case discussed in Section 2.2.3, the charge is stored in a planar array, so the double-layer voltage drop between this plane and, say, the solution appears in our formulas, just as does the Helmholtz voltage drop. However, if the oxide is of any reasonable thickness, the voltage drop will be very high with very little charge stored. Thus, for example, it has been suggested[2] that the Helmholtz double layer voltage on GaAs must be the order of 1.2 V. It seems much more likely that such a large voltage appears across the double layer of a thin oxide film rather that being associated with a Helmholtz double layer. Thus in this case (and it can be expected with many oxidizable semiconductors), the presence of an oxide layer and the (almost) inevitable high density of interface states will lead to very high apparent 'Helmholtz potentials" as determined by measurements of V_{fb}.

We will describe one example of the band model of an oxide film, specifically the SiO_2 film over Si. Undoubtedly this is the most studied oxide film from an electronic point of view. Its band diagram has been firmly established. As we will be discussing this material as an example of the possible defects in a film, and as an example for electronic conductance through a film, we will discuss its band diagram here as an example of band models.

Thermally grown SiO_2 is amorphous, it is an excellent insulator (at 1000 Å or so thickness), and it is of exceptional commercial importance because it is the main insulator used in integrated circuits and transistors. Thus extensive studies have been made of the position of the conduction and valence bands, the normal trapping levels, the normal space charges found in the oxide, and the interface states between the Si and the SiO_2. Also, as will be found very important in analyzing its electronic behavior, extensive studies have been made of ion movement in the material. Very few of these studies have been done using electrochemical techniques; most have been done with an evaporated metal layer replacing the electrolyte. A metal–oxide–semiconductor (MOS) configuration is extremely versatile in analyzing the electronic properties of the oxide film. An electrolyte–oxide–semiconductor (EOS) configuration is different, primarily in that the solution provides an infininte source of ions that can move from the electrolyte to the oxide. The MOS configuration has provided a detailed band model to use as background for interpretation of EOS studies of film conductance as is described in Section 8.2.

A band diagram showing the relative energy of the bands of SiO_2 to those of Si is shown in Figure 8.2. The experimental determination of the energy levels is summarized in the review by DiMaria.[(R49)] We will describe the various parameters, and how they are determined, but very briefly. The positions of the conduction band and valence band of SiO_2 are determined most directly by optically exciting electrons, from the silicon or the metal into the conduction band of SiO_2, and observing the wavelength required for the resulting current to flow through the silica. Such optical excitation into the conduction band is used to study electron trapping in the SiO_2. Also, using p-type silicon,[(R49)] hot electrons produced by avalanche breakdown can be made to enter the SiO_2 conduction band, permitting studies[(3)] of traps in the SiO_2. Hole injection into the valence band of the SiO_2 is accomplished by optically exciting hole–electron pairs in the SiO_2 with 10-eV illumination. By measuring the transit times, the electron and hole mobilities can be measured, and so forth.

We are here interested not only in the bands, but in the traps in the SiO_2. One in particular, the so-called "2.4-eV trap," which lies 2.4 eV below the SiO_2 conduction band, is observed by measurements of photoexcitation of the carriers out of the traps. These traps, as summarized by DiMaria,[(R49)] have an exceptionally high capture cross section for electrons (up to 10^{-16} m^2 at low electric fields). The traps are related to positive ions in the $SiO\text{-}_2$, in particular Na^+ ions. Sodium also provides another, deeper trap.

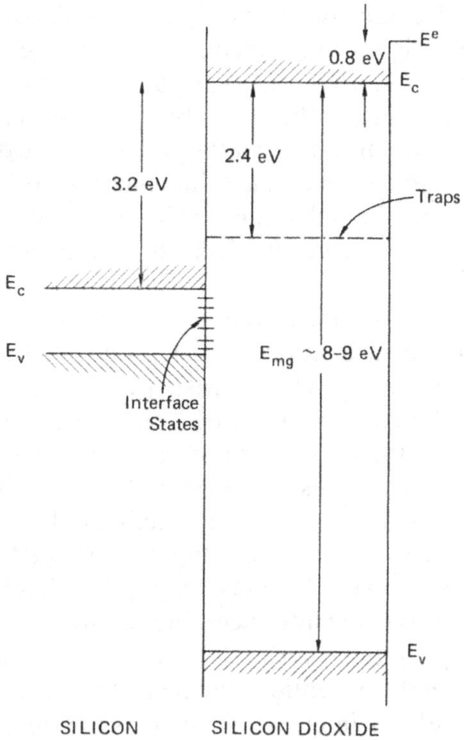

Figure 8.2. Band model of the Si/SiO_2 interface.

In addition to the attractive centers associated with positive impurities like sodium, there are neutral and repulsive traps. Some may be related to the incorporation of water into the SiO_2, although their origin is less clear than the origin of the 2.4-eV traps. Hole trapping occurs when holes are formed by uv or x-ray excitation of the SiO_2. Hole trapping is particularly effective near the Si/SiO_2 interface.[4] The traps may be associated with broken bonds or excess silicon near the interface. Finally, as a special case, traps introduced by ion implantation have been extensively surveyed. Tungsten, aluminum, arsenic, phosphorus, and boron all result in electron traps. References to individual studies of these trapping centers are found in the review article by DiMaria.[R49]

Nicollian *et al.*[5] described experiments where in water is diffused into the SiO_2 at temperatures less than 200°C. Water provides electron traps, able to capture electrons that are injected into the SiO_2 conduction band. When the water traps the electrons, apparently the protons from the water are reduced, for the neutral hydrogen is found to diffuse back to the surface and escape at room temperature.

Another center in the oxide is the positively charged Q_{ss} center, a permanently charged center in the SiO_2 near the silicon interface. The origin of this center (origin from a chemical point of view) is not known.

To continue our discussion of Figure 8.2, we note the presence of the so-called interface states—the states between the semiconductor silicon and the silicon oxide. Interface states probably arise dominantly from dangling bonds at the interface that are dangling because of poor registry between the oxide and the substrate.[6] The unpaired electrons on the dangling bonds can be detected by electron spin resonance, as investigated by Caplan *et al.*[7,8]

The interface states are measured usually by capacity/voltage characteristics as described in Section 5.2.2.1. Other methods of studying them include deep level transient spectroscopy (DLTS), as developed by Lang,[9] and used effectively by Johnson[10] on the Si/SiO_2 interface. At metal/oxide interfaces, the importance of interface states may well lie in their presence a few angstroms away from the actual interface, where they can become charged and provide a significant double-layer voltage. As interface states are in a highly defective region, the first layers of the oxide, they can accommodate probably a very high charge. For example, as will be discussed below, the interface states at the GaAs/oxide interface has trap densities up to the order of 10^{18} m^{-2} eV^{-1}. Such densities will be the rule rather than the exception; silicon is the exception with its very low (10^{14}–10^{15} m^{-2} eV^{-1}) interface state densities for a well-prepared oxide.

Interface states, especially on silicon, are reviewed in detail by Goetzberger *et al.*[11]

The above is intended as an introduction to the variability of electrically active defects in SiO_2 to illustrate how electronically complicated the material is. The discussion, hopefully, will give a reasonable background for the description in Section 8.2 of the silicon/silicon oxide/electrolyte measurements. Even more important, it will illustrate how defective, electronically speaking, oxide layers on conductors can be, even when the oxide is "perfect" in the sense of being a coherent, amorphous, uniform film on the surface of the conductor.

8.1.2. Thin Films

In this section, we will discuss types of very thin films in slightly more detail than was done in the introductory section. By very thin films we mean films of a few angstroms in thickness. The defect structure of these is less important than that of thick films, for tunneling directly through the film is expected to be the dominant carrier transfer mechanism. However, other features of the films, such as their double-layer voltage and their influence on the catalytic activity of the surface, are very important in defining electrode reactions.

Film growth is initiated, in general, as clusters at the surface—the clusters forming at particularly favorable centers, then expanding in circumference until they cover the surface. For example, Szpak[12] showed that an induction period during the anodization of GaAs is associated with nucleation, and at constant current the voltage starts to rise over 4 V only when the surface is completely covered. However, the important point from a "thin film" point of view is whether, after nucleation, the film grows laterally to form a monolayer of the second phase, or whether the film grows perpendicular to the surface to form crystallites of the new material.

With the oxidation of metals the film can, in general, be expected to grow laterally, because where the oxide is formed becomes a region of higher resistance, so the current flows more easily to the bare spots on the metal. On a semiconductor electrode, the deposition of reaction products may be either in the form of crystallites or in the form of uniform layers. Bindra *et al.*[13] have proposed a model for the growth of metal films on semiconductors that relates the growth habit with the work function. In the present section we are not very interested in cathodic deposition of metal layers on semiconductors, but the referenced approach to the problem is of interest and some features may be carried over to oxide films on semiconductors. They suggest that the form of the nucleation, whether monolayer or crystallite, depends on the difference in work function between the metal and the semiconductor. As the metal deposits, according to the model, the metal acts as an acceptor surface state if it has a higher work function than the semiconductor. As with any acceptor surface state, V_s, the surface barrier, can become high. With an n-type semiconductor a high V_s leads, according to Equation (2.10), to a low density of electrons available to reduce more of the metal. Thus the metal tends to deposit elsewhere than where the metal is already present—the metal tends to form a uniform layer. On the other hand, if the metal has a low work function—lower than the semiconductor—it tends to inject electrons into the semiconductor as the metallic phase forms. Then electrons are available to reduce more metal ions, and the metal layer grows large deposits where it is first formed—it forms nucleation centers. These concepts may possibly be extended to insulating films on semiconductors, but the case of insulating films may be complicated by interface states.

Oxide films that are clearly thin enough for tunneling are not common. The most important of such cases is the extremely thin films on noble metals, when used as electrodes. Even on the most noble metals, such oxide films do form, induced by the high fields at the surface when a moderate overvoltage is applied.[14] On all oxygen electrodes (especially electrodes used for the oxidation of water to evolve oxygen gas) at least a monolayer and sometimes several monolayers of oxide are formed.[R9]

Oxide layers on noble metals have a significant effect on the overvoltage for anodic evolution of oxygen. The effect can be due to the need to activate carriers over potential barriers with moderately thick films,[15] due to the resulting dipole formed because the oxygen on the surface results in a negative surface charge, or due to the removal of surface sites for adsorption, where catalytic effects are needed for a kinetically active reaction. Adsorption and catalytic behavior may have a strong effect on the exchange current. For example, Erdey-Gruz[R9] suggests that with a platinum electrode in a nitric acid solution the oxidation passivates the electrode but makes it a good catalyst for the decomposition of H_2O_2. Thus in this case the oxidation changes the rate-limiting step of the oxygen evolution. It is no longer peroxide decomposition, but electron capture by the oxygen molecule to form O_2^-. As another example we have the study of Faroque and Fahidy,[16] who studied the oxidation of H_2S on platinum. In this case it is found that if a monolayer of sulfur is formed it blocks surface sites, preventing the H_2S adsorption, and so in this case the layer causes a decrease in rate.

Thus, although relative to the thicker oxides discussed below, very thin oxides have a negligible effect on the electron transfer rate to species in solution, they do have three possible very important effects: first, the resulting double-layer potential shifts the pzc; second, the catalytic activity of the surface changes; and third, the density of adsorption sites is lowered. In Section 8.2.1 below, where we discuss current flow through thin oxides, we will be emphasizing tunneling as a rate-limiting step, and thus will be emphasizing the transition between the thin oxide and the thick oxide regimes. However, the double-layer and catalytic effects may be important at all oxide thicknesses.

Other forms of very thin layers are the chemically modified electrode and the deposited monomolecular layer. A chemically modified electrode is one with a chlorosilane or an equivalent molecule bonded to the surface, where the molecule is chosen to have a useful functional group. These layers were discussed in Section 7.3.3. In this case, clearly the primary function of the thin film is to change the chemical activity of the surface. A deposited monolayer, is one that is lifted onto the surface. The deposited monolayer is illustrated by Arden and Fromherz' study[17] discussed in Section 6.5.3. The use of the deposited monolayer provides an interesting approach to the study of the electrochemistry of thin layers that does not seem to have been emphasized. In this case the primary function of the film is probably to block adsorption sites, when a reasonably inert film is used.

8.1.3. The Structure of Thick Films

In this section we briefly discuss anodic oxide growth, then discuss the structure, emphasizing defects, of oxide layers in general.

Thick oxide films on semiconductors or metals are usually prepared by thermal oxidation or anodic oxidation. Although in principle the mechanism of growth of anodic oxides could be considered within the scope of the present work, only a very brief background discussion will be given. The reasons for this neglect are twofold: first there are excellent reviews available, for examply by Young[(R7)] and by Vermilyea[(R50)]; and second, the emphasis in this book has been on electron transfer mechanisms, and not on ion diffusion except where it affects the electronic properties of the solid.

Nucleation of an anodic film occurs when sufficient overvoltage (over the theoretically calculated open circuit potential for an equilibrium oxidation reaction) is applied. The nucleation occurs, as discussed above, at preferred sites, and the oxide broadens in area and then increases in thickness. If the growing film is noncoherent, breaking up as it grows, then the voltage necessary to continue the growth is the same as the initial voltage. If the growing oxide is coherent, forming a continuous film, then the voltage required to continue the growth increases with the thickness of the film.

During the growth of a thick coherent film, ions move between the solution and the substrate. In later sections we will be discussing electron movement in the same oxide films, and indeed during anodization electrons also can move. But usually, when discussing electronic conductance below, we will be considering electric fields almost an order of magnitude lower than those needed for anodizing. During the anodic growth of the oxide, typically the electric fields applied are the order of 5×10^8 V/m. Now if the activation energy for ion movement from a site to an adjacent site is E_0, and the sites are separated by a distance a, the activation energy under the applied field will become $E_0 - aq\, dV/dx$. Even if a is only the order of the distance between atoms in the solid, such high fields have a substantial effect. With a jump distance a of 3 Å, the activation energy becomes $(E_0 - 0.15)$ eV. Thus the rate varies exponentially with the field dV/dx.

The Cabrera and Mott[(18)] formulation for the rate of oxidation uses the above description to estimate the activation energy for motion of the ions, introduces an activation energy for the injection of metal ions into the oxide layer, and equates the two rates to determine the density of ions in the oxide at the interface. Further complications must be introduced into this picture if the moving ions represent a space charge in the oxide, because then the electric field through the oxide is not uniform.

As described above, we will not derive these fundamental relations, referring the reader to reviews of anodic oxidation. The main point of discussing the motion of ions is to be able qualitatively to relate it to the motion of electrons or holes through the oxide in later discussions, and the above qualitative description suffices.

The structure of oxide films varies from amorphous to polycrystalline, from hydrous to anhydrous. The composition will often change between the layers nearest the metal (or semiconductor) substrate and the layers nearest the outer surface of the film. And, of course, as mentioned earlier, there are many different types of flaws in the material. In the discussion below we will describe a few oxides that illustrate the possible complexity of the structure.

Iron oxide, for example, is amorphous and coherent for the first few layers,[19] than becomes porous. It is basically a gel-like structure, concluded by Revie *et al.*[19] to be hydrated Fe_2O_3. Others, for example Ord and DeSmet,[20] conclude the inner layers are Fe_3O_4, and only the outer layers are further oxidized to Fe_2O_3. It is a very complicated film.

The anodic oxidation of aluminum has been extensively studied and considerable is known about its structure as a function of oxidation parameters. Diggle *et al.*[R51] have reviewed the anodic oxidation of aluminum, offering more details than can be presented here. In some systems, such as sulfuric acid (a common electrolyte for commercial applications), an outer porous layer is formed, with a thin coherent "barrier" layer near the aluminum. In other electrolytes, such as citric acid, only the coherent barrier layer is formed, and it rapidly becomes thick enough to stop the oxidation. A recent review distinguishing the barrier layer on one hand and the porous layer on the other hand is given by Dell'Oca and Fleming.[R52] The porous layer is of great commercial interest in that it can be impregnated with dye molecules and sealed (by a high-temperature treatment) for colored protective coatings. For example, Cohen *et al.*[21] have determined that the tin dopant in anodized aluminum that gives a gold color is metallic, of zero oxidation number. Shimura,[22] in measurements of the anodic oxidation of aluminum in maleic acid, concludes the black color at high maleic acid concentrations arises from maleate ions coordinated to the Al^{3+} in complicated ways. Thus aluminum anodization can be cited also as an example of the incorporation of ions from the solution during growth.

The variation of composition in anodic oxides with distance from the metal surface is well documented; a case studied extensively is the anodic oxide on GaAs. At least two such features have been discussed in the literature. First, anodic oxides on gallium arsenide have given indications (in electrical measurements) of a "metamorphic" layer between the GaAs and the covering oxide, where the structure is different from either of the other phases.[23] Second, extensive Auger and radiotracer studies on the material indicate that beyond this metamorphic layer (if it exists) the composition of the oxide varies from the interface to the surface. The Ga:As ratio depends on the exact oxidation conditions, and indeed depends on heat treatment conditions following anodization.[24–26] Actually the variable composition is not unexpected, for both Ga and As are being oxidized in the anodic process, but arsenic oxides are quite soluble and (during heat treatment) quite volatile, so would tend to escape. On the other hand the gallium is much more electropositive than arsenic, and indeed there is evidence both from electrical measurements (to be discussed in Section 8.2.4) and from ESCA measurements, that a nonnegligible fraction of the As in the oxide is incompletely oxidized. Thus it is not always found that the oxide is gallium rich; the stoichiometry can be off in any direction.

Undoubtedly the most studied oxide (excepting perhaps alumina), with respect to passivating layers, is silica over silicon, because it is so important in semiconductor technology. In this case most of the work has been done on thermally grown oxide layers, and we will discuss mainly these layers, although some studies of silicon anodization are available in the literature.[27] In our review of thermally oxidized silicon we will first discuss laminar defect distributions, where the defects are in layers parallel to the surface, then we will discuss point defects, and finally we will discuss laterally distributed imperfections.

There are layers of defects both at the SiO_2/solution interface and at the Si/SiO_2 interface. The exact nature of the transition in stoichiometry from Si to SiO_2 at the Si/SiO_2 interface is not completely clarified, because the layer is apparently less than 20 Å or so in thickness. On the other hand, transition layers at an oxide/solution interface are common and the SiO_2/solution interface is no exception. Dignam[28] analyzes such cases. He is particularly interested in H^+ and OH^- from solution penetrating a short distance into an oxide. In the case of SiO_2 this definitely occurs—when exposed to aqueous solution the outer surface of the silicon becomes hydrated. The depth of the penetration by large concentrations of water seems small. Although some protons permeate the oxide, a real hydrous oxide layer, while present, penetrates silica very slowly. For example, Lansford [29] used the depth of penetration of the hydrous layer into glass to date archeological finds.

Point defects due to impurities and imperfections are found in the silica structure. In particular, one flaw known as the E' center is observed that is due to a dangling bond on the silicon ion. It is found particularly on silica subjected to nuclear radiation, both in glass and in oxidized silicon.[30]

Many different impurities can be introduced into silica when formed as glass, but only a few of these are of great importance in passive oxide layers on silicon. The most important are sodium, potassium, protons, and perhaps fluoride and chloride ions and alkaline earth cations. We will emphasize studies of sodium. Alkali ions in silica are known in the glass literature[R53] as network modifiers. Their action is to convert bridging oxygen ions (i.e., $\geq$Si—O—Si$\leq$) that connect two silicon ions in the network, to nonbridging (i.e., $\geq$Si—O—Na). If half the oxygen ions in the material become nonbridging, the material crystallizes, rather than forming amorphous glass, because the silica tetrahedra that form the glass network become too broken up. Sodium ions (and probably protons) are particularly important in the behavior of passive layers of oxide over silicon because of their high ionic mobility. Apparently there are two forms of the sodium in oxide layers, one highly mobile and one immobile. The work of Woods and Williams,[31] who deposited NaCl on the surface of an oxide layer and then electrostatically charged the oxide layer to drive the ions in, found a mobility for the rapidly diffusing ions estimated at 4×10^{-17} m^2/V at room temperature. This is the highly mobile form. Then with a layer the order of 1000 Å thick and applied voltage of 50 V, the ions can cross in the order of 5 sec. Such a high mobility of ions can have a dominating effect on the electrical properties, as discussed in Section 8.2.2. Woods and Williams' experiment is unique in that it applies (a) to movement from the surface to the Si/SiO_2 interface, and (b) without a metal overlayer. When either of these conditions is not met, the ions become trapped at flaws at the interface.[32–34] Then the motion of the sodium ions from the one interface to the other is limited by the rate at which the ions can be ejected from their traps at the interface region into the "bulk" of the oxide film. Normally a plot of concentration vs. position of sodium ions for a metal/oxide/semiconductor (MOS) structure will be U-shaped, with a minimum in the bulk oxide. It is not yet known how common this tendency for interface trapping of ions will be, for oxides that are formed on metals and semiconductors, but it may be common.[32] On silicon MOS structures, interface trapping of the impurities under aluminum metallization is an important technique for gettering the SiO_2 layer. The aluminum metallization is needed anyway for electrical connections and parts of active components of the circuit. Such gettering is very important in obtaining sodium-free oxide layers for integrated circuits.

The defects in thermally grown SiO_2 are not uniform laterally across the surface. This could, of course be expected; an oxide layer will presumably have weaker spots here and there, weaker in the sense that more flaws or defects are present than on the average.

Some of the flaws in the silica that show lateral inhomogeneity are associated with recrystallization of the layer. The silica is by nature amorphous, but tends to crystallize when impurities are present[35] and when heated in the presence of these impurities,[36,37] or in vacuum.[1] The crystallization, however, is local, and must be considered a defect, for, as will be discussed later, the boundary between the crystalline regions and the amorphous regions provides conductive paths through the oxide under certain conditions. Such paths may be related to the "channels" proposed by Revesz.[35] They are also perhaps responsible for the permeability of the oxide films to gases, as can be observed with the substrate silicon etched away.[38]

Studies by Madou *et al.*[1] of thermally grown silica illustrate substantial differences in the properties of the silica. The technique was simply to apply a cathodic voltage with the Si/SiO_2 sample used as an electrode. After a period of time consistent with the ability of Na^+ ions to move across the oxide under the electric field, current begins to flow and hydrogen is evolved at the surface. But the hydrogen evolution is not uniform. It tends to concentrate at points on the surface. Inspection of these points after moderate hydrogen evolution shows a slightly different optical density than the rest of the surface, and a slightly thicker layer, suggesting recrystallization. It is concluded that the "channels" of the original layer allow rapid in-diffusion of the ions into the material, and the presence of the ions leads to recrystallization. In more recent studies, it has been shown that mechanical damage to the silica surface, or impurities grown into the silica layer, generate such weak spots in the thermally grown oxide.

Zakzouk *et al.*[39] developed an interesting method to study the nonuniform distribution of defects on an SiO_2 layer, using a liquid crystal technique. Where the flaws were present, a voltage applied across the oxide and a thin layer of liquid crystal was found to cause a different electric field configuration at defective spots of the oxide. As the color of the liquid crystal depends on the field, such flaws are easily detected by the optical variations of the liquid crystal.

Another such study of such flaws is that of Irene,[40] who examined the flaws in silica using transmission electron microscopy. He found, depending on the preparation of the film, many different kinds of flaws. Significantly, he almost always found a very heterogenous film with flaws appearing at random over the surface. He shows the presence of different structures, for example, regions of excess silicon and particles of crystalline SiO_2 in the amorphous film.

The electrical properties of passivating layers of silica, emphasizing the injection of carriers into the material and the dependence of carrier transport on the imperfections found in the material, will be discussed in Section 8.2.

Other materials can be mentioned to illustrate the generality of the concepts developed above, where iron oxide, aluminum oxide, anodized gallium arsenide, and silicon oxide were used to provide examples of the structures to be expected. Vedel *et al.*[41] anodized cadmium to form CdS layers with variable stoichiometry using an aqueous solution of S^- ions. In this system excess cadmium is interstitial. They found that the stoichiometry could be varied by passing current through the CdS layer in the presence of Cd^{2+} ions in the solution. The presence of the interstitial cadmium in the cadmium sulfide was detected by the variation of the open circuit electrode potential as a function of the integrated current passed. Miller *et al.*,[42] also studying CdS formation on Cd, found that during anodization, hydroxide is incorporated into the layer. In fact, the competition between the formation of sulfide and the formation of hydroxide, and the competition between the cathodic removal of sulfide and the cathodic removal of hydroxide, provide a study of substantial interest in illustrating the complexity of the structure of anodic layers. For example, with increasing cathodic bias on their films, they could remove (strip) preferentially the hydroxide species, leading to a photosensitive semiconducting film, and then strip the sulfide film to return to the pure metal. They studied the system using the rotating ring-disk electrode method for corrosion studies.

Imperfections with a laminar or with a uniform distribution can be obtained on a wide variety of oxide layers by in-diffusion of impurities during electrochemical studies. A particularly interesting case of in-diffusion of impurities, as it relates to films on metals, is the effect of lithium on the overvoltage for the oxygen evolution at a nickel electrode. This was studied by Turyan and Tsinman, as reviewed by Erdey-Gruz.[R9] Lithium is the most popular dopant for NiO, making it strongly *p*-type, and so its introduction should have a marked effect on the properties of the layer. And, it is indeed found that lithium changes the electrode properties of nickel, increasing the overvoltage necessary for oxygen evolution. Presumably this effect can be traced back to the Li doping of the NiO layer on the nickel electrode. The general in-diffusion of ions into oxide semiconductors and oxide layers was reviewed in Section 5.2.2.2.

Finally, we can discuss some observations of lateral inhomogeneity on materials other than SiO_2. Corish and Warde[43] discuss the transfer of silver into silver sulfide as being preferential at grain boundaries. Flint and Varley[44] offer a particularly interesting study of ZrO_2 on Zr, where they conclude that electron injection into the oxide from the zirconium impedes the flow of anions through the oxide, because electron injection neutralizes the anion vacancies. They go on to suggest that electron injection is favored at dislocations in the zirconium oxide, because of a slightly different work function at the dislocations. Thus only cations can move where the substrate is dislocated, and the oxide characteristics change considerably. In the extreme case of mechanically polished zirconium, where the damage continues into the oxide, the whole oxide growth is attributable to cation motion.

To summarize this rather complicated picture, we have discussed three general types of defects in the oxide film: imperfections having a reasonably uniform defect density, those showing lateral nonuniformity in defect density, and those showing laminar nonuniformity. In the group of uniformly distributed defects we can include defects associated with a random variation in the degree of hydration. Also, there will be random variations due to the inclusion of point impurities during growth. The class of lateral nonuniformities would include boundaries between amorphous and crystalline regions, perhaps dislocations, and include in-diffused ions that find an easier pathway in these regions. The class of laminar imperfections includes variation in structure from amorphous to porous as the film grows thicker. It includes metamorphic layers at the conductor/oxide interface and hydrated layers at the oxide/solution interface. It also includes laminar stoichiometry variations in the "bulk" of the oxide that develop during the growth of the film.

8.2. Current Transport through Oxide Films

In this section we discuss the flow of electronic current through oxide films on the surface of semiconductors and metals. We will consider both semiconducting films (films that would be semiconductors if in the bulk crystalline form) and insulating films. We will consider tunneling as is expected for thin films, transport of carriers through a conduction or valence band, as is expected if the films are semiconductors, and transport due to flaws, as is expected if the material in the bulk crystalline form is an insulator. We will not, except as incidental to other discussions, analyze current flow by ionic motion. Thus, for example, the polarization capacity[R9] of electrodes due to film growth or removal will not be discussed, except as it relates to electronic transport through the oxide layer.

8.2.1. Thin Oxide Layers

As discussed earlier, the change in current flow, in going from a clean metal to an adsorbed monolayer, will probably not be dominated by tunneling limitations. Tunneling should be relatively easy through a monolayer. Other effects—the catalytic activity change of the surface, the occupation of adsorption sites, and the development of a surface double layer different from that on the oxide-free surface—will affect the electrode characteristics more than barrier thickness, while adsorbing the first monolayer. Only when the film thickness increases toward or above 10Å will tunneling be impeded and the current changes become dominated by tunneling probability.

Heusler and Yun[45] describe experiments and review other workers' experiments that show an exponential drop of current density with increasing film thickness while still in the "thin film" regime. The thin film regime is where carrier transfer by tunneling is still the dominant mechanism. The thin film regime ends, according to them, at about a few tens of angstroms—say, less than 30 Å, a reasonable distance for the cutoff of electron tunneling.

Schmickler and Ulstrup[46,47] have provided theoretical analyses in this "tunneling range" of oxide thickness. They analyzed the system in terms of Fowler–Nordheim tunneling (see Section 8.2.2) where the lowering (or narrowing of the barrier to tunneling by the electric field inside the oxide is considered. They conclude that electrons well above E_F (of a metal electrode) can make a substantial contribution to the tunneling current, for although there are few electrons with this high energy, both the barrier height and the barrier width are much smaller for the higher energy electrons.

Schmickler[46] also analyzes the case of "resonance tunneling," where the tunneling is aided by levels in the bandgap of the oxide. In other words, if tunneling through a barrier, say, 30 Å, is difficult, but a level at a convenient energy is available halfway through the barrier, it can act as sort of a stepping stone to aid the electron through.

A good quantitative comparison of theory and experiment is difficult, because at a metal/metal oxide/solution interface there are usually significant chemical reorganization processes accompanying any redox reaction, and it is difficult therefore to simplify the analysis. However, qualitatively Schmickler and Ulstrup[47] indicate an exponential dependence of current density on the oxide thickness. Schmickler and Schultze[48] report data that they feel indicate resonance tunneling through PtO_2 at high overvoltage. An interesting feature of these experiments is the transformation from metallic behavior to semiconductor behavior as the oxide thickens. Schmickler and Ulstrup report that the anodic transfer coefficient goes to zero as required for n-type semiconductors (Section 3.3.3.2).

Schultze and Vetter,[49] also analyzing reactions (oxygen evolution, CO oxidation, and Ce^{4+} reduction) on platinum, find that the current decreases exponentially with oxide thickness from 1.5 to 10 Å. They conclude the barrier to tunneling is about 2 eV high. Specifically they find

$$J = A(\exp\{-(E_a^o - \beta F\eta)/2.3RT\}) \exp(-x/x_0) \tag{8.1}$$

represents the dependence of the current on the thickness x and the overvoltage η. Here E_a^o and x_o are constants, and the other symbols have their usual meaning. Effectively we are rewriting Equation (3.17), where instead of tunneling through the solution, we are considering tunneling through the oxide film. From Equation (8.1) we observe that the oxide film has little effect on the form of the Tafel equation for metals except to decrease the exchange current by a tunneling probability, which decreases exponentially with oxide thickness. As long as the current flows dominantly by tunneling from the Fermi energy directly to ions in solution, the metal electrode will behave as a metal, as discussed at the beginning of Section 3.4.3. Damjanovic and co-workers[15,50] find similar results with oxide films on platinum, finding the current obeys Equation (8.1).

Photoeffects, usually considered typical of semiconductors, are observed with a Pt electrode, but there is some disagreement whether the photoeffects observed can be attributed to the production of holes[51] in the PtO_2, or whether they originate from solution processes. Kunugi *et al.*[52] find that the photoeffects are most marked for wavelengths longer than 420 nm, whereas PtO_2 absorbs light at wavelengths shorter than 400 nm. Thus Kunugi and co-workers suggest that the optical effects are associated with the excitation of molecules in solution, not the creation of hole–electron pairs. From a theoretical view, there are at least two ways that the absorption of light in the oxide could affect the process. First, there is trapping of carriers in the oxide leading to double-layer effects; and second, there is simple transport of current (photoproduction of holes). It seems unlikely, with a reasonable tunneling current density, that hole photoproduction would have a gross effect on the current. It could change the electric field by trapping, however, and thus, by the Fowler–Nordheim model, affect the tunneling.

Schultze and Stimming[53] have investigated tunneling through thin passive oxide layers on iron. Again the exchange current decreases rapidly with increasing oxide thickness. They suggest that in this case the carriers tunnel into the conduction band of the iron oxide, because the conduction band of the iron oxide is estimated to be near the Fermi energy of the iron. This makes a very complicated electrode, where the oxide is thin but has available energy levels.

Before leaving the topic of thin layers on metals, thin enough for tunneling, it should be emphasized that the analysis of the current/voltage characteristics of such layers is not so simple. In practice, one has great difficulty in distinguishing on the one hand currents due to the growth and cathodic removal of such films, from, on the other hand, currents due to tunneling from the metal to redox couples in solution. Additionally, as mentioned above, one has the difficulty in the first layers of oxide in separating barrier effects (tunneling) from changes in the catalytic activity of the surface or changes in the adsorption site density for the reactants. Erdey-Gruz[R9] suggests that most of the effects observed for very thin films can be explained satisfactorily by purely inner sphere effects (the electrode behaves like a metal and tunneling need not be considered). However, there seems to be substantial evidence that tunneling effects do dominate the systems with oxide films between, say, 5 and 25 Å.

8.2.2. Model of Electronic Conduction through Thick Coherent Layers

In this section and Sections 8.2.3 and 8.2.4 we will be interested in mechanisms of electronic conductivity through thicker films. We will base the discussion on a band model such as is illustrated in Figure 8.1. It will be found that there are many factors to consider[R54] that require modification of that simple model.

In order to conduct current through the bands the carrier must be sufficiently energetic to reach a band, and second, must move through the film without being trapped. For example, consider Figure 8.1 but assume there is no impurity band. With the conduction band and valence band of the oxide layer, as shown, far from the Fermi energy of the conductor, a reasonably thick film may act as an insulator owing to back-to-back diode effects: In either direction, with either carrier, there is a high activation energy to promote the carrier into the appropriate band of the oxide film. Such back-to-back diode effects will be discussed in detail below. If a carrier is by some means promoted to the conduction band or the valence band of the film, it may or may not be able to make its way across the film. If traps are present in the film, trapping may lead to the material being an insulator—the trapped charge builds up a space charge, neutralizing any applied voltage, so the carriers injected cannot cross the film. We will discuss these phenomena in more detail in the present section. First, we will discuss carrier injection into a "semiconducting layer," where the energy level of the bands in the film is not too far from the Fermi energy in the conducting substrate. We will assume there are no mobile ions in the oxide film. Then we will describe how injection into oxide layers can be so promoted by mobile ions that even insulating layers can be conducting. Finally, we will discuss trapping in the oxide, and how it blocks the current flow.

It should be realized, however, that the models developed will nonetheless be oversimplified, primarily because they do not satisfactorily describe the behavior with gross defects present. By gross defects we mean noncoherent oxides, oxides with voids and pinholes, etc. It is to be anticipated that most films on the surface of a semiconductor or a metal will not be well behaved in the sense of the diagram of Figure 8.1. With amorphous films, we may have crystallized grains where boundaries can be conductive.[1] With crystalline films, there may be grain boundaries and dislocations that can conduct current,[44] as discussed in Section 5.2.3.

There may be gross changes in stoichiometry that lead to highly conducting regions in what is presumed to be an insulating film. There may be complete disruption of the film (a noncoherent film) that permits the solution to penetrate to the substrate. And, even with a coherent amorphous film, there may be stoichiometry changes and gel structures leading to a high ionic mobility through the film that permits current flow. Thus, in order to suggest a realistic model, we should discuss in more detail the flaws present in an oxide film, and apply the concepts of Section 5.2 to suggest how these flaws may affect film conductance. Some of the effects of gross flaws will be discussed below, but the discussion will be limited because little is known. We can conclude, however, that gross flaws usually increase the conductance substantially over that expected by the simple models.

First, consider the case where the oxide is expected to be a simple semiconductor (e.g., one of the oxides shown in Figure 5.5), with its conduction or its valence band not too far from the Fermi energy of the conducting substrate.

Although evidence will be reported below that oxide layers may behave as semiconductors, there is one major problem—rectification—that must be recognized. The metal/semiconductor interface (and the semiconductor/electrolyte interface too, as was discussed in the introduction to Section 2.2.4) may not provide an "ohmic contact" to the semiconductor layer. Figure 8.3 indicates the normal band structure of a Schottky barrier diode. The terminology "Schottky barrier diode" was also intro-

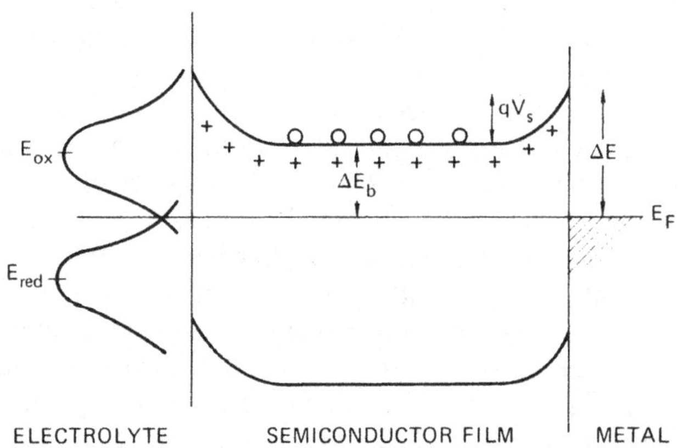

Figure 8.3. The band diagram for Schottky barrier diodes at the metal/semiconductor and at the semiconductor/solution interfaces. With this configuration electron transfer from the metal to the semiconductor is the reverse direction of that diode, and the reverse current is low. Unfortunately in the other direction of current flow, the Schottky diode is in the reverse direction at the semiconductor/electrolyte interface. Thus current is blocked in both directions.

duced in Section 2.2.4. At the metal/oxide interface, electron injection into the semiconductor oxide would be very difficult because the electron must be activated by an energy ΔE independent of the applied voltage. On the other hand, electrons can flow to the metal with very little positive voltage applied, for this is the forward direction, where V_s rapidly becomes small enough to pass current. Unfortunately, with a band model such as Figure 8.3, when the Schottky diode at the metal/semiconductor interface is in the forward direction (namely, when the metal is biased positively relative to the solution), the Schottky diode at the semiconductor/electrolyte interface will be biased in the reverse direction. Thus we have "back-to-back diodes": in either direction of applied voltage there will be a reverse-biased diode, according to Figure 8.3, preventing current flow. Such a configuration will allow reasonable Mott–Schottky plots (e.g., for estimation of carrier density in the oxide) where there is no current flow. However, the current flow itself will be blocked.

Two idealized mechanisms for electron injection into a band in an insulating oxide are usually compared. One is Schottky emission, dominant with the usual solid state Schottky diode; the other, Fowler–Nordheim (FN) tunneling. We will not develop the mathematical analyses here, but will simply indicate the models and their differences. The discussion by Osburn and Weitzman[54] provides a more mathematical treatment.

The route of electron injection by Schottky emission is shown in Figure 8.4. Clearly Schottky emission is a highly activated process with an activation energy ΔE, (or slightly less if the electrons can tunnel through the barrier at a lower energy than the top of the barrier).

To illustrate Fowler–Nordheim tunneling, consider the case of a cathodic voltage applied to a metal that is covered with an oxide layer. The metal is biased negative, and the applied voltage appears across the oxide layer. Consider further that there are positive mobile ions in the oxide that can move toward the metal/oxide interface. (The same effect is obtained with very heavy doping). Under the influence of the applied field, the ions will move to (or, with heavy doping, be in) a position shown in Figure 8.4. The applied voltage will appear between the negative charge on the metal and the equal and opposite positive charge in the oxide. If the applied voltage is sufficient to move the conduction band of the oxide below the Fermi energy of the metal, then it is clear from Figure 8.4 that it is energetically favorable to move electrons from the metal through the conduction band of the oxide to the electrolyte. If the barrier is thin enough, the electrons will tunnel. The FN tunneling depends very sensitively on the applied voltage, because as the voltage is increased, the field is increased, and the tunneling distance is shorter. The FN tunneling obviously depends also on the density of charges that are, or can be moved into, the space charge region. We can inject carriers even into the most

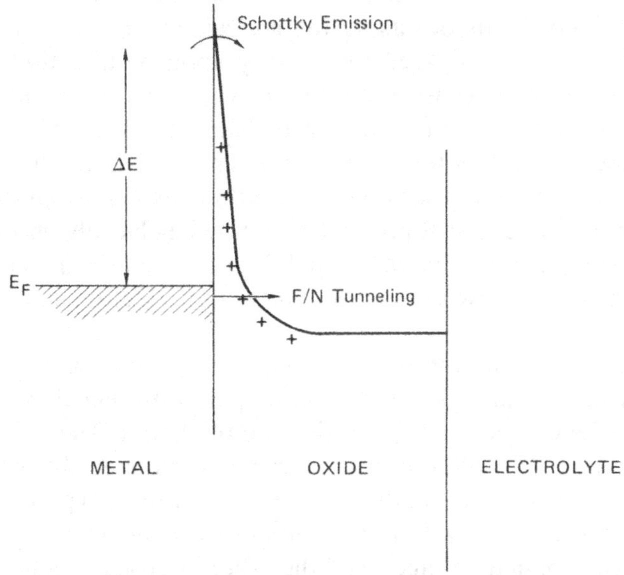

Figure 8.4. Electron injection into the oxide film, either by Schottky emission over the barrier, or by Fowler–Nordheim tunneling through the barrier.

insulating oxide (e.g., SiO_2, as discussed below) with a modest applied voltage, if the density of mobile ions is high enough.

There are four ways in which the rectifying properties at the conductor/oxide or the solution/oxide interfaces can be destroyed and carrier injection obtained into the conduction or valence band of a semiconducting film. There are four general ways to transform a rectifying contact into a quasiohmic contact. First, for a very low resistance contact at the metal/semiconductor interface, the Fermi energy of the metal can be above the conduction band (or below the valence band) edge of the semiconductor. Normally, when choosing metals to make electrical contacts to bulk semiconductors, various metals are tested until one is found to provide such an ohmic contact. Unfortunately, with an anodic film we have no choice in the metal to be used for the "contact." The metal substrate either naturally provides an ohmic contact or it does not. A second possibility to obtain carrier injection in at least one direction is to provide an "ohmic contact" at the semiconductor/solution interface. This means to use a redox couple with E^o_{redox} at or above the conduction band edge, or at or below the valence band edge.

The third and fourth possibilities to make a rectifying contact ohmic involve a defective or degenerate oxide layer. As discussed in Section 5.2.3.1, with many materials, dislocations or grain boundaries will provide paths for easy current flow through a Schottky diode, so the presence of dislocations or grain boundaries is the third possibility. In fact, in order to prepare Schottky diodes as useful devices, an almost defect-free semiconductor must be prepared to get very good rectification. In Section 5.2.3.1, results on ZnO were discussed where it was found that with a mechanically polished surface (one with dislocations) the diode effectively "broke down" at a few tenths of a volt. This was at the liquid/semiconductor contact. Such a low voltage breakdown is almost equivalent to an ohmic contact. The fourth possibility is Fowler–Nordheim tunneling, as discussed immediately above. If mobile ions are added that result in a narrow barrier, narrow enough for tunneling, electron injection will be rapid.

The above discussion describes carrier injection into "semiconductor" films in the dark. The situation is quite different if the film is illuminated, independent (almost) of the film thickness. There is, in general, no rectification with photoproduced carriers, as is easily seen in Figure 8.3. Photoproduced holes in the valence band of the *n*-type semiconductor indicated in Figure 8.3 will move readily to whichever interface they are drawn by the applied voltage. And the other interface will be the forward direction of the diode; in other words, electrons will be moving out of the oxide, and that is the forward direction. Thus it is to be expected that photoeffects will be observed with oxide films even if they are insulating in the dark. The photoeffects will occur (depending on crystallographic structure) at the same wavelength as will occur with the semiconductor in bulk.

We now turn to traps in the bulk film, and discuss how they block the current flow. In the discussions above we concerned ourselves with promoting the carrier into one of the bands of the semiconductor. However, even when the carrier can be promoted into one of the bands, there is no guarantee that the oxide will conduct. Trapping of carriers in the oxide may prevent carrier motion.

Before electronic conductivity can be realized through any of the bands, all appropriate traps must be filled or empty—all traps below the band of interest must be occupied by electrons for electron conduction; all traps above the band of interest must be unoccupied for hole conduction. If, for example, we wish to pass current from the solution to the metal through the conduction band E_{cf}, electrons must be in the conduction band. But, they can only be present in the conduction band if the traps below the conduction band are occupied, otherwise the injected electrons will become trapped in the lower energy levels. With many traps, it is difficult to make them all occupied, and this is why so many materials are good insulators.

As a specific illustration, consider the case where not only the conduction band but also many neutral traps at a level lower than the conduction band are empty, unoccupied by electrons. Then if one attempts to pass current through the material by injecting electrons into the conduction band, the electrons will become trapped and immobilized at the neutral traps. A space charge from the trapped electrons will build up, nullifying the electric field at the surface and preventing the injection of more electrons. Figure 8.5 illustrates the development of the space charge.

The "space-charge-limited current" associated with a layer dominated by such traps has been observed and analyzed in many studies of the electical properties of thin insulating layers.[R1–R3] With such trapping dominant, one finds a striking behavior for the thin films. Again we will refer to Figure 8.5. With a low voltage V_m the current is very small even with an injecting contact (or, in our case, it could be an injecting solution). The current corresponds to the density of carriers that are thermally present in the band with its low "Fermi energy," where the "Fermi energy" must be below the lowest-energy unoccupied trap (cf. Section 1.1.1.3). As the voltage is raised, charge is injected until the voltage across the space charge region equals the applied voltage V_m. The injected charge increases in accordance with the capacity law Equation (2.5), that is, $Q_2 = V_m/C$, where C is an effective capacity. The band model takes the form of Figure 8.5b. As indicated in Figure 8.5b, as V_m is increased, x_1 increases, because Q_2 must increase, and $Q_2 = N_{sc}x_1$, where N_{sc} is the density of traps. Finally, if the layer is thin enough and the trap density is low enough, when the voltage reaches some critical voltage the traps will all be filled through to the other side of the film. In Figure 8.5, x_1 becomes approximately equal to the film thickness. With the traps filled,

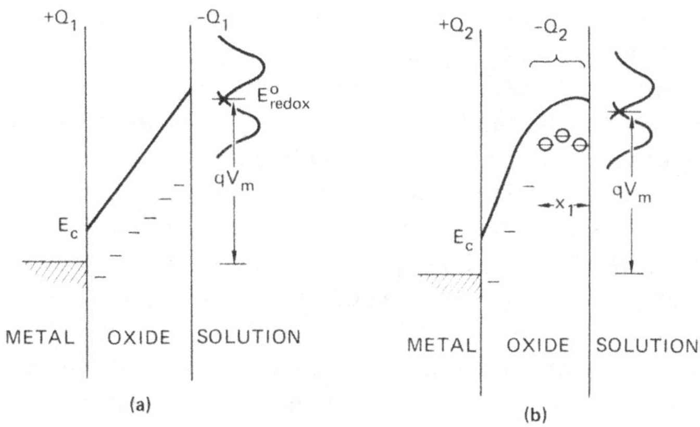

Figure 8.5. Space-charge-limited current, where traps in the insulator provide the space charge. (a) The band model before electron injection; $Q_1 = C_{ox}V_m$, where C_{ox} = oxide capacity. (b) The band model after electron injection, trapping having occurred until the electric field at the surface is nullified.

when an electron is injected from the contact, it can easily make its way through the film. Thus as we approach and pass this critical voltage, a high current suddenly starts to flow. We will not go through the mathematics. The reader is referred to the books on the subject.[R1,R54]

Such space-charge-limited current has apparently not been positively identified as such at the conductor/oxide/electrolyte interface, but with a good oxide it should be observed. Thus the above discussion of the qualitative features of the so-called space-charge-limited current associated with traps should be applicable to carefully designed studies.

8.2.3. Semiconducting Oxide Layers on Metal Electrodes

In this section and in Section 8.2.4 we will emphasize experimental observations; here we emphasize experiments on semiconducting oxides and list theoretical studies where the films are referred to as "semiconducting."

There are many cases where oxidation of a metal should lead to a semiconducting layer; in fact a glance at Table 5.1 or Figure 5.5 will suggest many. When such a semiconducting layer is formed, and it is thick, there is a possibility of a conduction mechanism involving injection of carriers to the conduction or valence band of the oxide and transport through the oxide with the oxide behaving as a simple semiconductor electrode.[R54,R55] This is particularly to be expected if the layer is crystalline, for then each grain will act as a small semiconducting crystal and the metal substrate will act as the metal contact to the crystal. Then, if the layer is thin enough that each grain is as thick as the layer, so the current does not have to cross grain boundaries, there is little difference between the oxidized metal electrode and the semiconductor electrode discussed in the preceding chapters.

In Section 8.2.2 it was pointed out that there are two problems: promoting carriers to the bands, and moving carriers through the (almost always defective) semiconductor. Depending on the location of the bands in the semiconductor relative to the Fermi energy of the metal, the perfection of the surface layer, the redox couple under study, and the presence of photoeffects, one can expect anything between very good transmission of current through the semiconductor to complete blocking of the current in both directions. As will be seen below, such practical problems in the analysis of semiconducting films on metal electrodes have not been given the attention they deserve, and the area promises to be a most interesting field of research for the future.

With the above concepts in mind, and particularly Figure 8.3, we will examine further the few theoretical papers and a few experimental studies describing the behavior of semiconducting films on metals. Dogonadze *et al.*[55] analyze the expected kinetics when a semiconductor film covers a metal electrode. They conclude that the transfer coefficient and the exchange current should depend on the thickness of the film, where the simplest semiconductor model is used, and quantum analysis is applied. Butler[56] also applies simple semiconductor theory to the system, discussing space charge effects in the film. Heusler and Yun[45] suggest more qualitatively that with thick oxides, semiconductor behavior is observed. As discussed above, such analyses are of interest, but it is to be expected that more fundamental questions must be resolved before a satisfactory theory can be developed. And, the fundamental questions involve impurity diffusion, defects in the layer, and diode effects.

We will first discuss the cases where the semiconductor behavior occurs in the dark, defying the theoretical qualifications of Section 8.2.2. Then we will discuss cases where the semiconductor effects are observed under illumination, where of course significant effects should be observed, as described above.

A substantial amount of work has been done interpreting TiO_2 films on Ti in terms of semiconductor theory. Heusler and Yun[45] report the semiconductor behavior of a TiO_2 anodic layer on Ti, and of a Nb_2O_5 layer on Nb. Figure 8.6 shows the current/voltage characteristics of TiO_2 on Ti from their study. In this case the ferricyanide/ferrocyanide couple were added to the solution to inject or extract electrons from the Ti/TiO_2

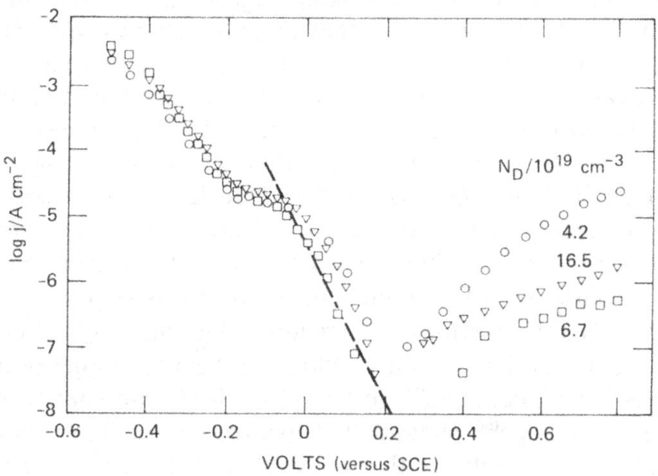

Figure 8.6. Polarization curves of 0.05 *M* Na_3 $Fe(CN)_6$ + $Na_4Fe(CN)_6$ in 1 *M* $NaClO_4$ containing 0.1 *M* sodium acetate–acetic acid buffer *p*H 4.6, 25°C, at titanium covered by anodic oxide films of different donor densities N_d and thicknesses. The high-field dielectric constant $\epsilon = 7$ was used to calculate donor densities from the Mott–Schottky plots. Dashed line reproduces results from (Reference 57) at rutile single crystals.

electrode. We note that cathodic current, electrons from Ti through TiO_2 to the ferricyanide, passes freely, but anodic current, requiring electron injection into the TiO_2 from ferrocyanide, is low. Such behavior is found with single crystals of TiO_2 as well. The dashed line in Figure 8.6 shows some results of the study of Dutoit *et al.*[57] for crystalline TiO_2. It is observed that at low currents the Ti/TiO_2 electrode shows the same characteristics as single-crystal TiO_2. At higher currents, specifically (according to Heusler and Yun) when the bands in the TiO_2 become flat, the characteristics become different from the straightforward semiconductor characteristics. The results suggest that at low currents the injection of charge from the metal to the semiconductor is not rate limiting, the charge transfer across the semiconductor/electrolyte interface is. The bands at the Ti/TiO_2 interface must be such that if there is a reverse biased diode present, it must have a reasonably high leakage current. But at high currents, presumably electron transfer across the Ti/TiO_2 interface becomes rate limiting.

In the studies of Heusler and Yun, both on oxidized Ti and on oxidized Nb, Mott–Schottky plots in an indifferent electrolyte show standard semiconductor behavior, as would be expected when the metal is effectively a "contact" to the covering semiconducting *n*-type oxide.

Laser *et al.*[58] and Quinn and Armstrong[59] also report that fairly thick oxides of titania on titanium act as semiconductors. Laser *et al.* find reduction of species, such as ferricyanide and quinone, that have their E^o_{redox} below the TiO_2 conduction band (Figure 5.5). The claim of Quinn and Armstrong in their abstract, that their oxides (up to 6000 Å thick) behave as semiconductors, is not clarified by specific examples in the text. However, they do show a very defective metal/oxide interface for Ti/TiO_2, with various suboxides identified. Such a defective structure may well explain the lack of diode effects there, and the "ohmic" electron injection from Ti into TiO_2 observed by Heusler and Yun and by Laser *et al.* The expected lack of diode effects with defects present was described in Section 8.2.2. Alternatively the cathodic current into the TiO_2 may be associated with FN tunneling, because it is well known (Section 5.2.2.2) that ions diffuse into TiO_2 readily. Alternatively, of course, the conduction band of TiO_2 could be close to the Fermi energy of Ti, and actually has been calculated[80] to be below the Fermi energy of Ti.

Wilde and Hodge[60] suggest that oxidized chromium acts as a semiconductor, based mainly on the low exchange current observed during cathodic hydrogen evolution. This would be an interesting example of a *p*-type semiconductor. However, the thickness of their layer is not indicated, so it is not known whether tunneling or hole transport through the valence band is the dominating mechanism.

Iron oxide films on iron are reported to behave similarly to iron oxide crystals, as described in Section 5.3. Wilhelm *et al.*[61] found an energy gap of about 2 eV for both anodically and thermally grown iron oxide films (the same as the bulk material), but found for the film a substantially more negative flat band potential, namely, about -0.25 V at pH 8.4.

Metikoš-Huković and Lovreček[62] studied Mott–Schottky plots on oxidized antimony, and found that it behaves as an n-type semiconductor. With a positive bias on the Sb metal, the solution/Sb_2O_3 interface is reverse biased, and they were able to make a Mott–Schottky capacity/voltage plot (similar to Figure 4.4), obtaining $N_D = 10^{23}$ m^{-3}. They found good agreement between these dark measurements and photovoltage measurements, suggesting recognizable semiconductor behavior of a thick ($\sim 10^{-7}$ m) film. Of course, as discussed in Section 8.2.2 above, a good "ohmic" contact is not required to obtain representative Mott–Schottky plots.

Peter[63] studied CdS on Cd, and found with a moderately thin layer (~50 Å) that a good Mott–Schottky plot is observed in this case also, until the space charge reaches the metal. With higher voltages, the surface layer is an insulator (the layer is depleted completely of mobile carriers) so a constant capacity representing the insulator thickness [Equation (2.5)] is measured.

As discussed above, photoeffects characteristic of the semiconductor film can be expected even if the interfaces block current flow in the dark. Thus most of the work where semiconductor behavior is clearly observed is work where the "semiconductor behavior" is photocurrent flow, occurring in the same spectral region that the bulk semiconductor is sensitive in—photons of bandgap energy. For example, Miller *et al.*[42] find that when they have the composition of CdS in their layers over cadmium, the material is photovoltaic. Wilhelm *et al.*[61] show that iron oxide has the same spectral response when present as a film or in the bulk semiconductor form. Dutoit *et al.*[64] observe both current doubling and photosensitivity on TiO_2 films over titanium, while Keeney *et al.*[65] report photovoltaic response and Clechet *et al.*[66] report photocatalyzed H_2O_2 generation over such films.

As can be discerned from the above discussion, the semiconductor surface layer over a metal is really very poorly understood. Measurements of photosensitivity show that the material on the surface will indeed be a semiconductor. This could have been anticipated. But there is much to learn about the behavior of the semiconducting layers in the dark, where the conductance mechanism becomes much more complicated. Some of the expected complications can be recognized when, in Section 8.2.4, we discuss the apparent behavior of insulating films on semiconductor material, and how defects and ion motion strongly affect this material. Actually, as will be discussed, in general there may not be too much difference between "semiconducting" layers and "insulating" layers insofar as ability to conduct cathodic current is concerned. This suggestion is made, bearing in mind that because of ion motion in the film, the back-to-back diode problem is not much less of a handicap with semiconductors than with insulating material.

8.2.4. Insulating Layers on Metal and Semiconductor Electrodes

By implication, one expects no electronic conductance through well-behaved insulating films over a conducting electrode. However, there are two conditions in which substantial currents are commonly observed with the best insulators. One is if sufficient electric field is applied to cause dielectric breakdown; the other is during cathodic bias, when for many systems substantial currents are found to flow for moderately thick films.[R56]

We will say little about breakdown that has not already been said in Section 2.2.4.3. Except for the influence of flaws and of ion movement, the concepts are similar in an insulating layer to those discussed in that section for a space charge layer in a semiconductor. There are, however, a few reports of interest that have bearing on the general theme of the present discussion. For example, Wood and Pearson[77] discuss anodic breakdown of oxide films of valve metals (metals that form a coherent resistive film). They investigated Nb, Ta, Zr, Hf, Al, Ti, W, Mo, and V. It is of interest to note, relevant to Section 8.2.3, that Ti and W, whose oxides should be semiconductors, did not break down, but acted as conductors leading to oxygen evolution. But the interesting observation for the present section is that breakdown on valve metals resulted in thicker films where the breakdown occurred, thus pinched itself off. The breakdown spot meandered over the surface. By analyzing their data they concluded that avalanche breakdown was being observed, where the electrons to initiate the avalanche are supplied by the electrolyte. The voltage required for this anodic breakdown increases with increasing ionicity of the oxide. In the discussions below, where we relate the passage of electronic current to the presence and mobility of ions in an insulator, this observation has considerable significance.

The high currents associated with cathodic bias may be related to cation in-diffusion, most often mobile protons. The influence of cathodic bias on diffusion of cations into oxides has been discussed with respect to bulk oxides in Section 5.2.2.2. There a few examples were cited of in-diffusion into oxide layers, for example, the work of Dyer and Leach.[78] These authors examined oxidized titanium and niobium, and found that hydrogen enters the films under a cathodic bias. And, it enters in non-negligible amounts. TiO_2, for example, converted to TiOOH, as concluded from observations of the change in the index of refraction. They observed up to 85% TiOOH. With niobium oxide, they found less change in the index of refraction, but a much larger change in the optical absorption constant when a cathodic stress was applied to the electrode. Thus cation in-diffusion into passivating films can be substantial under a cathodic bias.

Current passage through oxide layers when the metals are biased cathodically is common on the so-called valve metals, for example, zirconium, tantalum, and aluminum. Also silicon shows cathodic currents, as will be discussed in detail later. When the metals are biased anodically, however, the current is very low. Workers have found that this form of rectification has been difficult to explain, but there are many features that relate to our discussion of imperfections, and of tunneling associated with the imperfections.

For example, Vermilyea[79] and others[R50] showed that the cathodic currents are laterally inhomogenous, and correspond to flaws in the oxide. Schmidt[67,R56] suggests that the cathodic current is related to the in-diffusion of protons, for he found with tantalum and silicon that if an anhydrous electrolyte is used, there is no cathodic current. Young[R7] discusses experiments where, with a mechanically polished metal (many dislocations), the protons diffuse through quite homogeneously and are detected by the uniform hydrogen embrittlement of the substrate metal.

The effect of mobile ions on the current passage can be studied particularly well in studies of silicon because, in this case, the band model and the defect structure are known so well (Figure 8.2) and the observations can be directly interpreted. Specifically, a well-investigated case is that showing the effect on electron flow of mobile Na^+ in SiO_2, where electrons move from the silicon into and through the oxide. The Si/SiO_2 band model (Figure 8.2) shows that electron promotion to the conduction band of the oxide can be expected to be difficult. Many authors[39,54,68,69] have shown that a FN tunneling process occurs, aided by the motion of the Na^+ ions to the Si/SiO_2 interface, as in Figure 8.3. In all the experiments cited, the sample was a Si/SiO_2/metal or Si/SiO_2/organic coating structure, so the amount of sodium was limited. But with the very high applied fields (over 10^8 V/m) used, electrons from the silicon could, nonetheless, FN tunnel to the conduction band of the SiO_2. The current/voltage behavior followed the FN theory well, although the barrier height measured was lower than that anticipated from Figure 8.2. Below, we discuss the case of the Si/SiO_2/electrolyte system, where the source of positive ions is the electrolyte solution and, therefore, essentially infinite. Then the FN tunneling can be either to the Na^+ impurity band or to the silicon dioxide conduction band.

Madou *et al.*[1] have observed currents, both anodic and cathodic, with the Si/SiO_2 as the working electrode in an electrochemical cell. In order to induce electronic current flow in a thermally grown oxide film, it was found necessary to bias the electrode cathodically by approximately 2 V (SCE) for an extended period of time (the order of one-half hour). It is concluded that during this activation process a high density of positive ions, sodium or protons, diffuses into the SiO_2. With the high mobility of ions described in Section 8.2.3, there is ample time for such diffusion. The diffusion is most rapid in the most defective parts of the oxide.[70,36,37] Following this activation treatment, hydrogen evolution is observed at the SiO_2 surface, indicating the development of electronically conducting channels. It is concluded that at this stage tunneling as represented by Figure 8.4 is occurring, with tunneling to the conduction band of the SiO_2 or to an impurity band, depending on the applied potential. Visually one observes that the hydrogen evolution is not uniform but occurs only at localized points on the surface.

Once the SiO_2 is cathodically activated, conduction in both directions is possible, and, indeed, the electrode acts as a quasimetallic electrode in the sense of Section 3.4.3. The current to redox couples in solution begins as the electrode potential approaches E^o for the redox couple. The current densities observed are, however, much lower than those on a noble metal electrode, showing saturation on the order of 10^{-2}–10^{-1} A/m^2. However, the major point of interest is that there are apparently no significant voltage drops in the semiconductor, or in the oxide, that lead to overvoltages. As the current begins at E^o, the Fermi energy must be constant through the semiconductor and through the oxide to the solid/solution interface.

The above illustrates the dominating effect mobile ions can have if present in an insulating oxide film. To summarize, if mobile ions in the oxide film are available to move in response to the applied field, they may be able to move to the interface in high densities and effectively produce a breakdown condition or produce a tunneling condition. It would be anticipated that the observed elimination of the diode effects by a concentration of mobile ions is probably not possible at the solution/oxide interface, because in general one expects the ions to move right into the solution rather than pile up at the interface. Thus oxides on valve metals do not conduct current in the anodic direction.

In the theoretical discussions of Figure 8.1 the possibility of impurity band conduction was introduced, and SiO_2 may offer an example of such defect conduction. The band in the oxide that actually does the conducting in SiO_2 is not necessarily the conduction band. It was suggested by Madou *et al.*[1] that the band is associated with the moving positive ions, specifically that the band is the 2.4-eV trapping level illustrated in Figure 8.2. They observe that the cathodic applied voltage required is about 2 V, and this is sufficient to lower the 2.4-eV levels to the valence band of the silicon, so that electron transfer to these levels can readily occur. They argue that the positive ions can move into the oxide, capture an electron on their 2.4-eV levels (once the tunnelable barrier is formed), and become neutralized. Once neutralized, more ions can enter the oxide, until the oxide is heavily doped (by H^+ or Na^+, say). The dopant provides an impurity band at 2.4-eV below the SiO_2 conduction band.

Studies of the anodic oxide of GaAs, made by Frese and Morrison,[71] show that a conducting impurity band apparently exists in the anodic oxide under certain growth conditions. In the GaAs case, however, it is assumed that the band in the gap of the insulating oxide is associated with incompletely oxidized arsenic, not with ions diffusing in from the solution. Thus the "impurity band" is associated with nonstoichiometry. Again, when the sample is prepared so that the impurity band is present, electrons can flow from the GaAs to the solution through the oxide, assuming there is an appropriate oxidizing agent in the solution to capture the electrons. In this case, however, the band is found to be about 1 eV below the conduction band of the GaAs, and so rectification is observed: electrons can flow from the GaAs to the solution, but are blocked from flowing from the solution to the GaAs.

8.3. Deposition of Reaction Products on Semiconductor Electrodes

In Chapter 7, it was noted that the products of the electrode reaction on a semiconductor electrode can be insoluble and form a surface film. Three major classes of film formation can be expected—the formation of insoluble products after hole capture, the formation after electron capture, and the formation under photocatalytic conditions, with both hole and electron capture. The surface film can either be a new compound, or, as discussed in Section 7.3.4 for CdS, reprecipitation of the host material in another phase (in the cited case an amorphous phase).

Much of the material that could be considered for this section will not be presented because it is discussed elsewhere. For example, anodic oxidation of semiconductor electrodes by hole capture processes was described in Section 8.1.2. The formation of semiconductor films by the anodic oxidation of metals, as for example in the work of Miller *et al.*,[42] was also discussed above. In the present short section, we will describe the oxidation of sulfide ions to insoluble sulfur as an example of hole reactions, the reduction of metal ions as an example of electron reactions, and the photocatalytic deposition of noble metals as an example of combined hole and electron reactions.

The oxidation of CdS forming elemental sulfur provides an example of deposition following hole reactions. This system was investigated by Minoura *et al.*,[72] who were particularly interested in solar energy conversion, in particular the influence of corrosion on the photocell response. They conclude that the reaction of photoproduced holes h^+ could either be

$$CdS + 2h^+ \rightarrow Cd^{2+} + S \tag{8.2}$$

or

$$S^{2-} + 2h^+ \rightarrow S \tag{8.3}$$

and actually it is not clear that the distinction is meaningful. They find that SO_3^{2-} acts to remove the sulfur layer, as they find their photocurrent remains high in the presence of SO_3^{2-}. They suggest a reaction

$$SO_3^{2-} + S \rightarrow S_2O_3^{2-} \tag{8.4}$$

and this reaction of course is not preventing the corrosion, but simply preventing the buildup of the deposited surface layer. In the presence of Na_2S as a competitive agent for hole capture, they find that corrosion stops, to the limit of their measurement. Of course as discussed earlier, the use of Na_2S leads to an amorphous cadmium sulfide layer at the surface after long-term illumination, so presumably the polysulfide is also not preventing corrosion in the accurate sense of the word.

Deposition of metallic layers by reduction involving electron capture from the conduction band of a semiconductor is fairly common. Either the cation of the compound semiconductor can be reduced, or a metal ion from solution can be reduced.

The differences between metal deposition (plating) on metals and metal deposition on semiconductors or on thick oxide layers over metals should be emphasized, because it is particularly important for the understanding, not only of semiconductor electrochemistry, but also of plating (and of the reason that plating is often poor on oxidized metals). Again the difference in the electrode behavior between metals and semiconductors, specifically the difference in the origin of the overvoltage, leads to quite different behavior with respect to metal deposition. As discussed many times earlier (Section 3.3 in particular), most of the overvoltage applied to a metal (or to a degenerate semiconductor) electrode, appears across the Helmholtz double layer, so the energy levels of the metal ions in solution move below the Fermi energy, and the metal ions can be reduced. With a semiconductor, the change in voltage of the electrode appears across the semiconductor space charge region, and only changes the rate of a reaction, not whether it is energetically favorable. Thus with a nondegenerate semiconductor, any metal ion in solution that cannot be reduced at a low voltage (because its energy level is above E_{cs}) cannot be reduced at any voltage short of surface degeneracy. Thus it is of particular interest to identify which metals can be reduced and will form a film on a semiconductor electrode.

In general, in line with the above comments, only noble metal ions are easily reduced at nondegenerate semiconductor electrodes. Evidence that the reduction of the semiconductor ZnO can occur by electron flow to the surface was described in Section 7.3.1, but it probably happens only when the ZnO is made degenerate. With the very high conduction band energy of SiO_2, zinc reduction was observed by Madou *et al.*,[1] when conduction band electrons were induced to flow from the Si conduction band through the SiO_2 conduction band to the surface. But there seems to be little evidence that in aqueous solutions cations more electropositive than zinc can be reduced to the metal by electron flow. Presumably hydrogen is reduced first.

The deposition of noble metals from solution has been observed both cathodically and photocatalytically for many systems. We list a few examples. Möllers *et al.*[73] report the photocatalytic deposition of Ag and Pd on TiO_2. In photocatalysis we must always consider the reactions of both carriers (Section 6.3.3). In TiO_2, of course, the holes are such strong oxidizing agents that they can oxidize OH^- groups to oxygen, as described in Sections 6.6.3 and 9.1, so the fate of the holes is clear. Kelly and Vondeling[74] give a review of early work in this area. They showed similar results, and by ellipsometry measurements were able to show that about 10^{-2} monolayers of noble metal were deposited. They were particularly interested in the coverage obtained, because they wanted to use the technique to deposit nucleation centers for later electroless deposition of various metals for printed circuits.

Bindra *et al.*[13] describe a series of possible systems for the electrolytic deposition of such thin metal films on semiconductors. They studied Pb, Ag, and Pd, with the semiconductors ZnO, CdS, *n*-GaP, *n*-GaAs, and SnO_2. Morrison[75] also showed Ag^+ is reduced cathodically, and Hada *et al.*[76] studied the photocatalytic reduction of Ag^+ on ZnO.

9

Applications of Semiconductor Electrodes

The applications of semiconductor electrodes are, to date, confined to special areas. Electrodes of non degenerate semiconductors will never replace metal electrodes in most electrochemical power applications, any more than transistors can replace bus bars. In batteries, fuel cells, and other high-current devices, semiconductors will only be dominant when they are degenerate and behave as metals. The future of nondegenerate semiconductor electrodes for practical exploitation is in electrocatalysis, sensors, and possibly solar cells. Such possibilities will be explored in this chapter. Their future impact in basic knowledge, however, as has been indicated in earlier chapters, is an exciting new insight into surface science: corrosion, anodic oxidation, and a general understanding of reactions at the solid/liquid interface.

9.1. Solar Energy Conversion

9.1.1 Introduction

The possibility of practical application of semiconductor electrodes in the area of solar energy conversion has undoubtedly been the main impetus in the last few years for the rapid development seen in semiconductor electrochemistry. There are two dominant forms of photoelectrochemical (PEC) solar cells[1]. In the first form, the photogenerated hole is used to oxidize one species and the electron is used to reduce another. The reaction is chosen such that the products can be stored. When at some later time they are rereacted, the energy is recovered. This is optical to chemical energy conversion. Such a process is typified by the first PEC cells attempted, where the photoproduced carriers were used to decompose water. In that case, the holes oxidized OH^- ions, the electrons reduced H^+ ions; the solar energy was stored as the fuel hydrogen. In the second form of the PEC cell, the photovoltaic PEC cell, the photoproduced hole electron pair is used to generate electricity directly, in the same way as it is done in the solid state solar cells. The devices are very similar to the Schottky barrier type of solid state solar cells. The operation is based on the separation of charge by the surface barrier; the details of the mechanism will be discussed further below. This form of cell has, at times, been termed a "liquid junction solar cell".[R57]

Four possible advantages can be considered for the PEC cells over solid state solar cells. The first advantage is insensitivity to gross defects; the PEC cells should be less sensitive than the Schottky barrier solid state solar cells to grain boundaries in the material. A good rectifying metal contact is needed for a Schottky barrier cell, and on a badly defective semiconductor a metal contact will be much more ohmic than an electrolytic contact (see Section 8.2.2). Now, in order to obtain low cost solar cells, defective polycrystalline material must be used, so this advantage of the PEC cell is important. However, as discussed in Section 5.2.3.1, the rectification characteristics of the Schottky barrier will be much less than ideal when dislocations and grain boundaries are present. The topic of dislocation and grain boundary effects in the electrode behavior of semiconductors was described in Section 5.2.3.1, so the topic will not be discussed further in this section. However, the work of several authors who successfully used polycrystalline electrodes (albeit with modest conversion efficiency) should be mentioned. Examples are Kim and Laitinen[2] with SnO_2, Kennedy and Frese[3] with Fe_2O_3 electrodes, Miller *et al.*[4] with sulfide semiconductors, Okuda *et al.*[5] with titanates, and Frank and Bard[6] and Hardee and Bard[7] with TiO_2.

A second advantage of PEC solar cells may be the cost of manufacture. If we compare them to the *p–n* junction version of solid state solar cells, it is clear that the costly diffusion process is avoided with the PEC cells. A third advantage of PEC cells is the possibility of direct chemical storage, assuming a suitable reactant/semiconductor system can be selected. As will be seen below, the photolysis of water, with the object of storing hydrogen, seems not to be the ideal reaction.

Finally the fourth advantage of PEC cells is the possibility, when using the PEC cell in a chemical storage mode, of using a powder rather than an electrochemical cell. When used as a photocatalytic powder, both the reduction and the oxidation occur on the same particle, although one may arrange for a second material deposited on the powder grains so that the reduction occurs on one material, the oxidation on the other. In other words the hole–electron pair formed in a particular particle reacts at its surface, the hole-oxidizing a species and the electron-reducing a species. Thus the products are formed in the same volume. For a useful system we require that the products (a) store energy, and (b) be unable to recombine in the system provided. If the products can react spontaneously, they will do so of course, and there will be no chemical storage. A system has to be conceived where the reverse reaction is highly activated, so the energy remains stored until needed. To recover the energy at a later date, the reaction can be catalyzed. Now the great advantage of such a system is the extremely low cost—the solar convertor is prepared simply by sprinkling the powder in a tray and making arrangements to add reactants and collect products.

On the other hand, to compensate for the above potential advantages, there are great difficulties with the PEC cells as compared to solid state cells. The two major difficulties are (a) corrosion and (b) practical problems of maintaining an electrolyte in the liquid form when in a field or a rooftop under a hot sun. A great effort has been directed at the first problem, and indeed, much of the work described in Section 7.4 stems from this effort. Very little effort has been directed toward the more mundane problem (b). These problems, while great, are not impossible to solve, in a compromising way, and so interest in PEC solar cells has continued.

The history of interest in PEC cells is very short. Following the work of Boddy,[8,R34] who observed that holes from TiO_2 could oxidize water, and concurrent with the work of Möllers and Memming,[9] who also observed oxidation of water by holes, this time using holes from SnO_2, Fujishima and Honda[10] suggested that such oxidation of water by holes, combined with the reduction of protons by the conduction band electrons, is a method of directly converting optical to chemical energy. This work, as mentioned above, initiated a plethora of publications examining all possible systems, without necessarily a regard for practicality, but aiming at an improved understanding of the requirements. In the discussions to follow, we shall refer to only a few of these publications. Here we will list a few review articles that provide a much more detailed description of the studies than we will attempt. For example, there is the proceedings of a meeting on the topic: *Semiconductor Liquid-Junction Solar Cells*, edited by Heller.[R57] Additionally, reviews by Archer,[R58] Gerischer,[R59,R60] Haneman,[R61] Gerrard and Rouse,[R62] Wrighton *et al.*,[R45] Murphy,[R63] Nozik,[R25,R73] Harris and Wilson,[R27] Morrison,[R4] Manassen *et al.*,[11] Butler and Ginley,[R75] and Memming[R64] are of great value for a summary of the field.

In the discussions to follow, we will only present a qualitative theory of the behavior of the PEC solar cells. A quantitative analysis leading to an expression for the efficiency possible for these cells would follow the form of analyses for solid state solar cells,[12] and actually an analysis along this line has been presented by Wilson.[13] In Wilson's calculation there was no attempt to describe realistically the complicated carrier transfer steps between the semiconductor and the solution, as described in Chapters 3, 6, and 7. As discussed in the introductory paragraphs of Section 6.4, Wilson simply defines a simple parameter S to describe charge transfer to the solution. The parameter is more or less equivalent to the surface recombination velocity, described in Section 6.4.2, leading the Equation (6.33), and so essentially bypasses any chemical complications.

In the present description, we will neither attempt a quantitative mathematical theory describing a general model for solar cell behavior, nor will we repeat Wilson's model. We will simply describe the concepts qualitatively, indicating in words the important features. We feel most of the problems are in the electron transfer reactions, and these have been dealt with individually in Chapters 3, 6, and 7 (including the background for Wilson's model). At the present time a "general" expression that described all possible important steps would be impossible. Perhaps for some simple systems, where the surface reactions are well defined, a mathematical model based on the concepts of Chapter 3 and Chapter 7 can be developed, but this is for the future, when better electronic–chemical models are developed.

9.1.2. Photovoltaic Cells

In this section we describe the action of PEC photovoltaic cells qualitatively, attempting to bring out the various processes that must be accounted for in a rigorous theory.

In Figure 9.1 we show a band model for a semiconductor with a bandgap of the order of 1.4 eV, and a redox couple in solution, suitable for photovoltaic conversion of energy. For illustration, we use an *n*-type semiconductor.

The bandgap of the semiconductor should be the order of 1.4 eV because a semiconductor with this energy gap will absorb the dominant part of the solar spectrum. Materials with energy gaps larger than this will not absorb efficiently (because the solar spectrum peaks in the green), so the rate of hole production (the short circuit current, I_{sc}), will be low. Materials with bandgaps much smaller than this cannot efficiently convert the photons (say, of the order of 2 eV on the average) into voltage — the energy will be lost as the photoproduced holes and electrons are de-excited to the band edges. When the carriers reach the surface for reaction, they only have an energy available, at most, equal to the bandgap. Thus, if E_g is low, the open circuit voltage, V_{oc}, will be low. For the most power output—power being a product of current and voltage—it is desirable to start with V_{oc} and I_{sc} high.

The redox couple in solution in Figure 9.1 is indicated with a λ slightly

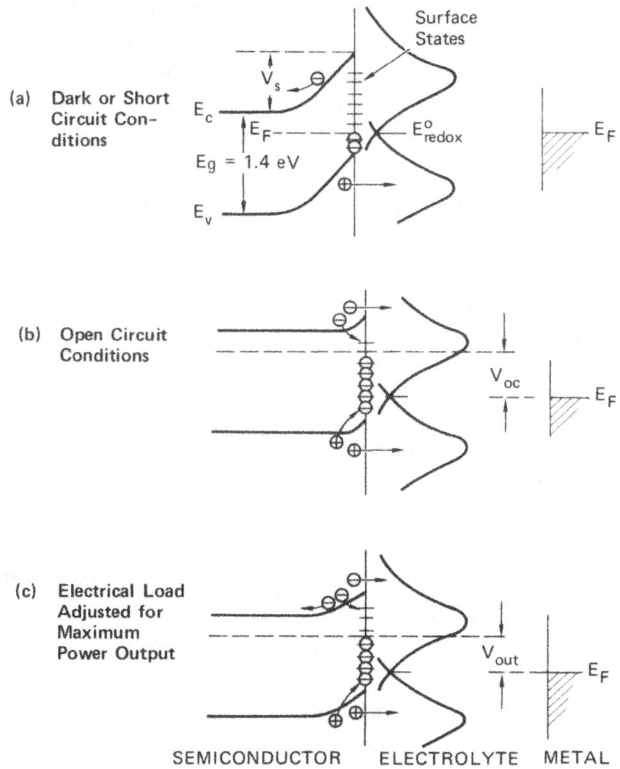

Figure 9.1. Band model for PEC photovoltaic cell.

less than 1 eV, as expected for most redox couples, and with an E^o_{redox} near the valence band edge. Because λ is so large, one observes that the Gaussians for the redox couple extend well into the band regions. The reason it is desirable to have E^o_{redox} low is to encourage band bending. As indicated in Figure 9.1a, in the dark, with E^o_{redox} near the valence band, the bands bend up, because electrons from the semiconductor move to the oxidizing agent in solution. It is this band bending that leads to a good high open circuit photovoltage V_{oc} upon illumination. Thus we want the E^o_{redox} to be near the valence band edge, so there will be as much band bending in the dark as possible. However, the value of E^o_{redox} cannot be below the valence band edge because the reducing agent of the couple must be able to capture photoproduced holes. Thus the desirable position of E^o_{redox}, relative to the bands in the semiconductor, is firmly defined —near the valence band edge. For example, Tributsch *et al.*[14] find that V_{oc} on WSe_2 increases steadily as E^o_{redox} moves from the value for Ti (III/IV) to the value for Ce(III/IV).

Finally, in Figure 9.1 we indicate surface states, or, more accurately, recombination centers, at the surface of the semiconductor. Such recombination centers can play a very important role in the determination of conversion efficiency, as will be discussed below.

The relation of E_F, E^o_{redox}, and V_s in Figure 9.1 follows reasonable assumptions. At equilibrium in the dark, or with an external short circuit, we have Figure 9.1a where E_F of the metal counterelectrode equals E_F of the semiconductor. We assume in all the diagrams of Figure 9.1 that there is no overvoltage at the metal counterelectrode, that $E^o_{\text{redox}} = E_F$ for the metal. For simplicity, we assume that at the semiconductor electrode also, equilibrium obtains in the dark—that in the dark $E^o_{\text{redox}} = E_F$ of the semiconductor. It is to be noted, as an unimportant detail, that we are using E^o_{redox} instead of E_{redox} to describe the redox couple. In principle, this means that the reducing agent and the oxidizing agent of the couple are there in equal concentration; in practice it means, within a few kT error [Equation (1.45)], only that there is a significant quantity of both the oxidizing agent and the reducing agent present. The reason for using E^o_{redox} in our qualitative discussion is the conceptual simplicity of obtaining its value from the standard electrode potential (redox potential) of the couple from handbooks [Equation (1.68)].

If the sample is illuminated, but the semiconductor and the counterelectrode are short circuited, the arrows in Figure 9.la indicate the fate of the photoproduced electrons and holes. Because of the high surface barrier, the electrons move away from the surface, through the short circuit and to the metal. The electrons flowing from the metal reduce (are captured by) the oxidizing agent of the couple. The holes move toward the surface, and oxidize the reducing agent of the couple. The only effect is current flow. No external power is generated, of course, because there is no voltage drop, V, in the external circuit and external power is given by $I \cdot V$. The energy of the hole–electron pair is lost in two places, first in the solid, and second, in the solution. In the solid, the hole loses energy as it moves through the space charge region to the surface, and the electron loses any energy it has (relative to E_F) as it moves through the space charge region to the bulk semiconductor and to the metal. In the solution, the hole loses its remaining energy after oxidizing the reducing agent, because the free energy of the redox system is given by E^o_{redox}. The lost energy is dissipated by the readjustment of the outer solvation sphere around the ion.

In Figure 9.1b we show the band diagram under open circuit conditions. Here the fate of the photoproduced carriers is again shown by arrows. The net flow of carriers to the surface at open circuit must be zero—the electron flow must equal the hole flow. When the sample is first illuminated, there is a transient response because initially electron flow to the surface is low, and holes dominate the flow. This builds a positive charge (lowers the negative charge N_s in Equation 2.25) that lowers the barrier V_s until the electrons can reach the surface at the same rate as the holes. At the metal/electrolyte interface there is no change, and E^o_{redox} still equals E_F of the metal. But because of the lowering of the surface barrier E_F of the semiconductor moves up in energy relative to E^o_{redox}. Thus an external voltage is measured, given by the difference of the two Fermi energies, and denoted by V_{oc}, the open circuit voltage. This is the maximum external voltage obtainable from the system. It is noted that this voltage is somewhat less than the bandgap. Three small voltage losses add up to the difference: the residual V_s, the separation of the bulk conduction band from the bulk Fermi energy, and the separation of E^o_{redox} from the valence band edge.

Again we note that in this open circuit case there is no external power generation, because although now a voltage is produced, no current flows. This time the energy of the hole–electron pair is lost as heat during the recombination process, either through the recombination centers as shown, or through the redox couple as shown.

Figure 9.1c shows the band model under conditions approaching optimum power conversion. Conditions are adjusted so that some of the photoproduced carriers go through the external circuit, but a load (an external resistor, say) is provided so that external work is done. For maximum power, the load is adjusted. Clearly, from Figures 9.1a and 9.1b, as the load is increased from zero, we move from a case of no external voltage and hence no power, through a case of maximizing the product $I \cdot V$ (the power), to a case of no external current and no power. Describing this by a band model picture, we see that moving from case 9.1a toward 9.1c, the surface barrier is lowered (the amount depending on the load) and as it is lowered, some of the electrons start to flow to the oxidizing agent and to the recombination centers. The energy from these processes is lost, just as in Figure 9.1b. Thus, the barrier cannot be lowered too far. However, the more the barrier is lowered, the greater the voltage produced by the cell, and that determines (by power $= I \cdot V$) the power obtained from the carriers that do go through the external circuit. The barrier is lowered (by adjusting the external load) to maximize the power delivered to the external load.

Clearly, for maximum efficiency, it is desirable to block majority carrier flow to the surface. It would be desirable to have no recombination centers associated with the semiconductor. It would be desirable also, to have a redox couple that obeys the fluctuating energy level model ideally, for then electron capture by oxidizing agents would be low, because the conduction band edge is near the tail of the Gaussian distribution. Thus majority carrier capture due to surface states, to dislocations (as in Figures 3.5 or 7.1), or due to the presence of different complexes of the oxidizing agent (as in Figure 7.1), will all seriously degrade the solar cell characteristics. We note, of course, that we cannot dispense with the oxidizing agent itself, for that is needed in the solution to capture electrons from the metal. Also, we need the oxidizing agent to provide the barrier in the first place.

Many studies have been made of PEC systems designed to be operated in the photovoltaic mode. Ohashi *et al.*[15] combined *p*-type and *n*-type semiconductors, instead of using a metal for one of the electrodes, studying TiO_2 and $SrTiO_3$ as the *n*-type, and CdTe or GaP as the *p*-type electrodes. They were attempting, not only to generate electrical power, but at the same time to photolyze water (Section 9.1.3). The efficiencies of solar energy conversion were low, however, with less than 1% of the incident power appearing on the external load. With CdSe, Chai *et al.*[16] managed to obtain slightly greater than 1% efficiency, using CH_3OH as the electrolyte, and limiting the wavelength of the illumination. But they found the current doubled (Section 6.3.1), so concluded the electrolyte was not stable. Kohl and Bard[17] used GaAs in acetonitrile but found that traces of water in the acetonitrile lead to corrosion and inefficient conversion. Chang *et al.*[18] describe a system based on *n*-GaAs with Se^{2-}/Se_2^{2-} as the electrolyte, where they claim 9% efficiency, a value approaching that of solid state solar cells. Parkinson *et al.*[19] and Johnson *et al.*[95] using the same semiconductor and electrolyte, but treating the GaAs surface with rithenium to displace a surface state, and perhaps block surface recombination, report efficiencies to 12%. They also claim negligible corrosion of the electrode, a problem, as will be discussed in Section 9.1.4, that is difficult to overcome and difficult to test. Other systems have been reported, as are described in the review papers listed earlier.

9.1.3. Conversion of Optical to Chemical Energy

The other form of PEC solar cells is that designed to convert optical to chemical energy. In this case, the requirements on the system are somewhat different. The bands should be almost flat, but bent just enough to avoid too great a loss due to recombination. The desired picture is shown in Figure 9.2.

In this case the external circuit is short circuited, because we do not wish to draw electrical power, but want the current to flow freely. At the *n*-type semiconductor electrode, we want to oxidize the reducing agent of a redox couple #1 that has a low E^o_{redox}. At the metal electrode we want to reduce the oxidizing agent of a redox couple #2 that has a high E^o_{redox}. The desired processes are shown by arrows. Then the available energy that can be recovered from the stored chemicals is the difference, indicated in the diagram as ΔE.

There are, however, a few complications. After the reducing agent #1 is oxidized it should not be rereduced by either the conduction band

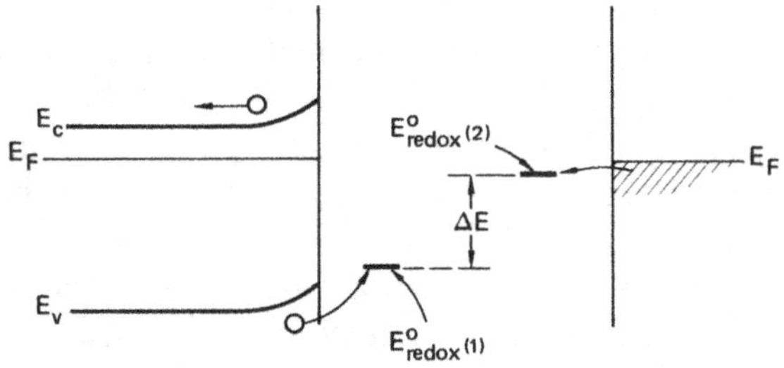

Figure 9.2. Band model for PEC cell for optical-to-chemical energy conversion.

of the semiconductor or by the metal. Thus it is preferrable to have it automatically removed. For example, in the photolysis of water, the oxidized OH^- ions convert to oxygen:

$$2h^+ + 2OH^- \rightarrow H_2O_2 \rightarrow H_2O + (1/2)O_2 \uparrow \tag{9.1}$$

and the gaseous oxygen is no longer available for rereduction (at least in principle). Similarly, after the oxidizing agent #2 is reduced at the metal electrode, it should not move to the semiconductor electrode and be available for oxidation by holes. This is also more or less taken care of in the photolysis of water because when the protons are reduced at the metal, they move into the gas phase. Unfortunately, as discussed below, the photolysis of water does not appear to have promise as an efficient system. For another system to be useful in storing chemical energy, the separation of the products, to avoid a backreaction, must be arranged.

Now that the diagram of Figure 9.2 represents the ideal case. The semiconductor and the redox systems must be chosen to match as closely as possible the case illustrated. The Fermi energy of the semiconductor (and hence of the metal) must be very close to, or slightly above E^o_{redox} (2). Then the electrons from the semiconductor, after traversing the external circuit, will have sufficient energy to reduce the species in solution, but not excess energy that would be wasted. The energy of the valence band edge must be just below the E^o_{redox} (1), in the ideal case, so that the holes can oxidize the reducing agent #1, but just barely. Then the optimum amount of energy is stored. It should be recognized, however, that we have indicated thermodynamic energy levels in solution, not real energy levels. In terms of real energy levels, the levels of the semiconductor must be such that carriers can transfer to the real energy levels, and the real energy level E^o_{redox} for the reducing agent will, if anything, be lower than the thermodynamic energy level; and the real energy level E^o_{redox} for the oxidizing agent will, if anything, be higher than the thermodynamic energy level. For this and for other reasons, there will be an overvoltage for the reactions, and the Fermi energy of the semiconductor may, unfortunately, have to be chosen substantially higher than E^o_{redox} (2), and the valence band edge may have to be chosen substantially below E^o_{redox} (1). Any overvoltage necessary means that that amount of energy (in electron volts) is lost, and it means, of course, that the bandgap of the semiconductor used must be greater than ΔE by that amount.

There are many compromises that must be made, so that the high-efficiency system of Figure 9.2 cannot be realized, and a much lower efficiency is expected. For example, the photolysis of water stores energy at 1.23 eV per equivalent (per hole–electron pair). Thus, a semiconductor with an energy gap in the desired range, near 1.4 eV, should, in principle, be able to photolyze the water. However, the reduction of protons, in practice, almost always requires an overvoltage (except, say, on platinized platinum), and the rapid oxidation of water on a metal always requires a polarization overvoltage. When a reaction on a metal requires a polarization overvoltage, we anticipate, as discussed in Section 3.4.2, that on a semiconductor a band edge farther from E^o_{redox} will be required. Further, for kinetically very rapid reactions (so that the desired reaction is kinetically favored, compared to competing reactions such as corrosion or simple recombination) the levels of the band edge should not be just shifted to the E^o_{redox}'s of the active complex, but substantially beyond them, where Gaussians of the fluctuating energy levels of the desired reactants are more favorable. In other words, E_{cs} should be closer to E_{ox} to improve the kinetics; E_{vs} should be closer to E_{red} (see for example Figure 3.2). And finally, we must have that modest value of V_s as shown in Figure 9.2, to prevent surface recombination. And the latter is hard to control experimentally without adding recombination centers to the system. Putting all these problems together, we find that bandgaps of more than, say, 2.5 eV, are required to photolyze water, and, indeed, most of the studies have been done on materials with bandgaps of 3 eV or larger. Nozik[20] estimates that bandgaps the order of 3.3–3.5 eV are needed for efficient water photolysis. With such bandgaps, very little of the solar spectrum is utilized in the photolysis of water, and, of course, the efficiency of solar conversion is negligible. Other optical-to-chemical conversion systems will have similar efficiency problems.

A glance at Figure 5.5. will indicate which semiconductors are possible candidates to photolyze water. The first requisite is that the conduction band electrons be able to reduce protons to hydrogen. For this, the value of E_{cs} must be above zero on the hydrogen scale, by the definition of the hydrogen scale. There are only a few oxides suitable. In addition, the holes from the valence band must be able to oxidize water; E_{vs} must be below 1.3 eV, say, on the hydrogen scale. Except for CdS, clearly only the oxides have a low enough valence band. And unfortunately, of those listed, only those oxides are suitable (for reducing protons) that have an energy gap greater than 2.6 eV.

Several authors[R59,R77] have reviewed semiconductors suitable for H_2O photolysis. Mavroides *et al.*[21] have discussed photolysis at TiO_2 anodes. Kim and Laitenin[22] describe work on SnO_2. Yeh and Hackerman[23] and Kennedy and Frese[24,25], discuss photo-oxidation of water at Fe_2O_3 electrodes. Schleich *et al.*[26] describe their work on $BaTiO_3$ and mixtures of barium titanate and iron titanate (where, in the second case, the iron addition quenched the photolysis). In some of these cases the conduction band is high enough to reduce protons, in other cases it is not. The attractive feature in all these cases is that the semiconductor is not subject to corrosion by holes (see Section 9.1.4). But, in all these cases the energy gap is so large that there is no possibility of a practical device. The scientists were investigating the photolysis process, not aiming directly for a commercially useful device.

Two semiconductor electrodes in series can be used to photolyze water[R25]—one where electrons are the minority carrier and are suitable to reduce protons, and the other where holes are the minority carrier, and are suitable to oxidize water. Then two photons are required for a "hole–electron pair." Such a configuration, in principle, makes the required bandgap smaller,[15] but decreases the ultimate efficiency immediately by a factor of 1/2. Thus, as discussed by Memming,[27] *p*-GaAs can be the one-half of the cell, irradiated so the minority electrons will reduce protons. The conduction band is amply high in GaAs and in many other semiconductors, as observed in Figure 5.5 and as reviewed by Bockris and Uosaki.[28] However, there is a problem finding suitable narrow bandgap *n*-type semiconductors that can be used for the other half of the cell, to oxidize the water. With this two-semiconductor system, where half the conversion efficiency is given up before we begin, there are many narrow bandgap semiconductors such that the photoproduced holes *could* oxidize water, but the problem is their corrosion stability, as will be discussed further below. The most desirable semiconductors for solar energy conversion, those with a narrow bandgap, seem to be the least stable against self-oxidation by holes.

9.1.4. Corrosion of PEC cells

Corrosion represents the major difficulty in PEC solar cell development.[R27,R74] And, the corrosion that must be considered is not only dissolution of the electrode, but also the formation of a new phase on the electrode, such as an oxide, as holes reach the surface.[29]

Four ways have been proposed to avoid the corrosion problem. First is the use of corrosion-resistant oxide semiconductors (TiO_2 and titanates mostly), but this, in practice, seems to mean wide bandgap materials that have too low a conversion efficiency for practical interest (it requires greater than 10% efficiency before a solar cell becomes of practical interest, and even the narrow bandgap PEC cells have just barely made that figure). Second is the use of corrosion-resistant *d*-band semiconductors, an approach pursued primarily by Tributsch and co-workers—an approach that will be discussed further below. Third is the use of species in the solution with a very high capture cross section for minority carriers, so that the minority carriers are captured before they have an opportunity to decompose the electrode. And fourth is the use of a coating of a thin film of some noncorroding material to protect the semiconducting substrate, while permitting the minority carriers to move through to the surface.

The use of *d*-band semiconductors[30,31,14] is based on the concept that dissolution occurs because the holes, when they reach the surface, represent a broken bond—a missing electron in a bonding orbital. And by so removing the electrons from bonding orbitals, the holes decompose the crystal. If the holes can be in nonbonding orbitals, then the missing electron from this orbital does not represent a broken bond. As *d* orbitals can be nonbonding, semiconductors will be stable against this type of corrosion if they have a filled nonbonding *d* band lying energetically above the band occupied by the bonding electrons. The holes will be in the *d* band, not the band occupied by the bonding electrons. Transition metal compounds are candidates for such systems.[R73] Tributsch and co-workers[30,31,14] have examined MoS_2, $MoSe_2$, and WSe_2 as possible solar energy converters. With WSe_2, for example, they obtain 2% efficiency for photolysis of water,[31] which, of course, is much too low for a practical system, but suggests that the approach may be useful. On the other hand, in some cases the oxygen produced can oxidize the sulfide or selenide anions of the crystal and make the electrode unstable.[30]

The introduction into solution of species with a very high capture cross section for minority carriers has been the approach used by most workers (Section 7.4.3). Because of their unusually high value of E^o_{redox}, the polychalcogenides (polysulfide, polyselenide, or polytelluride ions) have been the most popular candidate for capture of holes from *n*-type electrodes.[32,R65] As described in Section 6.3.2, and in contradiction to the simple fluctuating energy level model, in all cases studied the hole capture cross section is higher for higher E^o_{redox}. However, as Gerrard and Rouse[R62] comment, "the extent to which the polysulfide-type electrolytes stabilize compound semiconductors to photocorrosion has yet to be accurately established." The point is, that for an electrode to be practical, its lifetime must be measured in terms of years, not hours, so any small residual corrosion, barely measureable by the techniques now available, will make the electrode impractical. And, if the action of these strong reducing agents is to rereduce the electrode when it becomes oxidized (by holes), this action may not suffice for stability. This may be what is happening with CdS electrodes in a polysulfide electrolyte. Here, as discussed at the end of Section 7.3, after a lengthly exposure it is found that the surface layer, while still cadmium sulfide, is now amorphous cadmium sulfide. Perhaps the surface lattice sulfide ions are oxidized to sulfur, breaking up the crystal, but then are promptly rereduced to sulfide ions by the electrolyte. But, if the newly reduced cadmium sulfide does not follow the original crystal structure,[33] the overall process represents corrosion.

Other redox couples, in addition to polychalcogenides, to compete with corrosion for the photoproduced minority carriers, have been investigated. For example, Eu^{3+}/Eu^{2+} have been suggested as a stabilizing couple for use with GaAs[34]; ferrocene/ferricenium has been suggested for silicon[32,35]; ferrocyanide/ferricyanide ions have been suggested with CdS electrodes.[36] But because of the reasons discussed above, there seems no reason to believe, until proven, that these will provide the long-term stability necessary for a practical device.

The fourth approach to prevent corrosion, as listed above, is to use a coating of a stable material over the semiconductor, hopefully such that minority carriers can penetrate the coating but ions cannot. Nakato *et al.*[37] have emphasized the use of metal coatings over the semiconductors. Unfortunately, it is found that if the metal layer is thin, it is in the form of small crystallites, and thus does not protect the semiconductor substrate[39] (although it is of value as a catalyst, as will be described in Section 9.2). On the other hand, if the metal layer is thick enough to protect the semiconductor substrate, it may be too thick for the "hot" holes to penetrate without being scattered, so the holes assume the energy of the Fermi energy of the metal. Then the solar cell becomes a simple metal Schottky barrier solar cell and the only role of the electrolyte is to make contact to the metal.

Finally, wide bandgap semiconductors have been used[40–42] as coatings over the desirable narrow bandgap semiconductors in an effort to protect the latter against corrosion. MoO_3, TiO_2, and many other sputtered materials have been used over the various semiconductors, such as Si, GaAs, CdS, InP, etc. It is found that one of two problems remains—either the carriers cannot make their way through the film, or corrosion still occurs through the film. A clearly successful coating/semiconductor system has not yet been identified.

9.1.5. The Future Potential of PEC Solar Cells

From the above discussions, it is clear that at present, the technology developed for PEC cells is insufficient to produce a practical cell. Were it not for the fact that conversion of solar energy into high-grade electrical energy (as opposed to low-grade heat) is so desperately desired, there is no doubt that the current high rate of activity in this field would be much diminished. At the present time, strictly solid state solar cells come much closer to being a practical device than do PEC solar cells. However, solid state solar cells are still not practical as a low cost approach to energy generation, and several serious problems can be listed that must be solved before they will ever be practical, probably problems as serious as those faced by the PEC designers. Thus, because of the strong motivation, and the ever-present possibility of a breakthrough that will lead to a practical device, work is still continuing at a reasonable level to develop a workable PEC solar energy converter.

Other modifications of PEC solar cells than those described above are under study. For an important example, we have the configuration of Nakato and co-workers,[37] who found that platinum, deposited directly on grains of the semiconductor, would catalyze the reduction of protons without requiring an external connection to a platinum electrode compartment. This approach, where the "electrochemical cell" is one grain of powder, was mentioned as a potential advantage of PEC cells in the introduction. The approach, if it can be made successful, will lead to very-low-cost optical-to-chemical energy conversion, and is an approach not accessible to the solid state sensor developers. The approach is reviewed by Bard.[R37] A similar approach[38] involves microspheres of silicon, all in contact with a metal support. A thin gold corrosion-resistant coating is in contact with an electrolytic solution. All are prepared as *p–n* junctions, half with the *p* region, half with the *n* region in contact with the support. The photocurrent associated with the incident light is then circular, with Br^- or I^- oxidized on the *p* layers exposed to the solution, and H^+ reduced to H_2 as a fuel on the *n* layers exposed to the solution. However, there seems a problem with corrosion resistance despite the gold coating.

Another approach, which has been relatively unsuccessful, has been the use of organic dyes in the electrolyte, where the organic dyes are selected to inject carriers into the bands of the semiconductor when the dye is excited (Section 6.5.3).[43,44,R66] Unfortunately this approach has, to date, yielded negligible conversion efficiency. First, it is difficult to shine a light at the interface because the dye is opaque at the active wavelength, and second, the dye only absorbs a small part of the solar spectrum. This approach has been reviewed by Rajeshwar *et al.*[R66] and Archer.[R58] Undoubtedly, new ideas are needed for practical PEC solar cells.

9.2. Electrocatalysis on Semiconductors

9.2.1. General

In this section we discuss the formation of valuable products by the electrochemical oxidation or reduction of low-cost reactants, and discuss the use of semiconductor electrodes for this purpose. We use the term electrocatalysis to represent the process. Often such processes on metal electrodes are catalytic, requiring adsorption of the reactant on the metal or on a deposited "catalyst" to force the reaction toward the desired product.[R55] But because there does not seem to be a better term for the process without a catalyst, and because the term is a natural extension of the generally used term "photocatalysis" when both electrons and holes reach the surface of a semiconductor (with no deposited catalyst), we are using the term as a general expression for desirable electrode reactions. Thus by electrocatalysis we will mean the movement of holes or electrons to the surface under an anodic or cathodic bias, respectively, to form desired products in the solution.

There seems to be a close correspondence between electrocatalysis and catalysis, which suggests that electrocatalysis on semiconductors will be of value. Successful metal catalysts for gas phase reactions have been compared in extensive literature[45,46,R67] with successful electrocatalysis on metals.[47] Now many of the highly selective catalysts for gas phase oxidation/reduction reactions are semiconductors.[R4] So, it can be anticipated that these successful semiconductor catalysts for gas phase reactions will be paralleled by successful semiconductors for electrocatalysis when semiconductor electrodes are somewhat better understood.

The purpose of this section is to indicate why a high selectivity in semiconductor electrocatalysis is to be expected. A few cases of electrocatalysis on semiconductor electrodes have been studied, and will be discussed, but in general (not always) equivalent results would have been obtained with metal electrodes. And if there is no other difference, metal electrodes must always be preferred because of their much higher current-handling capability. But, in this section we will not only report available studies, but first will examine the general theories of semiconductor electrochemistry to suggest cases where special and desirable characteristics in electrocatalysis can be anticipated for semiconductor electrodes.

The advantages of semiconductor electrodes must, of course, lie in the differences in behavior between semiconductor and metal electrodes. Such phenomena as described in earlier chapters: the possible low capture cross section for highly exothermic reactions, current doubling, hole injection, radical generation, hot carriers, and controlled tunneling, all have the potential of making reaction routes occur on semiconductor electrodes that would be abnormal for metal electrodes.

The low capture cross section for highly exothermic reactions is clearly of interest, for it would mean that of two oxidizing agents, or two complexes in solution, the semiconductor could be designed to reduce the unfavorable (thermodynamically) one, whereas the metal will usually reduce the more obvious species. Of course, as discussed in Chapter 6, especially in Section 6.2, it is clear that for such thermodynamically unfavorable reactions to be kinetically faster, the electrode must be free of imperfections, and even then there are only a few examples indicating the feasibility of this route to selectivity, even though the fluctuating energy level model predicts it.

Current doubling is of interest in that it makes two-equivalent reactions virtually irreversible,[R31] but is of less interest because it usually is only important when high-energy (and thus expensive) illumination is provided. The general irreversibility of semiconductor photoelectrolytic reactions has also been pointed out by Williams and Nozik.[48]

Hole injection is one type of process that makes semiconductors so useful in gas-phase catalysis.[R4] An oxidizing agent injects a hole into the semiconductor catalyst and the hole in turn oxidizes another species in solution. If such a process is favorable when the two species would not directly react, then the semiconductor is acting as a catalyst. Similarly at an electrode, if a semiconductor can be chosen so the oxidizing agent injects holes and then those holes oxidize a reducing agent, we have catalysis. If the reducing agent is two-equivalent, and the electrode is biased anodically, then we will be taking advantage of current doubling to make the reaction irreversible. Such a system could provide a valuable electrocatalytic process. Incidentally, although we have used hole injection as the example, the reverse process on a p-type semiconductor, electron injection followed by current doubling, is equally possible, leading to the reduction of a two-equivalent oxidizing agent by an electron-injecting reducing agent with the help of an applied cathodic voltage. Unfortunately, to date the only example of such a reaction is the catalyzed oxidative corrosion of Ge or GaAs by oxidizing agents such as ferricyanide (Section 6.5.1).

Hot electron, or hot hole effects, are of particular interest for electrocatalysis. Hot electrons are known to be active in gas phase catalysis.[49] Hot carriers are carriers at an energy far from the Fermi energy produced by breakdown or by illumination. A known example of the generation of hot electrons applicable to electrodes was discussed in Section 8.2.3, namely, hot electrons generated in silicon. Either by biasing a p-type silicon electrode into deep depletion, or by providing a p–n junction near the silicon surface, one can produce hot electrons at the silicon surface.[R49] In the studies described both in Section 8.2.3, and in DiMaria's review, the hot electrons were produced with an SiO_2 layer at the surface of the silicon, and the electrons were so hot that they were able to enter the conduction band of the SiO_2 (only 0.8 eV below the free electron energy, as shown in Figure 8.2!). Such hot electrons could reduce most species in solution, and indeed in the study of Madou *et al.*[50] they were found to reduce zinc ions in aqueous solution, and (judging from the emission of visible light) probably reduced Na^+ ions as an intermediate before the Na reacted with protons from the aqueous solvent. The point is, of course, that such hot carriers can be obtained easily on semiconductors and can be made to react with species otherwise difficult to reduce or oxidize.

The potential of controlled tunneling has been demonstrated in the work of Pettinger *et al.*[51] as was described near the end of Section 6.5.2, using Figure 6.9. Here electron tunneling from the conduction band of ZnO to $IrCl_6^{-3}$ was found to occur when the value of E_{ox} was lined up with the conduction band of the ZnO. A process similar to this could have very high selectivity, if there is no interference by surface states. Of course, one disadvantage would be the low current density.

As has been mentioned, there are a few experimental studies of reactions on semiconductor electrodes that can be described as electrocatalysis, but most of these have already been reviewed in Chapter 6, or in the preceding section (Section 9.1). Studies where the production of reaction products were considered of most interest (as opposed to basic electrode reaction studies, or to solar energy conversion) include such studies as Miyake *et al.*[52,53] They examined the oxidation of alcohols on TiO_2 to produce aldehydes or acetone, and examined the reduction of methylene blue to a leuco form. They suggest that electrocatalytic reactions on semiconductors may be of more practical interest than photocatalytic reactions (Section 6.3.3) because with photocatalytic reactions one always has the competition of direct recombination. Another example is given by the studies of Frank and Bard,[6] who also studied oxidation of various species on illuminated TiO_2 and who coined the term "photoelectrosynthesis" to describe the process. They obtained the same products without the added difficulty of illuminating the sample by simply using a platinum anode. Bard[R37] reviews "photoelectrosynthesis" on TiO_2. Landsberg *et al.*[54] reduced aromatic hydrocarbons with electrons from GaAs and GaP, but needed an accumulation layer for the reduction. The study was interesting in that the reduction was found irreversible on GaP.

Inoue *et al.*[55] show that various hydrocarbons can be generated by reduction of CO_2 using semiconductors with E_{cs} higher than 0.25 eV, the E^o_{redox} for the CH_3OH/H_2CO_3 couple. This observation may lead to important use of semiconductor electrodes in hydrcarbon synthesis.

As described, electrocatalysis using semiconductor electrodes without special characteristics has great promise for controlled chemical synthesis, but to date has been neglected, probably because our understanding of the electrode characteristics of semiconductors has been weak. Hopefully, now that, in principle, band edges can be "designed" for a catalytic reaction (i.e., a semiconductor with the desired band edge can be chosen) this field will begin to develop rapidly.

9.2.2. Surface State Additives and Narrow Bands in Electrocatalysis

In this short section we indicate two variations on the preceding simple semiconductor model that can be effective in promoting electrocatalytic activity. The first is the addition of active species to act as surface states to catalyze the reactions; the second is to use *d*-band semiconductors for their natural catalytic activity.

Surface additives have been discussed earlier. For example, the action of deposited metals in catalyzing the reduction of protons was discussed in Section 7.3.3. A photocatalytic process to deposit the metals was discussed in Section 8.3. Then we have surface states in the form of "chemically modified electrodes" as discussed at the end of Section 7.3.3. The active groups deposited as surface states on the electrode in this process can be chosen to be catalytically active: they can be chosen to adsorb the reactant of interest, to react rapidly with the reactant of interest, or simply to prevent undesirable reactions with other species in solution. Evans *et al.*[(56)] have explored the use of chemically modified electrodes for electrocatalysis.

Another form of surface group that can be involved in electrocatalysis is the intrinsic surface state. In gas phase catalysis, for example, Bickley and Jayantz[(57)] find that the OH groups intrinsic to the surface of TiO_2 act as traps for holes, and are involved in the oxidation of isopropyl alcohol to formic acid or acetaldehyde. And, as discussed at the end of Section 6.3.2, undoubtedly intrinsic surface states also play a dominant role in hole capture by ions in solution. However, so little is known about the intrinsic surface state, especially in solution, that it seems it will be far in the future before we can use them in a constructive way in electrocatalysis.

Finally, it is important to discuss narrow bands, although here the exact role of semiconductor theory is not clear.

The basis for the activity of narrow, partially filled bands, in catalysis and in electrocatalysis, is clear from the studies of Wolfram and co-workers,[(58)] who describe the formation of bulk and surface state levels on solids in simple terms, starting from Madelung, covalent, crystal field forces, followed by the influence of band formation, and the influence of irregularities such as edges, corners, and grain boundaries. They, in their discussion, are interested primarily in perovskites, but the concepts are reasonably general. The point is, that with the perovskites, with a conduction band of *d* levels, one gets interesting surface state orbitals and interesting bonding possibilities.

When considering such narrow, partly filled bands, it is not clear whether the band model ever provides the best description of the reactions.[(R4)] If space charge regions are negligible, because the density of states at the Fermi energy is very high (a high density of cations of both valences are present), then the material acts as a highly degenerate semiconductor—essentially acting as a metal. Then it may well be that there is no advantage or purpose to using the band model description, and one should turn, say, to Marcus' model of electron transfer between ions in a homogeneous solution, to provide a more useful picture of reactions at an electrode.[59]

It is also not clear that the "semiconductor properties," listed in Section 9.2.1 as the advantages of semiconductors in electrocatalysis, will be realized with narrow band semiconductors.

Some examples that suggest possible semiconductor-type behavior with narrow band materials can, however, be cited. We have the case of silver oxide as an electrode, where it is concluded that holes reaching the surface act in all respects as Ag^{2+} ions.[60] The reaction on these surface states or bands (however, one wants to define them) with ethylene glycol, results in the oxidation of the ethylene glycol and the observation of subsequent current doubling. Also we can describe NiO as a semiconductor.

Yoke *et al.*[61] consider holes arriving at the surface of NiO as being in a Ni^{2+}/Ni^{3+} band. Large densities of holes become localized at the surface (become Ni^{3+} ions) at about a 0.9 V (vs. NHE) anodic electrode potential. They discuss the conductance in this material as being "polaron" conductance, a concept we have not discussed in detail in this book, but which may be of interest as a description of the solid state properties of many solids used as electrodes, particularly solids with a high dielectric constant. Now NiO is of interest as an oxygen electrode. Tseung[R68] has reviewed such electrodes, discussing the chemical processes that lead to overvoltages and how they can be bypassed for an efficient electrode. However, in this case, the space charge region is definitely present at the surface,[62] so clearly the material is best described by semiconductor models. This system was briefly reviewed in Section 6.6. Another group of materials that has led to recent interest as an oxygen electrode is the rare earth cobaltites,[63] such as $La_{0.5}Sr_{0.5}CoO_3$.[64] It should be pointed out that in the sense of Chapter 8, all materials used for an oxygen electrode will have a layer of oxide on their surfaces, so can be considered nonmetallic electrodes.

Another example of a commercially important narrow band electrode, where again it is not entirely clear whether the band model can be helpful, is that of RuO_2, an electrode used for chlorine generation from brine. Here[65] the material is characterized by a narrow conduction band of *d* electrons, specifically a half-filled Ru–O antibonding π orbital. In fact, Doblhofer *et al.*[66] conclude that the surface ruthenium atoms are so active that they can be oxidized or reduced from Ru^{+2} to Ru^{+6} as one changes the applied voltage between the limits of hydrogen evolution and oxygen evolution in aqueous solution. However, according to Pizzini *et al.*,[67] the incorporation of chlorine increases the resistance of the RuO_2, because the chlorine affects the *d* orbitals. And, because it is as a chlorine electrode that we are interested in RuO_2, there is plenty of chlorine available. Thus Kuhn and Mortimer[68] suggest that the material can best be considered as an *n*-type semiconductor. On the other hand, Kuhn agrees with Trasatti[69] that RuO_2 does not show semiconductor behavior, in terms of temperature dependence of resistance, for example, and a semiconductor description may not be best. In summary, it appears probable that RuO_2 is certainly not a simple semiconductor, but in order to describe the important bonding orbitals at the surface when chlorine[67] or even water[68] is present, it may be that a band model may be required to describe the high-resistance space charge region.

9.3. New Devices

As indicated in the first few words of this chapter, to date few new devices have arisen from studies on semiconductor electrochemistry, but then, to date there have been few studies of semiconductor electrochemistry aimed at producing devices. Indeed, we have just reached that level of understanding of the processes occurring at the semiconductor/electrolyte interface where one can expect ideas for applications to develop. In the present section, we will discuss a few additional (in addition to energy conversion and electrocatalysis) applications and potential applications of the concepts developed in the preceding chapters.

Printed circuits and photography are areas where the photocatalytic reduction of noble metals (Section 8.3) on materials such as TiO_2 has been used. Kelly and Vondeling[70] discuss the deposition of palladium and silver on TiO_2 when the material is illuminated. By illuminating selectively, the printed circuit or the photograph is exposed. The resulting nuclei of silver are then "developed" by electroless plating, using an electroless process that deposits further metal on the nuclei but not on the bare TiO_2. Kelly and Vondeling[70] and Möllers *et al.*[71] each provide a list of references to earlier work on this process.

Semiconductor electrodes can also, in principle, be used as a source of luminescence. Faulkner and Bard[72] discuss luminescence on semiconductor electrodes. Noufi *et al.*[73] describe luminescence associated with the reduction of $S_2O_8^{2-}$ on TiO_2. Madou *et al.*[50] discuss luminescence associated with electrons flowing from the conduction band of an SiO_2 overlayer on Si into aqueous solution. And Young[R7] reports many cases of luminescence upon application of an anodic voltage to oxide films. None of these systems seems practical, but more interesting systems may be found in the future.

Nicholson[74] suggested that image conversion could be realized, using one of a group of aromatic hydrocarbon dyes that emit light in solution.

If placed between two metal electrodes, these dyes are oxidized to D^+ upon anodic bias, reduced to D^- upon cathodic bias, and the reaction

$$D^+ + D^- \rightarrow 2D + h\nu$$

emits light. If one of the electrodes is replaced by a reverse-biased semiconductor electrode, such that the reaction at the semiconductor will only be possible by illuminating the semiconductor, the luminescence in solution will only result when photons strike the semiconductor — image conversion. Luttmer and Bard[75] have demonstrated the effect.

A somewhat different optical effect is "electrochromism," where use is made of the observation that under the high fields of the space charge region at the surface of a semiconductor electrode, ions can move in by field-aided diffusion. Thus Gissler and Memming[76] and Mohapatra[77] report studies of tungstates whose color is changed by the in-diffusion of H^+ or Li^+ in an electrochemical cell. Both these authors cite extensive literature on the subject.

Finally, we will mention ISFETs and OSFETs, sensors using a silicon electrode with a silica layer on the surface. The silicon substrate is prepared as a field effect transistor (FET), a design where the conductance through the silicon is extremely sensitive to the potential on the outer face of the silica. In the normal FET, the MOSFET, the outer face of the silica is metallized, and a voltage to be amplified is applied to this outer face of the silica. In an ISFET or an OSFET, the outer face of the silica is exposed to the solution, and the voltage appearing there is determined by the adsorption of ions from the solution. This makes an extremely sensitive device for the determination of the presence of such species. The term ISFET means Ion-Sensitive (or Ion-Selective) Field Effect Transistor, and the term OSFET means Oxide/Semiconductor FET (to compare to the normal MOSFET, the Metal/Oxide/Semiconductor FET).

We have not described ISFETs in detail in this book because, in principle, they do not involve electron transfer, which has been the main emphasis here. Reviews of these devices are available by Buck and Hackleman,[78] Zemel,[79,80] Janata and Moss,[81] together with other authors,[82–85] and a book edited by Cheung.[R71] The topic has been very well studied in recent years, but there seems no practical device on the market as yet. As discussed in Section 8.2.3, silica is not extremely stable in aqueous solution.

9.4. Electropolishing of Semiconductors

When germanium transistors were dominant, in the 1950s, electrochemical polishing of transistors was an important part of the processing. Indium was alloyed into the n-type germanium to form a p–n junction, and current passing anodically through the indium would etch selectively the region of indium-doped germanium. The *n*-type germanium would not etch (except near the junction) because there were no holes and, as discussed in Section 7.4.1, holes reaching the surface are required for the dissolution of the germanium as Ge^{+4}, as described also by Equation (6.27). And the indium did not etch. Thus the region of the junction could be selectively etched, removing enough material so that flaws in the remaining material were minimum.

However, when silicon transistors became dominant, in the 1960s, and in turn gave way to integrated circuitry, electropolishing of transistors became not only unnecessary (because of the design of the transistors), but almost impossible, because silicon does not easily electropolish. Silicon dioxide forms rather than the silicon dissolving, and even with fluoride ions present in solution to dissolve the silicon dioxide, a satisfactory polish is much more easily obtained with a chemical etch, rather than an electrochemical one.

Studies of the electropolishing of silicon have, however, been made and are available for special effects. For example, Pham[(86)] combines electropolishing with ferricyanide hole injection and electrode potential measurements, to determine the thickness of damaged silicon. Reviews are presented by Pankove in the book edited by Holms,[(R69)] and in Myamlin and Pleskov's book,[(R17)] for example. More recent work on the electropolishing of silicon is reported by Memming and Schwandt,[(87)] and by Meek.[(88)]

Other materials are now becoming important in solid state technology, for example, GaAs, and electrochemical polishing may again become of interest, because these materials do not have the natural advantage of silicon in its near-perfect oxide. Studies of the electropolishing of GaAs by Hollan *et al.*[(89)] and by Faktor *et al.*,[(90)] and Rode *et al.*[(91)] do not show promise of desirable characteristics, but special problems in preparing GaAs for FETs were found to be solved by using the technique of electropolishing under breakdown conditions. The procedure was discussed at the end of Section 7.4.2.2.

The electrochemical removal of material from semiconductors can be used for three purposes. First, of course, is the general etching or electropolishing of a uniform slice. With *n*-type material, usually the sample must be illuminated to bring the necessary holes to the surface. With *p*-type material the holes are present, and the sample simply needs to be biased anodically in a solvent in which the cation is soluble. Second is electrochemical processes that take advantage of the difference in etch rate between the *p*- and *n*-type material. This can be used to display *p–n* junctions, to etch down to *p–n* junctions (etching the *p* region only), and to etch *n* regions, which are close enough to a *p–n* junction that holes from the junction reach the surface. And, finally, as discussed below, *n*-

type material can be etched in patterns, the etching only occurring where light is allowed to strike. These various features of electrochemical etching can make it of technical importance in some cases.

An example process where the special characteristics of semiconductor electropolishing are utilized is that of hologram recording, as discussed by Tyagai *et al.*,[92] who used CdS as the semiconductor, Baliga and Ghandi,[93] who were interested in shaping SnO_2, and Dalisa *et al.*,[94] who used silicon. In the CdS work, etching was induced where the sample was illuminated. Using this simple technique, they were able to record one-dimensional diffraction gratings with up to 6×10^3 lines/mm by using intersecting laser beams to produce the interference patterns. In the silicon work,[94] an SiO_2 film is produced on *n*-silicon where it is illuminated, and the film can be removed, leaving a pattern, using HF. SnO_2 patterns, useful for transparent conductors, can also be produced by electrochemical etching, where the pattern desired is retained by masking.[93]

To conclude, it is well to refer to the excellent review by Schnable and Schmidt,[R70] which covers the use of semiconductor electrochemistry in the fabrication of semiconductor devices in far more detail than we do here.

References

References with the prefix R are listed separately under "References for Review Articles and Books" on page 374.

References for Chapter 1

1. J. N. Hobstetter, in *Semiconductors*, ed. N. B. Hannay (Reinhold, New York, 1959), p. 508.
2. J. M. Thomas, *Adv. Catal.* **19,** 293 (1969).
3. S. R. Morrison and T. Freund, in NBS Spec. Publ. No. 455, "Electrocatalysis on Non-metallic Surfaces" (1976).
4. H. D. Schulte, thesis, University of Göttingen (1971).
5. T. Wolfram, F. J. Morin, and R. Hurst, in NBS Spec. Publ. No. 455, "Electrocatalysis on Non-metallic Surfaces" (1976).
6. S. R. Morrison, *J. Catal.* **47,** 69 (1977).
7. J. D. Levine and P. Mark, *Phys. Rev.* **144,** 751 (1966).
8. H. Noller and W. Kladnig, *Cat. Rev.-Sci. Eng.* **13,** 149 (1976).
9. S. R. Morrison, *Surf. Sci.* **50,** 329 (1975).
10. B. Bazraktarogler, S. J. Hannah, and H. L. Hartnagel, *Electron. Lett.* **13,** 45 (1977).
11. D. Haneman and F. Steenbeeke, *J. Electrochem. Soc.* **124,** 861 (1977).
12. M. Madou, thesis, University of Ghent (1979).
13. R. M. Noyes, *J. Am. Chem. Soc.* **84,** 513 (1962).
14. F. Lohmann, *Z. Naturforsch.* **22a,** 843 (1967).
15. S. Trasatti, *J. Electroanal. Chem.* **52,** 313 (1974).
16. B. Case, N. S. Hush, R. Parsons, and M. E. Peover, *J. Electroanal. Chem.* **10,** 360 (1965).
17. W. N. Hansen and D. M. Kolb, *J. Electroanal. Chem.* **100,** 493 (1979).
18. R. Gomer and G. Tryson, *J. Chem. Phys.* **66,** 4413 (1977).
19. R. R. Dogonadze, A. M. Kuznetsov, and A. A. Chernenko, *Russ. Chem. Rev.* **34,** 759 (1965).
20. R. A. Marcus, *J. Chem. Phys.* **24,** 966 (1956).
21. R. A. Marcus, *Can. J. Chem.* **37,** 155 (1959).
22. R. A. Marcus, *J. Chem. Phys.* **43,** 679 (1965).

23. S. G. Christov, *Ber. Bunsenges. Phys. Chem.* **79,** 357 (1975).
24. V. G. Levich, in *Advances in Electrochemistry and Electrochemical Engineering,* Vol. 4 eds. P. H. Delahay and C. W. Tobias (Interscience, New York, 1966), p. 249.
25. S. Efrima and M. Bixon, *Chem. Phys.* **36,** 161 (1979).
26. H. Gerischer, *Surf. Sci.* **18,** 97 (1969).
27. H. Gerischer, *Z. Phys. Chem. (Frankfurt)* **26,** 233 (1960); **27,** 40 (1961).
28. S. R. Morrison, *Surf. Sci.* **10,** 459 (1968).
29. K. Hauffe, in *Electrochemistry the Past Thirty and Next Thirty Years,* eds. H. Bloom and F. Gutmann (Plenum Press, New York, 1968).

References for Chapter 2

1. H. P. Boehm, *Discuss. Faraday Soc.* **52,** 264 (1971).
2. B. Pettinger, H. R. Schöppel, T. Yokoyama, and H. Gerischer, *Ber. Bunsenges. Phys. Chem.* **78,** 1024 (1974).
3. H. Gerischer, *Surf. Sci.* 18, 97 (1969).
4. R. O. James and T. W. Healy, *J. Colloid Interface Sci.* **40,** 42, 53, 65 (1972).
5. K. Micka and H. Gerischer, *J. Electroanal. Chem.* **38,** 397 (1972).
6. P. Lemasson, A. M. Baticle, and P. Vennereau, *Surf. Sci.* **59,** 177 (1976).
7. P. J. Boddy, *Surf. Sci.* **13,** 52 (1969).
8. F. Cardon and W. P. Gomes, *Z. Phys. Chem. N.F.* 78, 205 (1972).
9. H. Gerischer, A. Maurer, and W. Mindt, *Surf. Sci.* **4,** 431 (1966).
10. R. Memming, *Surf. Sci.* **2,** 436 (1964).
11. R. M. Cornell, A. M. Posner, and J. P. Quick, *J. Colloid Interface Sci.* **53,** 6 (1975).
12. I. Uchida, H. Akahoshi, and S. Toshima, *J. Electroanal. Chem.* **88,** 79 (1978).
13. M. Tschapek, C. Wasowski, and R. M. Torres-Sanchez, *J. Electroanal. Chem.* **74,** 167 (1976).
14. H. S. Ritter, *Pigment Handbook,* Vol. III, ed. T. C. Patton (Wiley Interscience, New York, 1973).
15. P. Roy and D. W. Fuerstenau, *Surf. Sci.* **30,** 487 (1972).
16. I. Uchida, H. Urushibata, and S. Toshima, *J. Electroanal. Chem.* **96,** 45 (1979).
17. B. Timmer, M. Sluyters-Rehback, and J. H. Sluyters, *Surf. Sci.* **18,** 44 (1969).
18. S. R. Morrison and T. Freund, *J. Chem. Phys.* **47,** 1543 (1967).
19. T. Freund and S. R. Morrison, *Surf. Sci.* **9,** 119 (1968).
20. T. W. Healy, D. E. Yates, L. R. White, and D. Chan, *J. Electroanal. Chem.* **80,** 57 (1977).
21. M. Hofman-Perez and H. Gerischer, *Z. Elektrochem.* **65,** 771 (1961).
22. S. R. Morrison and T. Freund, *Electrochim. Acta* **13,** 1343 (1968).
23. A. R. Elsharkawi and K. C. Kao, *Solid State Electron.* **19,** 939 (1976).
24. W. H. Laflere, R. L. van Meirhaeghe, F. Cardon, and W. P. Gomes, *Surf. Sci.* **59,** 401 (1976).
25. Yu. V. Pleskov, *Croat. Chem. Acta* **44,** 179 (1972).
26. P. F. Schmidt, *J. Electrochem. Soc.* **115,** 168 (1968).
27. M. A. Butler and D. S. Ginley, *J. Electrochem. Soc.* **125,** 228(1978).
28. K. Frese, *J. Vac. Sci. Technol.* **16,** 1042 (1979).
29. L. Pauling, *The Nature of the Chemical Bond* (Cornell University Press, Ithaca, 1940).
30. R. T. Sanderson, *Chemical Periodicity* (Reinhold, New York, 1960).

References for Chapter 3

1. M. R. deRiepen, J. Range, and K. Hauffe, *Z. Phys. Chem. N. F.* **99,** 257 (1976).
2. H. Kokado, T. Nakayama, and E. Inouye, *J. Phys. Chem. Solids* **31,** 2785 (1971).
3. S. R. Morrison, *Surf. Sci.* **15,** 363 (1969).
4. H. H. Bauer, *J. Electroanal. Chem.* **16,** 419 (1968).
5. P. van Rysselberghe, *Electrochim. Acta* **8,** 583, 709 (1963); **9,** 1547 (1964); **10,** 107 (1965).
6. R. R. Dogonadze, A. M. Kuznetsov, and A. A. Cherenko, *Russ. Chem. Rev.* **34,** 759 (1965).
7. S. G. Christov, *Ber. Bunsenges. Phys. Chem.* **79,** 357 (1975).
8. R. A. Marcus, *J. Chem. Phys.* **24,** 966 (1956).
9. R. A. Marcus, *Can. J. Chem.* **37,** 155 (1959).
10. R. A. Marcus, *J. Chem. Phys.* **43,** 679 (1965).
11. R. R. Dogonadze, A. M. Kuznetsov, and Yu. A. Chizmadzhev, *Russ. J. Phys. Chem.* **38,** 652 (1964).
12. V. G. Levich, *Electrochem. Electrochem. Eng.* **4,** 249 (1966).
13. J. M. Hale, in *Reactions of Molecules at Electrodes*, ed. N. S. Hush (Wiley Interscience, London, 1971).
14. D. M. Tench and E. Yeager, *J. Electrochem. Soc.* **121,** 318 (1974).
15. R. Memming and F. Möllers, *Ber. Bunsenges. Phys. Chem.* **76,** 475 (1972).
16. T. Freund and S. R. Morrison, *Surf. Sci.* **9,** 119 (1968).
17. H. Gerischer, *Z. Phys. Chem. N. F.* **26,** 223 (1960).
18. H. Gerischer, *Z. Phys. Chem. N. F.* **27,** 48 (1961).
19. R. A. L. Vanden Berghe, F. Cardon, and W. P. Gomes, *Surf. Sci.* **39,** 368 (1973).

References for Chapter 4

1. M. Tomkiewicz, in *Semiconductor Liquid-Junction Solar Cells*, ed. A. Heller (Electrochemical Society, Princeton, New Jersey, 1977), p. 92.
2. T. Freund and S. R. Morrison, *Surf. Sci.* **9,** 119 (1968).
3. J. F. Dewald, *J. Phys. Chem. Solids* **14,** 155 (1960).
4. C. G. B. Garrett and W. H. Brattain, *Phys. Rev.* **99,** 376 (1955).
5. S. R. Morrison and T. Freund, *J. Chem. Phys.* **47,** 1543 (1967).
6. W. H. Brattain and P. J. Boddy, *J. Electrochem. Soc.* **109,** 574 (1962).
7. A. Goetzberger, E. Klausmann, and M. J. Schulz, *CRC Crit. Rev. Solid State Phys.* **6,** 1 (1976).
8. M. Kuhn, *Solid State Electron.* **13,** 873 (1970).
9. R. De Gryse, W. P. Gomes, F. Cardon, and J. Vennik, *J. Electrochem. Soc.* **122,** 711 (1975).
10. D. M. Tench and E. Yeager, *J. Electrochem. Soc.* **120,** 164 (1973).
11. S. Toshima and I. Uchida, *Electrochim. Acta.* **15,** 1717 (1970).
12. R. L. Van Meirhaeghe, E. C. Dutoit, F. Cardon, and W. P. Gomes, *Electrochim. Acta.* **20,** 995 (1975).
13. M. J. Madou, F. Cardon, and W. P. Gomes, *J. Electrochem. Soc.* **124,** 1623 (1977).
14. E. C. Dutoit, R. L. van Meirhaeghe, F. Cardon, and W. P. Gomes, *Ber. Bunsenges. Phys. Chem.* **79,** 1206 (1975).
15. P. A. Kohl and A. J. Bard, *J. Am. Chem. Soc.* **99,** 7531 (1977).
16. S. N. Frank and A. J. Bard, *J. Am. Chem. Soc.* **97,** 7427 (1975).

17. N. R. Armstrong, A. W. C. Lin, M. Fujihara, and T. Kuwana, *Anal. Chem.* **48,** 741 (1976).
18. R. K. Jain, H. C. Gain, and B. J. Welch, *J. Electroanal. Chem.* **79,** 211 (1977).
19. W. W. Hansen, R. A. Osteryoung, and T. Kuwana, *J. Am. Chem. Soc.* **88,** 1062 (1966).
20. R. Memming and F. Möllers, *Ber. Bunsenges. Phys. Chem.* **76,** 475 (1972).
21. W. Schmickler, *Ber. Bunsenges. Phys. Chem.* **82,** 477 (1978).
22. S. R. Morrison, *Surf. Sci.* **15,** 363 (1969).
23. T. Freund, *J. Phys. Chem.* **73,** 468 (1969).
24. A. C. Riddiford, *Adv. Electrochem. Electrochem. Eng.* **4,** 47 (1966).
25. F. Opekar and P. Beran, *J. Electroanal. Chem.* **69,** 1 (1976).
26. S. Bruckenstein and B. Miller, *Acc. Chem. Res.* **10,** 54 (1977).
27. G. Ritzler and M. Gross, *J. Electroanal. Chem.* **94,** 209 (1978).
28. A. Caprani, C. Deslouis, M. Kaddan, Ph. Morel, and B. Tribollet, *Electrochim. Acta.* **22,** 1231 (1977).
29. K. Aoki and H. Matsuda, *J. Electroanal. Chem.* **94,** 157 (1978).
30. S. R.Morrison and T. Freund, NBS Spec. Publ. No. 455, "Electrocatalysis on Nonmetallic Surfaces" (1976).
31. H. Kiess, *J. Phys. Chem. Solids* **31,** 2379 (1970).
32. W. W. Gärtner, *Phys. Rev.* **116,** 84 (1959).
33. M. A. Butler, *J. Appl. Phys.* **48,** 1914 (1977).
34. D. S. Ginley and M. A. Butler, *J. Appl. Phys.* **48,** 2019 (1977).
35. Yu. V. Pleskov, *Progr. Surf. Membrane Sci.* **7,** 57 (1973).
36. G. W. Simmons, E. Kellerman, and H. Leidheiser, Jr., *J. Electrochem. Soc.* **123,** 1276 (1976).
37. J. E. Dubois, P. C. Lacaze, R. Courtel, C. C. Herrmann, and D. Mangis, *J. Electrochem. Soc.* **122,** 1454 (1975).
38. A. K. Zakzouk, R. A. Stuart, and W. Eccleston, *J. Electrochem. Soc.* **123,** 1551 (1976).
39. D. E. Aspnes and A. Frova, *Phys. Rev.* B **2,** 1037 (1970).
40. H. Gobrecht and R. Thull, *Ber. Bunsenges. Phys. Chem.* **74,** 1234 (1970).
41. H. Gobrecht, R. Thull, F. Hein, and M. Schaldach, *Ber. Bunsenges. Phys. Chem.* **73,** 68 (1969).
42. K. Vos and H. J. Krusemeyer, *J. Phys. C* **10,** 3893 (1977).
43. J. F. Pond, *Surf. Sci.* **37,** 596 (1973).
44. L.-S. R. Yeh and N. Hackerman, *J. Phys. Chem.* **82,** 2719 (1978).
45. N. J. Harrick, *Ann. N.Y. Acad. Sci.* **101,** 928 (1963).
46. F. Möllers and R. Memming, *Ber. Bunsenges Phys. Chem.* **77,** 879 (1973).
47. E. Yeager and J. Kuta, in *Physical Chemistry,* Vol. IX A, ed. H. Eyring (Academic Press, New York, 1970).
48. A. H. Reed and E. Yeager, *Electrochim. Acta.* **15,** 1345 (1970).
49. W. H. Brattain and C. G. B. Garrett, *Bell Sys. Tech. J.* **34,** 129 (1955).
50. F. Cardon, *Physica* **52,** 144 (1971).
51. W. P. Gomes and F. Cardon, *J. Solid-State Chem.* 3, 125 (1971).
52. F. Cardon and W. P. Gomes, *Ber. Gunsenges Phys. Chem.* **74,** 436 (1970).
53. V. A. Tyagai and N. B. Lukyanchikova, *Surf. Sci.* **12,** 331 (1968).
54. M. Gleria and R. Memming, *Z. Phys. Chem. N.F.* **101,** 171 (1976).
55. E. W. Grabner, *Electrochim. Acta.* **20,** 7 (1975).
56. M. R. de Riepen, J. Range, and K. Hauffe, *Z. Phys. Chem. N.F.* **99,** 257 (1976).
57. A. B. Ellis and B. R. Kares, *J. Am. Chem. Soc.* **101,** 236 (1979).
58. G. Petermann, H. Tributsch, and R. Bogomolni, *J. Chem. Phys.* **57,** 1026 (1972).
59. D. J. Bernard and P. Handler, *Surf. Sci.* 40, 141 (1973).
60. B. Pettinger, H.-R. Schöppel, T. Yokoyama, and H. Gerischer, *Ber. Bunsenges. Phys. Chem.* **78,** 1024 (1974).

61. H. Morisaki and K. Yazawa, *Appl. Phys. Lett.* **33,** 1013 (1978).
62. K. Hauffe and H. Volz, *Ber. Bunsenges. Phys. Chem.* **77,** 967 (1973).
63. K. W. Frese, Jr. and S. R. Morrison, *J. Electrochem. Soc.* **126,** 1235 (1979).

References for Chapter 5

1. J. F. Dewald, *Proceedings of the Second Conference on Semiconducting Surfaces*, ed. J. Zemel (Pergamon Press, New York, 1960).
2. F. Lohmann, *Ber. Bunsenges. Phys. Chem.* **70,** 428 (1966).
3. A. Fujishima, A. Sakamoto, and K. Honda, *Seisan-Kenkyu* **21,** 450 (1969).
4. E. C. Dutoit, F. Cardon, and W. P. Gomes, *Ber. Bunsenges. Phys. Chem.* **80,** 475 (1976).
5. M. J. Madou, F. Cardon, and W. P. Gomes, *Ber. Bunsenges. Phys. Chem.* **81,** 1186 (1977).
6. W. H. Laflere, F. Cardon, and W. P. Gomes, *Surf. Sci.* **44,** 541 (1974).
7. J. H. Kennedy and K. W. Frese, Jr., *J. Electrochem. Soc.* **125,** 723 (1978).
8. F. El. Halonani, H. Traore, G. Allais, and A. Deschanvres, *Surf. Sci.* **79,** 498 (1979).
9. F. Möllers and R. Memming, *Ber. Bunsenges. Phys. Chem.* **76,** 469 (1972).
10. P. J. Boddy, *Surf. Sci.* **13,** 52 (1969); *J. Electrochem. Soc.* **115,** 199 (1968).
11. D. J. Bernard and P. Handler, *Surf. Sci.* **40,** 141 (1973).
12. D. S. Ginley and M. A. Butler, *J. Electrochem. Soc.* **125,** 1968 (1978).
13. W. H. Brattain and P. J. Boddy, *J. Electrochem. Soc.* **109,** 574 (1962).
14. H. Gerischer, A. Maurer and W. Mindt, *Surf. Sci.* **4,** 431 (1966); H. Gerischer and W. Mindt, *ibid.* **4,** 440 (1966).
15. H. Gerischer and I. Mattes, *Z. Phys. Chem. N.F.* **49,** 112 (1966).
16. S. R. Morrison, *Surf. Sci.* **34,** 462 (1975).
17. H. Gerischer, *Surf. Sci.* **18,** 97 (1969).
18. P. A. Kohl and A. J. Bard, *J. Am. Chem. Soc.* **99,** 7531 (1977).
19. V. A. Tyagai and G. Ya. Kolbasov, *Surf. Sci.* **28,** 423 (1971).
20. I. Uchida, H. Akahoshi, and S. Toshima, *J. Electroanal. Chem.* **88,** 79 (1978).
21. I. Uchida, H. Urushibata, and S. Toshima, *J. Electroanal. Chem.* **96,** 45 (1979).
22. R. O. James and T. W. Healy, *J. Colloid Interface Sci.* **40,** 42, 53, 65 (1972).
23. W. W. Harvey, *J. Electrochem. Soc.* **114,** 472 (1967).
24. T. Freund and S. R. Morrison, *Surf. Sci.* **9,** 119 (1968).
25. R. A. L. Vanden Berghe, W. P. Gomes, and F. Cardon, *Ber. Bunsenges. Phys. Chem.* **77,** 289 (1973).
26. F. Williams and A. J. Nozik, *Nature* **271,** 137 (1978).
27. S. R. Morrison, in *Current Problems in Electrophotography*, eds. C. Berg and K. Hauffe (de Gruyter, Berlin, 1972).
28. S. R. Morrison and T. Freund, in NBS Spec. Publ. No. 455, "Electrocatalysis on Nonmetallic Surfaces" (1976).
29. W. P. Gomes, T. Freund, and S. R. Morrison, *J. Electrochem. Soc.* **115,** 818 (1968).
30. M. A. Butler, *J. Electrochem. Soc.* **126,** 338 (1979).
31. T. Ambridge, *Electron. Lett. (G.B.)* **10,** 204 (1974).
32. H. Gerischer and J. Gobrecht, *Ber. Bunsenges. Phys. Chem.* **82,** 520 (1978).
33. K. W. Frese, Jr. and S. R. Morrison, *J. Electrochem. Soc.* **126,** 1235 (1979).
34. L. M. Terman, *Solid State Electron.* **5,** 285 (1962).
35. M. T. Pham and J. Hueller, *J. Appl. Electrochem.* **7,** 531 (1977).
36. W. T. Read, Jr., *Phil. Mag.* **45,** 775, 1119 (1954).
37. D. J. Thomas and J. J. Lander, *J. Chem. Phys.* **25,** 1136 (1956).
38. W. D. Mackintosh, F. Brown, and H. H. Plattner, *J. Electrochem. Soc.* **124,** 1168 (1977).

39. F. Kornfeil, *J. Electrochem. Soc.* **109,** 349 (1962).
40. C. K. Dyer and J. S. L. Leach, *J. Electrochem. Soc.* **125,** 23 (1978).
41. D. Haneman and F. Steenbeeke, *J. Electrochem. Soc.* **124,** 861 (1977).
42. L. A. Harris and R. H. Wilson, *J. Electrochem. Soc.* **124,** 1010 (1976).
43. M. A. Butler, *J. Electrochem. Soc.* **126,** 338 (1979).
44. R. N. Noufi, P. A. Kohl, S. N. Frank, and A. J. Bard, *J. Electrochem. Soc.* **125,** 246 (1978).
45. C. N. Sayers and N. R. Armstrong, *Surf. Sci.* **77,** 301 (1978).
46. S. K. Mohapatra, *J. Electrochem. Soc.* **125,** 285 (1978).
47. H. Yoneyama, S. Mayumi, and H. Tamura, *J. Electrochem. Soc.* **125,** 68 (1978).
48. S. Gottesfeld and J. D. E. McIntrye, *J. Electrochem. Soc.* **126,** 742 (1979).
49. J. Corish and C. J. Warde, *Ber. Bunsenges. Phys. Chem.* **82,** 282 (1978).
50. C. T. Sah and J. W. Walker, *Appl. Phys. Lett.* **22,** 384 (1973).
51. M. G. Buehler, *Solid State Electron.* **15,** 69 (1972).
52. D. V. Lang, *J. Appl. Phys.* **45,** 3014, 3023 (1974).
53. N. M. Johnson, *Appl. Phys. Lett.* **34,** 802 (1979).
54. A. R. Elsharkawi and K. C. Kao, *Solid-State Electron.* **19,** 939 (1976).
55. N. Johnson (private communication).
56. H. Gerischer, F. Hein, M. Lübke, E. Meyer, B. Pettinger, and H.-R. Schöppel, *Ber. Bunsenges. Phys. Chem.* **77,** 284 (1973).
57. F. Vanden Kerchove, J. Vandermolen, W. P. Gomes, and F. Cardon, *Ber. Bunsenges. Phys. Chem.* **83,** 230 (1979).
58. R. A. L. Vanden Berghe, W. P. Gomes, and F. Cardon, *Ber. Bunsenges. Phys. Chem.* **77,** 289 (1973).
59. J. Hueller and M. T. Pham, *J. Appl. Electrochem.* **7,** 531 (1977).
60. W. Schmickler, *Ber. Bunsenges. Phys. Chem.* **82,** 477 (1978).
61. J. C. Tranchart, L. Hollan, and R. Memming, *J. Electrochem. Soc.* **125,** 1185 (1978).
62. W. Shockley and W. T. Read, Jr., *Phys. Rev.* **87,** 835 (1952).
63. S. R. Morrison, *Phys. Rev.* **104,** 619 (1956).
64. T. Figielski, *Solid State Electron.* **21,** 1403 (1978).
65. D. M. Tench and H. Gerischer, *J. Electrochem. Soc.* **124,** 1612 (1977).
66. A. B. Ellis and B. R. Karas, *J. Am. Chem. Soc.* **101,** 236 (1979).
67. J. R. Maltby, C. E. Reed, and C. G. Scott, *Surf. Sci.* **51,** 89 (1975).
68. A. Heller, K. C. Chang, and B. Miller, *J. Electrochem. Soc.* **124,** 697 (1977).
69. W. Gissler and R. Memming, *J. Electrochem. Soc.* **124,** 1710 (1977).

References for Chapter 6

1. R. Landsberg, P. Janietz, and R. Dehmlow, *Z. Phys. Chem. (Leipzig)* **257,** 657 (1976).
2. S. R. Morrison, *Surf. Sci.* **10,** 459 (1968).
3. R. A. L. Vanden Berghe, F. Cardon, and W. P. Gomes, *Surf. Sci.* **39,** 368 (1973).
4. R. Memming and F. Möllers, *Ber. Bunsenges. Phys. Chem.* **73,** 475 (1972).
5. J. M. Hale, in *Reactions of Molecules at Electrodes*, ed. N. S. Hush (Wiley, London, 1971).
6. R. Memming and F. Möllers, *Ber. Bunsenges. Phys. Chem.* **76,** 609 (1972).
7. S. R. Morrison and T. Freund, *J. Chem. Phys.* **47,** 1543 (1967).
8. K. M. Sancier and S. R. Morrison, *Surf. Sci.* **36,** 622 (1973).
9. M. J. Madou, F. Cardon, and W. P. Gomes, *Ber. Bunsenges. Phys. Chem.* **81,** 1186 (1977).
10. M. Gleria and R. Memming, *J. Electroanal. Chem.* **65,** 163 (1975).

11. T. Freund and S. R. Morrison, *Surf. Sci.* **9,** 119 (1968).
12. S. R. Morrison, *Surf. Sci.* **15,** 363 (1969).
13. F. Cardon and W.P. Gomes, *Z. Phys. Chem. N.F.* **78,** 205 (1972).
14. H. Gerischer and I. Mattes, *Z. Phys. Chem. N.F.* **49,** 112 (1966).
15. H. Gerischer and E. Meyer, *Z. Phys. Chem. N.F.* **74,** 302 (1971).
16. V. A. Tyagai and G. Ya. Kolbasov, *Surf. Sci.* **28,** 423 (1971).
17. E. W. Grabner, *Electrochim. Acta* **20,** 7 (1975).
18. M. Gleria and R. Memming, *Z. Phys. Chem. N.F.* **101,** 171 (1976).
19. B. Pettinger, H.-R. Schöppel, and H. Gerischer, *Ber. Bunsenges. Phys. Chem.* **77,** 960 (1973); **78,** 450 (1974).
20. I. Uchida, K. Niki, and H. A. Laitinen, *J. Electrochem. Soc.* **125,** 1759 (1978).
21. P. A. Kohl and A. J. Bard, *J. Am. Chem. Soc.* **99,** 7531 (1977).
22. S. N. Frank and A. J. Bard, *J. Am. Chem. Soc.* **97,** 7427 (1975).
23. L.-S. R. Yeh and N. Hackerman, *J. Phys. Chem.* **82,** 2719 (1978).
24. P. Kofstad, *Nonstoichiometry, Diffusion and Electrical Conductivity in Binary Crystals* (Wiley Interscience, New York, 1972).
25. J. Vandermolen, W. P. Gomes, and F. Cardon, *J. Electrochem. Soc.* **127,** 324 (1980).
26. R. Memming, in *The Semiconductor Liquid-junction Solar Cell,* ed. A. Heller, (Electrochemical Society, Princeton, New Jersey, 1977).
27. A. Fujishima, (private communication).
28. W. P. Gomes. T. Freund, and S. R. Morrison, *Surf. Sci.* **13,** 201 (1968); *J. Electrochem. Soc.* **115,** 818 (1968).
29. T. Freund, in *Effect of Environment on Thermal Control Coatings* Final Report, Jet Propulsion Lab., NASA, Contract 951522, October 1969, by S. R. Morrison and T. Freund (for details see Reference R4).
30. R. A. L. Vanden Berghe, W. P. Gomes, and F. Cardon, *Z. Phys. Chem. N.F.* **92,** 91 (1974).
31. K. M. Sancier and S. R. Morrison, AIAA Paper No. 72–264, *Proceedings of AIAA 7th Thermophysics Conference, San Antonio,* 1972 (American Institute of Aeronautics and Astronautics Library, New York); *Surf. Sci.* **36,** 622 (1973).
32. K. Micka and H. Gerischer, *J. Electroanal. Chem.* **38,** 397 (1972).
33. W. P. Gomes and F. Cardon, *J. Solid State Chem.* **3,** 125 (1971).
34. F. Cardon and W. P. Gomes, *Surf. Sci.* **27,** 286 (1971).
35. J. Cunningham and H. Zainal, *J. Phys. Chem.* **76,** 2362 (1972).
36. F. Vanden Kerchove, J. Vandermolen, W. P. Gomes, and F. Cardon, *Ber. Bunsenges. Phys. Chem.* **83,** 230 (1979).
37. H. Gerischer, *Surf. Sci.* **18,** 97 (1969).
38. R. Memming, *J. Electrochem. Soc.* **116,** 785 (1969).
39. R. N. Noufi. P. A. Kohl, S. N. Frank, and A. J. Bard, *J. Electrochem. Soc.* **125,** 246 (1978).
40. B. Pettinger, H.-R. Schöppel, and H. Gerischer, *Ber. Bunsenges. Phys. Chem.* **80,** 849 (1976).
41. Y. G. Chai, W. W. Anderson, and L. B. Anderson, *IEEE Trans. Electron. Devices* **ED24,** 492 (1977).
42. A. Kunugi, S. Muyoshi, and S. Nagaura, *Bull. Chem. Soc. Japan* **51,** 700 (1978).
43. E. C. Dutoit, F. Cardon, and W. P. Gomes, *Ber. Bunsenges Phys. Chem.* **80,** 1285 (1976).
44. S. N. Frank and A. J. Bard, *J. Am. Chem. Soc.* **99,** 4667 (1977).
45. M. Miyake, H. Yoneyama, and H. Tamura, *Denki. Kagaku* **45,** 411 (1977).
46. H. G. Völz, G. Kämpf, and H. G. Fitzky, *Farbe Lack* **78,** 1037 (1972).
47. M. J. Madou, K. W. Frese, Jr., and S. R. Morrison, *J. Electrochem. Soc.* **126,** 1827

(1979); K. W. Frese, Jr., M. J. Madou, and S. R. Morrison, *Phys. Status Solidi A* **57**, 705 (1980).
48. T. Kobayashi, H. Yoneyama, and H. Tamura, *Chem. Lett.* 457 (1979).
49. F. Williams and A. J. Nozik, *Nature* **271**, 137 (1978).
50. R. H. Wilson, *J. Appl. Phys.* **48**, 4292 (1977).
51. G. Irick, Jr., *J. Appl. Polym. Sci.* **16**, 2387 (1972).
52. R. B. Cundall, B. Hulme, R. Rudham, and M. S. Salim, *J. Oil Col. Chem.* **61**, 351 (1978).
53. K. Hauffe and H. Volz, *Ber. Bunsenges. Phys. Chem.* **77**, 967(1973).
54. P. Clechet, C. Martelet, J. R. Martin, and R. Olsen, *Electrochim. Acta* **24**, 457 (1979).
55. M. Miyake, H. Yoneyama, and H. Tamura, *J. Catal.* **38**, 22 (1979).
56. B. Krautler and A. J. Bard, *J. Am. Chem. Soc.* **100**, 598 (1978).
57. H. Yoneyama, Y. Toyoguchi, and H. Tamura, *J. Phys. Chem.* **76**, 3460 (1972).
58. M. Miyake, H. Yoneyama, and H. Tamura, *Electrochim. Acta* **21**, 1065 (1976).
59. A. H. Boonstra and C. A. H. A. Mutsaers, *J. Phys. Chem.* **79**, 1940 (1975).
60. S. R. Morrison, *Surf. Sci.* **50**, 329 (1975).
61. J. Cunningham and S. Corkery, *J. Phys. Chem.* **79**, 933 (1975).
62. R. A. L. Vanden Berghe, F. Cardon, and W. P. Gomes, *Ber. Bunsenges. Phys. Chem.* **78**, 331 (1974).
63. P. J. Boddy, *J. Electrochem. Soc.* **115**, 199 (1968).
64. H. Morisaki, M. Hariya, and K. Yazawa, *Appl. Phys. Lett.* **30**, 7 (1977).
65. F. Decker, J. F. Juliao, and M. Abramovich, *Appl. Phys. Lett.* **35**, 397 (1979).
66. R. A. L. Vanden Berghe, W. P. Gomes, and F. Cardon, *Ber. Bunsenges. Phys. Chem.* **77**, 289 (1973).
67. S. Toshima and I. Uchida, *Electrochim. Acta* **15**, 1717 (1970).
68. W. H. Brattain and C. G. B. Garrett, *Bell System Technical Journal* **34**, 129 (1955).
69. Y. V. Pleskov, *Dokl. Akad. Nauk. S.S.S.R.* **132**, 1360 (1960).
70. A. Wolkenberg, *Electrochim. Acta* **21**, 1165 (1976).
71. S. R. Morrison, *J. Vac. Sci. Tech.* **7**, 84 (1970).
72. K. M. Sancier and S. R. Morrison, *Surf. Sci.* **83**, 29 (1979).
73. R. Memming and G. Neumann, *Surf. Sci.* **10**, 1 (1968).
74. H. Gerischer and W. Mindt, *Surf. Sci.* **4**, 440 (1966).
75. H.-D. Schulte, dissertation, University of Göttingen (1971).
76. U. Plitt, H.-D. Schulte, and H. Seliger, Phys. Lett. **30A**, 266 (1969).
77. K. W. Frese, Jr. and S. R. Morrison, *J. Vac. Sci. Tech.* **17**, 609 (1980).
78. E. C. Dutoit, F. Cardon, F. Vanden Kerchove, and W. P. Gomes, *J. Appl. Electrochem* **8**, 247 (1978).
79. W. P. Gomes, *Surf. Sci.* **19**, 172 (1970).
80. R. A. L. Vanden Berghe and W. P. Gomes, *Ber. Bunsenges. Phys. Chem.* **76**, 481 (1972).
81. F. Beck and H. Gerischer, *Z. Elektrochem.* **63**, 943 (1959).
82. H. Gerischer and F. Beck, *Z. Phys. Chem.* **13**, 389 (1957).
83. T. Freund, *J. Phys. Chem.* **73**, 468 (1969).
84. D. J. Benard and P. Handler, *Surf. Sci.* **40**, 141 (1973).
85. C. F. Jones, R. L. Segall, R. St. C. Smart, and P. S. Turner, *J. Chem. Soc. Faraday Trans.* 1 **74**, 1615, 1624 (1978).
86. W. Schmickler, *Ber. Bunsenges. Phys. Chem.* **82**, 477(1978).
87. W. Franz, *Ergeb. Exakten Naturwiss.* **27**, 1 (1953).
88. N. R. Armstrong, A. W. C. Lin, M. Fujihara, and T. Kuwana, *Anal. Chem.* **48**, 741 (1976).

89. B. Pettinger, H.-R. Schöppel, T. Yokoyama, and H. Gerischer, *Ber. Bunsenges. Phys. Chem.* **78,** 1024 (1974).
90. T. Freund, *Surf. Sci.* **33,** 295 (1972).
91. M. Gleria and R. Memming, *Z. Phys. Chem. N.F.* **98,** 303(1975); **101,** 171 (1976).
92. T. Watanabe, A. Fujishima, and K. Honda, *Ber. Bunsenges. Phys. Chem.* **79,** 1213 (1975).
93. W. Arden and P. Fromherz, *Ber. Bunsenges. Phys. Chem.* **82,** 868 (1978).
94. H. J. Danzmann and K. Hauffe, *Ber. Bunsenges. Phys. Chem.* **79,** 438 (1975).
95. U. Bode and K. Hauffe, *J. Electrochem Soc.* **125,** 51 (1978).
96. A. Fujishima, T. Iwase, and K. Honda, *J. Am. Chem. Soc.* **98,** 1625 (1976).
97. E. Daltrozzo and H. Tributsch, *Photogr. Sci. Eng.* **19,** 308 (1975).
98. H. J. Danzmann, thesis, University of Göttingen (1973).
99. M. Vodenicharova and G. H. Jensen, *J. Phys. Chem. Solids* **36,** 1241 (1975).
100. R. R. Dogonadze, A. M. Kuznetsov, and Yu. A. Chizmadzhev, *Russ. J. Phys. Chem.* **38,** 652 (1964).
101. H. Gerischer, *J. Electroanal. Chem.* **82,** 133 (1977).
102. M. J. Madou, F. Cardon, and W. P. Gomes, *Ber. Bunsenges. Phys. Chem.* **82,** 819 (1978).
103. K. D. N. Brummer, *J. Electrochem. Soc.* **114,** 1274 (1967).
104. T. Freund, W. Gomes, and S. R. Morrison, in Proceedings of the IVth International Congress on Catalysis, Moscow, 1968.
105. Yu. V. Pleskov, *Dokl. Akad. Nauk. SSSR* **149,** 1399 (1962).
106. R. Landsberg, P. Janietz, and R. Dehmlow, *Z. Phys. Chem. Leipzig* **257,** 657 (1976).
107. P. A. Kohl and A. J. Bard, *J. Am. Chem. Soc.* **99,** 7531 (1977).
108. E. C. Dutoit, F. Cardon, and W. P. Gomes, *Ber. Bunsenges. Phys. Chem.* **80,** 475 (1976).
109. M. T. Pham and J. Hueller, *Phys. Status Solidi A* **43,** 505 (1977).
110. H. Marisaki and K. Yazawa, *Appl. Phys. Lett.* **33,** 1013 (1978).
111. A. Fujishima and K. Honda, *Nature* **238,** 37 (1972).
112. J. C. Tranchart, L. Hollan, and R. Memming, *J. Electrochem. Soc.* **125,** 1185 (1978).
113. S. R. Morrison and T. Freund, in NBS Spec. Publ. No. 455, "Electrocatalysis on Non-metallic Surfaces (1975).
114. W. N. Hansen, R. A. Osteryoung, and T. Kuwana, *J. Am. Chem. Soc.* **88,** 1062 (1966).
115. F. Möllers and R. Memming, *Ber. Bunsenges. Phys. Chem.* **77,** 879 (1973).
116. L. D. Burke, O. S. Murphy, J. F. O'Neill, and S. Venkatesan, *J. Chem. Soc. Faraday Trans.* **1,** 73 (1977).
117. E. J. Fraser and R. Woods, *J. Electroanal. Chem.* **102,** 127 (1979).
118. K. Doblhofer, M. Metikos, Z. Ogumi, and H. Gerischer, *Ber. Bunsenges. Phys. Chem.* **82,** 1046 (1978).
119. B. Koiller and L. M. Falicov, *J. Phys. C Solid State Phys.* **7,** 299 (1974).
120. D. M. Tench and E. Yeager, *J. Electrochem. Soc.* **120,** 164 (1973); **121,** 318 (1974).
121. R. DeGryse, W. P. Gomes, F. Cardon, and J. Vennick, *J. Electrochem. Soc.* **122,** 711 (1975).
122. M. C. Markham, J. Barry, M. Iava, and J. Haddad, *J. Electrochem. Soc.* **61,** 1665 (1957).
123. J. C. Kuriacose, V. Ramakrishnan, and E. P. Yesodharan, *Ind. J. Chem.* **16A,** 754 (1978).
124. H. Gerischer and E. Meyer, *Z. Phys. Chem. N.F.* **74,** 302 (1971).
125. V. A. Tyagai and N. B. Lukyanchikova, *Surf. Sci.* **12,** 331 (1968).
126. E. W. Grabner, private communication (1980).

References for Chapter 7

1. H. Gerischer and H. Rösler, *Chem. Ing. Tech.* **4,** 176 (1970).
2. M. Gleria and R. Memming, *Z. Phys. Chem. N.F.* **98,** 303 (1975).
3. R. Memming and F. Möllers, *Ber. Bunsenges. Phys. Chem.* **76,** 609 (1972).
4. S. R. Morrison, *Surf. Sci.* **15,** 363 (1969).
5. U. Bode, K. Hauffe, Y. Ishikawa, and H. Pusch, *Z. Phys. Chem. N.F.* **85,** 144 (1973).
6. H. J. Danzmann and K. Hauffe, *Ber. Bunsenges. Phys. Chem.* **79,** 438 (1975).
7. K. Hauffe and O. Haeggquist, *Z. Phys. Chem. N.F.* **85,** 181 (1973).
8. H. Gerischer, *Surf. Sci.* **18,** 97 (1969).
9. T. Sakata, K. Tanimura, and T. Kawai, *Annual Review of the Institute for Molecular Sciences*, Okazaki, Japan, (1978), p.62.
10. H. G. Völz, G. Kämpf, and H. G. Fitzky, *Farbe Lack* **78,** 1037 (1972).
11. R. I. Bickley and R. K. M. Jayantz, *Discuss Faraday Soc.* **58,** 194 (1974).
12. K. Micka and H. Gerischer, *Electroanal. Chem. Interf. Electrochem.* **38,** 397 (1972).
13. M. Miyake, H. Yoneyama, and H. Tamura, *Denki Kagaku* **45,** 411 (1977).
14. A. Caprani, C. Deslonis, M. Keddam, Ph. Morel, and B. Tribollet, *Electrochim. Acta* **22,** 1231 (1977).
15. R.A.L. Vanden Berghe, F. Cardon, and W. P. Gomes, *Surf. Sci.* **39,** 368 (1973).
16. R. Memming and F. Möllers, *Ber. Bunsenges. Phys. Chem.* **76,** 475 (1972).
17. R. O. James and T. W. Healy, *J. Coll. Interf. Sci.* **40,** 42, 53, 65 (1972).
18. D. M. Tench and E. Yeager, *J. Electrochem. Soc.* **121,** 318 (1974).
19. K. H. Hauffe, *Photogr. Sci. Eng.* **20,** 124 (1976).
20. P. Lemasson, A. M. Baticle, and P. Vennereau, *Surf. Sci.* **59,** 177 (1976).
21. H. Yoneyama, Y. Matsumoto, and H. Tamura, *J. Phys. Chem.* (in press).
22. H. Yoneyama and H. A. Laitinen, *NBS Spec. Publ.* 455 (1976).
23. B. A. Parkinson, A. Heller, and B. Miller, *J. Electrochem. Soc.* **126,** 954 (1979).
24. S. Toshima and I Uchida, *Electrochim. Acta.* **15,** 1717 (1970).
25. A. N. Gupta, *Ind. J. Chem.* **15A,** 181 (1977).
26. K. Hauffe and H. Volz, *Ber. Bunsenges. Phys. Chem.* **77,** 967(1973).
27. W. H. Brattain and P. J. Boddy, *J. Electrochem. Soc.* **109,** 574, 812 (1962).
28. R. Memming, *Surf. Sci.* **2,** 436 (1964).
29. H. Yoneyama, S. Mayami, and H. Tamura, *J. Electrochem. Soc.* **125,** 68 (1978).
30. K. M. Sancier, *J. Catal.* **20,** 106 (1971).
31. G. E. Batley, A. Ekstrom, and D. A. Johnson, *J. Catal.* **36**, 285 (1975).
32. Y. Nakato, S.Tonomura, and H. Tsubomura, *Ber. Bunsenges. Phys. Chem.* **80,** 1289 (1976).
33. Y. Nakato, K. Abe, and H. Tsubomura, *Ber. Bunsenges. Phys. Chem.* **80,** 1002 (1976).
34. G. N. Schrauzer and T. D. Guth, *J. Am. Chem. Soc.* **99,** 7189 (1977).
35. B. E. Frith and L. L. Miller, *J. Am. Chem. Soc.* **98,** 8272 (1976).
36. R. J. Burt, G. J. Leigh, and C. J. Pickett, *J. Chem. Soc. Chem. Commun.* **1976,** 940.
37. N. R. Armstrong, A. W. C. Lin, M. Fujihara, and T. Kuwana, *Anal. Chem.* **48,** 741 (1976).
38. D. G. Davis, and R. W. Murray, *Anal. Chem.* **49,** 194 (1977).
39. D. F. Smith, K. Willman, K. Kuo, and R. W. Murray, *J. Electroanal. Chem.* **95,** 217 (1979).
40. P. R. Moses, L. Wier, and R. W. Murray, *Anal. Chem.* **47,** 1882 (1975).
41. P. R. Moses and R. W. Murray, *J. Am. Chem. Soc.* **98,** 7435 (1976).
42. T. Osa and M. Fujihira, *Nature (London)* **264,** 349 (1976).
43. H. O. Finklea and R. W. Murray, *J. Phys. Chem.* **83,** 353 (1979).
44. A. B. Bocarsly, E. G. Walton, M. G. Bradley, and M. S. Wrighton, *J. Electroanal. Chem.* **100,** 283 (1979).

45. M. S. Wrighton, R. G. Austin, A. B. Bocarsly, J. M. Bolts, O. Haas, K. D. Legg, L. Nadjo, and M. C. Palazzotto, *J. Am. Chem. Soc.* **100,** 1602 (1978).
46. M. O'Keefe, Y. Ebisuzaki, and W. J. Moore, *J. Phys. Soc. Japan* **18,** Suppl. 2, 131 (1963).
47. W. P. Gomes, *Surf. Sci.* **19,** 172 (1970).
48. C. N. Sayers and N. R. Armstrong, *Surf. Sci.* **77,** 301 (1978).
49. Y. G. Chai, W. W. Anderson, and L. B. Anderson, *IEEE Trans. Electron Devices,* ED24, 492 (1977).
50. D. Cahen, G. Hodes, and J. Manassen, *J. Electrochem. Soc.* **125,** 1623 (1978).
51. H. Gerischer and J. Gobrecht, *Ber. Bunsenges. Phys. Chem.* **82,** 520 (1978).
52. H. Gerischer, A. Maurer, and W. Mindt, *Surf. Sci.* **4,** 431 (1966).
53. G. Brouwer, *J. Electrochem. Soc.* **114,** 743 (1967).
54. C. F. Jones, R. L. Segall, R. St. C. Smart, and P. S. Turner, *J. Chem. Soc. Faraday Trans.* **174,** 1615, 1624 (1978).
55. K. Hauffe and K. Reinhold, *Ber. Bunsenges. Phys. Chem.* **76,** 616 (1972).
56. H. Tributsch and H. Gerischer, *J. Appl. Chem. Biotechnol.* **26,** 747 (1976).
57. B. MacDougal and M. Cohen, *J. Electrochem. Soc.* **124,** 1185 (1977).
58. K. Dyrek, R. Schumacher, and R. N. Schindler, *Z. Naturforsch.* **32a,** 1157 (1977).
59. M. C. Markham, J. Barry, M. Iava, and J. Haddad, *J. Electrochem. Soc.* **61,** 1665 (1957).
60. H. Gerischer, *J. Electroanal. Chem.* **82,** 133 (1977).
61. D. Eger, A. Many, and Y. Goldstein, *Phys. Lett* **55A,** 197 (1975).
62. F. Stern, *J. Comp. Phys.* **6,** 56 (1970).
63. M. J. Madou, F. Cardon, and W. P. Gomes, *J. Electrochem. Soc.* **124**, 1623 (1977).
64. M. J. Madou, K. W. Frese, Jr., and S. R. Morrison, *J. Electrochem. Soc.* **126,** 1827 (1979).
65. R. L. Meek, *J. Electrochem. Soc.* **118,** 437 (1971).
66. R. Memming and G. Schwandt, *Surf. Sci.* **4,** 109 (1966).
67. L. A. Harris, D. R. Cross, and M. E. Gerstner, *J. Electrochem. Soc.* **124,** 839 (1977).
68. M. E. Gerstner, *J. Electrochem. Soc.* **126,** 944 (1979).
69. A. J. Bard and M. S. Wrighton, *J. Electrochem. Soc.* **124,** 1706 (1977).
70. F. Williams and A. J. Nozik, *Nature* **271,** 137 (1978).
71. S. Park and M. E. Barber, *J. Electroanal. Chem.* **99,** 67 (1979).
72. D. D. Justice and R. M. Hard, *J. Electrochem. Soc.* **118,** 1417 (1971).
73. B. Pettinger, H.-R. Schöppel, T. Yokoyama, and H. Gerischer, *Ber. Bunsenges. Phys. Chem.* **78,** 1024 (1974).
74. H. Erbse, K. Hauffe, and J. Range, *Z. Phys. Chem. N.F.* **74,** 248 (1971).
75. R. Baumeler, P. Rys. and H. Zollinger, *Helv. Chem. Acta* **56,** 2450 (1973).
76. A. Wolkenberg, *Electrochim. Acta* **21,** 1165 (1975).
77. M. M. Faktor, D. G. Fiddyment, and M. R. Taylor, *J. Electrochem. Soc.* **122,** 1566 (1975).
78. A. Yamamoto and S. Yano, *J. Electrochem. Soc.* **122,** 260 (1975).
79. J. C. Tranchart, L. Hollan, and R. Memming, *J. Electrochem. Soc.* **125,** 1185 (1978).
80. S. R. Morrison and T. Freund, in NBS Spec. Publ. No. 455, *Electrocatalysis on Non-metallic Surfaces* (1976).
81. D. L. Rode, B. Schwartz, and J. V. DiLorenzo, *Solid State Electron.* **17,** 1119 (1974).
82. R.A.L. Vanden Berghe, W. P. Gomes, and F. Cardon, *Ber. Bunsenges. Phys. Chem.* **77,** 289 (1973).
83. S. R. Morrison, *J. Vac. Sci. Tech.* **7,** 84 (1970).
84. K. M. Sancier, *Surf. Sci.* **21,** 1 (1970).
85. A. Fujishima (private communication).
86. W. P. Gomes (private communication).
87. T. Kobayashi, H. Yoneyama, and H. Tamura, *Chem. Lett.* 457 (1979).

88. R. Memming, *J. Electrochem. Soc.* **125,** 117 (1978).
89. W. P. Gomes, T. Freund, and S. R. Morrison, *J. Electrochem. Soc.* **115,** 818 (1968); *Surf. Sci.* **13,** 201 (1968).
90. K. Nakatani, S. Matsudaira, and H. Tsubomura, *J. Electrochem. Soc.* **125,** 406 (1978).
91. R. Memming, *Ber. Bunsenges. Phys. Chem.* **81,** 733 (1977).

References for Chapter 8

1. M. J. Madou, K. W. Frese, Jr., and S. R. Morrison, *J. Electrochem. Soc.* **126,** 1827 (1979); *Phys. Status Solidi A* **57,** 705 (1980).
2. J. O'M. Bockris and K. Uosaki, *J. Electrochem. Soc.* **124,** 1348 (1977).
3. E. H. Nicollian and C. N. Berglund, *J. Appl. Phys.* **41,** 3052 (1970).
4. M. H. Woods and R. Williams, *J. Appl. Phys.* **47,** 1082 (1976).
5. E. H. Nicollian, C. N. Berglund, P. F. Schmidt, and J. M. Andrews, *J. Appl. Phys.* **42,** 5654 (1971).
6. J. J. Simonne, *Solid State Electron.* **16,** 121 (1973).
7. P. J. Caplan, J. N. Helbert, B. E. Wagner, and E. H. Poindexter, *Surf. Sci.* **54,** 33 (1976).
8. P. J. Caplan, E. H. Poindexter, B. E. Deal, and R. R. Razouk, *J. Appl. Phys.* **50,** 5847 (1979).
9. D. V. Lang, *J. Appl. Phys.* **45,** 3014, 3023 (1974).
10. N. M.Johnson, *Appl. Phys. Lett.* **34,** 802 (1979).
11. A. Goetzberger, E. Klausmann, and M. J. Schulz, *CRC Crit. Rev. Solid State Sci.* 1, (1976).
12. S. Szpak, *J. Electrochem. Soc.* **124,** 107 (1977).
13. P. Bindra, H. Gerischer, and D. M. Kolb, *J. Electrochem. Soc.* **124,** 1012 (1977).
14. M. J. Weaver, *J. Electroanal. Chem.* **51,** 231 (1974).
15. A. Damjanovic and B. Jovanovic, *J. Electrochem. Soc.* **123,** 374 (1976).
16. M. Faroque and T. Z. Fahidy, *J. Electrochem. Soc.* **124,** 1191 (1977).
17. W. Arden and P. Fromherz, *Ber. Bunsenges. Phys. Chem.* **82,** 868 (1978).
18. N. Cabrera and N. F. Mott, Repts. *Prog. Phys.* **12,** 163 (1948–1949).
19. R. W. Revie, B. J. Baker, and J. O'M. Bockris, *J. Electrochem. Soc.* **122,** 1460 (1975).
20. J. L. Ord and D. J. DeSmet, *J. Electrochem. Soc.* **123,** 1876 (1976).
21. R. L. Cohen, Ch. J. Raub and T. Muramaki, *J. Electrochem. Soc.* **125,** 34 (1978).
22. M. Shimura, *J. Electrochem. Soc.* **125,** 190 (1978).
23. E. Kohn and H. L. Hartnagel, *Solid State Electron.* **21,** 409 (1978).
24. T. Ishii and B. Jeppson, *J. Electrochem. Soc.* **124,** 1785 (1977).
25. C. C. Chang, P. H. Citrin, and B. Schwartz, Proceedings of the 4th Conference on the Physics of Semiconductor Interfaces (1977).
26. J. C. Verplanke and R. P. Tijburg, *J. Electrochem. Soc.* **124,** 802 (1977).
27. J. D. E. Beynon, G. G. Bloodworth, and I. M. McLeod, *Solid State Electron.* **16,** 309 (1973).
28. M. J. Dignam, *Can. J. Chem.* **56,** 595 (1978).
29. W. A. Lansford, *Science,* **196,** 975 (1977).
30. E. H. Poindexter (private communication).
31. M. H. Woods and R. Williams, *J. Appl. Phys.* **44,** 5506 (1973).
32. W. Marciniak and H. M. Przewlocki, *J. Electrochem. Soc.* **123,** 1207 (1976).
33. M. R. Boudry and J. P. Stagg, *J. Appl. Phys.* **50,** 942 (1979).
34. E. H. Snow, A. S. Grove, B. E. Deal, and C. T. Sah, *J. Appl. Phys.* **36,** 1664 (1965).
35. A. G. Revesz, *J. Electrochem. Soc.* **126,** 122 (1979).

36. T. Sugano, K. Hoh, K. Kuno, and H. Hishinuma, *Jpn. J. Appl. Phys.* **7,** 715 (1968).
37. N. Nagasima and H. Enari, *Jpn. J. Appl. Phys.* **10,** 441 (1971).
38. S. W. Ing, Jr., R. E. Morrison, and J. E. Sandor, *J. Electrochem. Soc.* **109,** 221 (1962).
39. A. M. Zakzouk, *J. Electrochem. Soc.* **126,** 1772 (1979); A. M. Zakzouk, R. A. Stuart, and W. Eccleston, *J. Electrochem. Soc.* **123,** 1551 (1976).
40. E. A. Irene, in *The Physics of SiO_2 and its Interfaces*, ed. S. T. Pantelides (Pergamon, New York, 1978).
41. J. Vedel, M. Soubeyrand, and E. Castel, *J. Electrochem. Soc.* **124,** 177 (1977).
42. B. Miller, S. Menezes, and A. Heller, *J. Electroanal. Chem.* **94,** 85 (1978).
43. J. Corish and C. J. Warde, *Ber. Bunsenges. Phys. Chem.* **82,** 282 (1978).
44. O. Flint and J. H. O. Varley, *J. Phys. Chem. Solids* **6,** 213 (1958).
45. K. E. Heusler and K. S. Yun, *Electrochim. Acta* **27,** 977 (1977).
46. W. Schmickler, *Ber. Bunsenges. Phys. Chem.* **82,** 477 (1978).
47. W. Schmickler and J. Ulstrup, *Chem. Phys.* **19,** 217 (1977).
48. W. Schmickler and J. W. Schultze, *Z. Phys. Chem. N.F.* **110,** 277 (1978).
49. J. W. Schultze and K. J. Vetter, *Electrochim. Acta* **18,** 889 (1973).
50. A. Ward, A. Damjanovic, E. Gray, and M. O'Jea, *J. Electrochem. Soc.* **123,** 1599 (1976).
51. H. Gerischer, E. Meyer, and J. K. Sass, *Ber. Bunsenges. Phys. Chem.* **76,** 1191 (1972).
52. A. Kunugi, T. Shmizu, S. Miyosha, and S. Nagaura, *Denki Kagaku* **45,** 598 (1977); *Bull. Chem. Soc. Japan* **51,** 700 (1978).
53. J. W. Schultze and U. Stimming, *Z. Phys. Chem. N.F.* **98,** 285 (1975).
54. C. M. Osburn and E. J. Weitzman, *J. Electrochem. Soc.* **119,** 603 (1972); C. M. Osburn and D. W. Ormond, *J. Electrochem. Soc.* **121,** 1195 (1974).
55. R. R. Dogonadze, A. M. Kuznetsov, and A. A. Chernenko, *Russ. Chem. Rev.* **34,** 759 (1965).
56. J. N. Butler, *J. Chem. Phys.* **35,** 636 (1961).
57. E. C. Dutoit, F. Cardon, and W. P. Gomes, *Ber. Bunsenges. Phys. Chem.* **80,** 1285 (1976).
58. D. Laser, M. Yaniv, and S. Gottesfeld, *J. Electrochem. Soc.* **125,** 358 (1978).
59. R. K. Quinn and N. R. Armstrong, *J. Electrochem. Soc.* **125,** 1790 (1978).
60. B. E. Wilde and F. G. Hodge, *Electrochim. Acta* **14,** 619 (1969).
61. S. M. Wilhelm, K. S. Yun, L. W. Ballenger, and N. Hackerman, *J. Electrochem. Soc.* **126,** 419 (1979).
62. M. Metikoš-Huković and B. Lovreček, *Electrochim. Acta* **23,** 1371 (1978).
63. L. M. Peter, *Electrochim. Acta.* **23,** 165 (1978).
64. E. C. Dutoit, F. Cardon, F. Vanden Kerchove, and W. P. Gomes, *J. Appl. Electrochem.* **8,** 247 (1978).
65. J. Keeney, D. H. Weinstein, and G. M. Haas, *Nature* **253,** 719 (1975).
66. P. Clechet, C.Martelet, J. R. Martin, and R. Olsen, *Electrochim. Acta* **24,** 457 (1979).
67. P. F. Schmidt, *J. Appl. Phys.* **28,** 278 (1957).
68. R. Williams and M. H. Woods, *Appl. Phys. Lett.* **22,** 459 (1973).
69. M. Lenzinger and E. H. Snow, *J. Appl. Phys.* **40,** 278 (1969).
70. A. G. Revesz, *Phys. Status Solidi* **24,** 115 (1967).
71. K. W. Frese, Jr., and S. R. Morrison, *J. Vac. Sci. Tech.* **17,** 699 (1980).
72. H. Minoura, T. Oki, and M. Tsuiki, *Chem. Lett.* 1279 (1976).
73. F. Möllers, H. J. Tolle, and R. Memming, *J. Electrochem. Soc.* **121,** 1160 (1974).
74. J. J. Kelly and J. K. Vondeling, *J. Electrochem. Soc.* **122,** 1103 (1975).
75. S. R. Morrison, *Surf. Sci.* **15,** 363 (1969).
76. H. Hada, H. Tanemura, and Y. Yonezawa, *Bull. Chem. Soc. Japan* **51,** 3154 (1978.
77. G. C. Wood and C. Pearson, *Corrosion Sci.* **7,** 119 (1967).
78. C. K. Dyer and J. S. L. Leach, *J. Electrochem. Soc.* **125,** 23 (1978).
79. D. A. Vermilyea, *J. Appl. Phys.* **27,** 963 (1956).
80. K. W. Frese, *J. Vac. Sci. Tech.* **16,** 1042 (1979).

References for Chapter 9

1. A. J. Nozik, in *Semiconductor Liquid-junction Solar Cells*, ed. A. Heller (Electrochemical Society, Princeton, New Jersey, 1977).
2. H. Kim and H. A. Laitinen, *J. Electrochem. Soc.* **122,** 53 (1975).
3. J. H. Kennedy and K. W. Frese, Jr., *J. Electrochem. Soc.* **125,** 709 (1978).
4. B. Miller, S. Menezes, and A. Heller, *J. Electroanal. Chem.* **94,** 85 (1978).
5. M. Okuda, K. Yoshida, and N. Tanaka, *Jpn. J. Appl. Phys.* **15,** 1599 (1976).
6. S. N. Frank and A. J. Bard, *J. Am. Chem. Soc.* **99,** 4667 (1977).
7. K. L. Hardee and A. J. Bard, *J. Electrochem. Soc.* **122,** 739 (1975).
8. P. J. Boddy, *J. Electrochem. Soc.* **115,** 199 (1968).
9. F. Möllers and R. Memming, *Ber. Bunsenges. Phys. Chem.* **76,** 469 (1972).
10. A. Fujishima and K. Honda, *Nature* **238,** 37 (1972).
11. J. Manassen, D. Cahen, G. Hodes, and A. Sofen, *Nature* **263,** 97 (1976).
12. J. Loferski, Proceedings of the IEEE 1973 Photovoltaic Specialists' Conference, Palo Alto, 1973, p. 1.
13. R. H. Wilson, in *Semiconductor Liquid-junction Solar Cells*, ed. A. Heller (Electrochemical Society, Princeton, New Jersey, 1977), p. 67.
14. H. Tributsch, H. Gerischer, C. Clemen, and E. Bucher, *Ber. Bunsenges. Phys. Chem.* **83,** 655 (1979).
15. K. Ohashi, J. McCann, and J. O'M. Bockris, *Energy Research* **1,** 259 (1977).
16. Y. G. Chai, W. W. Anderson, and L. B. Anderson, *IEEE Trans. Electron Devices* **ED24,** 492 (1977).
17. P. A. Kohl and A. J. Bard, *J. Electrochem. Soc.* **126,** 603 (1979).
18. K. C. Chang, A. Heller, B. Schwartz, S. Menezes, and B. Miller, *Science* **196,** 1097 (1977); see also *Semiconductor Liquid-junction Solar Cells*, ed. A. Heller (Electrochemical Society, Princeton, New Jersey, 1977), p. 132.
19. B. A. Parkinson, A. Heller, and B. Miller, *J. Electrochem. Soc.* **126,** 954 (1979).
20. A. J. Nozik, *Nature* **257,** 383 (1975).
21. J. G. Mavroides, D. I. Tchernev, J. A. Kafalas, and D. F. Kolesar, *Mat. Res. Bull.* **10,** 1023 (1978).
22. H. Kim and H. A. Laitinen, *J. Electrochem. Soc.* **122,** 53 (1975).
23. L.-S. R. Yeh and N. Hackerman, *J. Electrochem. Soc.* **124,** 833 (1977).
24. J. H. Kennedy and K. W. Frese, Jr., *J. Electrochem. Soc.* **123,** 1683 (1976).
25. J. H. Kennedy and K. W. Frese, Jr., *J. Electrochem. Soc.* **125,** 709 (1978).
26. D. M. Schleich, C. Derrington, W. Godek, D. Weisberg, and A. Wold, *Mat. Res. Bull.* **12,** 321 (1977).
27. R. Memming, *J. Electrochem. Soc.* **125,** 117 (1978).
28. J. O'M. Bockris and K. Uosaki, *J. Electrochem. Soc.* **124,** 1348 (1977).
29. M. J. Madou, F. Cardon, and W. P. Gomes, *Ber. Bunsenges. Phys. Chem.* **81,** 1186 (1977).
30. H. Tributsch, *Ber. Bunsenges, Phys. Chem.* **82,** 169 (1978); **81,** 361 (1977).
31. J. Gobrecht, H. Gerischer, and H. Tributsch, *Ber. Bunsenges. Phys. Chem.* **82,** 1331 (1978).
32. M. S. Wrighton, J. M. Bolts, A. B. Bocarsly, M. C. Palazzato, and E. G. Walton, *J. Vac. Sci. Tech* **15,** 1429 (1978).
33. D. Cahen, G. Hodes, and J. Manassen, *J. Electrochem. Soc.* **125,** 1623 (1978).
34. R. Memming, *J. Electrochem. Soc.* **125,** 117 (1978).
35. K. D. Legg, A. B. Ellis, J. M. Bolts, and M. S. Wrighton, *Proc. Nat. Acad. Sci. USA* **74,** 4116 (1977).
36. R. Memming, *Ber. Bunsenges. Phys. Chem.* **81,** 733 (1977).

37. Y. Nakato, S. Tonomura, and H. Tsubomura, *Ber. Bunsenges. Phys. Chem.* **80,** 1289 (1976); Y. Nakato, K. Abe, and H. Tsubomura, *Ber. Bunsenges. Phys. Chem.* **80,** 1002 (1976).
38. J. S. Kilby, J. W. Lathrop, and W. A. Porter, U.S. Patent 4,021,323 (1977).
39. P. Bindra, H. Gerischer, and D. M. Kolb, *J. Electrochem. Soc.* **124,** 1012 (1977).
40. P. A. Kohl, S. N. Frank, and A. J. Bard, *J. Electrochem. Soc.* **124,** 225 (1977).
41. S. Gourgaud and D. Elliot, *J. Electrochem. Soc.* **124,** 102 (1977).
42. M. Tomkiewicz and J. M. Woodall, *J. Electrochem. Soc.* **124,** 1437 (1977).
43. H. Tsubomura, M. Matsumura, K. Nakatani, K. Yamamoto, and K. Maeda, *Solar Energy* **21,** 93 (1978).
44. V. R. Shepard, Jr.,and N. R. Armstrong, *J. Phys. Chem.* **83,** 1268 (1979).
45. K. F. Blurton and J. R. Stetter, *J. Catal.* **46,** 230 (1977).
46. J. Bett, K. Kinoshita, K. Routses, and P. Stonehart, *J. Catal.* **29,** 160 (1973).
47. G. P. Sakellaropoulos and S. H. Langer, *J. Catal.* **44,** 25 (1976).
48. F. Williams and A. J. Nozik, *Nature* **271,** 137 (1978).
49. J. Figar and W. Haidinger, *Chem. Phys. Lett.* **19,** 564 (1973).
50. M. J. Madou, K. W. Frese, Jr., and S. R. Morrison, *J. Electrochem. Soc.* **126,** 1827 (1979); *Phys. Status Solidi* **57,** (1980).
51. B. Pettinger, H.-R. Schöppel, and H. Gerischer, *Ber. Bunsenges. Phys. Chem.* **78,** 450 (1974).
52. M. Miyake, H. Yoneyama, and H. Tamura, *Bull. Chem. Soc. Jpn.* **50,** 1492 (1977); *Electrochim. Acta* **21,** 1065 (1976).
53. M. Miyake, H. Yoneyama, and H. Tamura, *J. Catal.* **58,** 22 (1979).
54. R. Landsberg, P. Janietz, and R. Dehmlow, *Z. Phys. Chem. Leipzig* **257,** 657 (1976).
55. T. Inoue, A. Fujishima, S. Konishi, and K. Honda, *Nature* **277,** 637 (1979).
56. J. F. Evans, T. Kuwana, M. T. Henne, and G. P. Royer, *J. Electroanal. Chem.* **80,** 409 (1977).
57. R. I. Bickley and R. K. M. Jayantz, *Discuss. Faraday Soc.* **58,** 194 (1974).
58. T. Wolfram, F. J. Morin, and R. Hurst, in NBS Spec. Publ. No. 455,"Electrocatalysis on Non-metallic Surfaces" (1976).
59. R. A. Marcus, *Can. J. Chem.* **37,** 155 (1959).
60. A. Kunuge, S. Miyoshi, and S. Nagaura, *Bull. Chem. Soc. Jpn.* **51,** 700 (1978).
61. D. Yoke, A. Riga, R. Greef, and E. Yeager, *Electrochim. Acta* **13,** 1351 (1968).
62. D. M. Tench and E. Yeager, *J. Electrochem. Soc.* **120,** 164 (1973).
63. T. Kudo, H. Obayashi, and M. Yoshida, *J. Electrochem. Soc.* **124,** 321 (1977).
64. D. B. Hibbert and A. C. C. Tseung, *J. Electrochem. Soc.* **125,** 74 (1978).
65. D. B. Rogers, R. D. Shannon, A. W. Sleight, and J. L. Gillson, *Inorg. Chem.* **8,** 841 (1969).
66. K. Doblhofer, M. Metikos, Z. Ogumi, and H. Gerischer, *Ber. Bunsenges. Phys. Chem.* **82,** 1046 (1978).
67. S. Pizzini, G. Buzzanca, C. Mari, L. Rossi, and S. Torchio, *Mat. Res. Bull.* **7,** 449 (1972).
68. A. T. Kuhn and C. J. Mortimer, *J. Electrochem. Soc.* **120,** 231 (1973).
69. S. Trasatti, *J. Electrochem. Soc.* **120,** 1703 (1973).
70. J. J. Kelly and J. K. Vondeling, *J. Electrochem. Soc.* **122,** 1103 (1975).
71. F. Möllers, H. J. Tolle, and R. Memming, *J. Electrochem. Soc.* **121,** 1160 (1974).
72. L. R. Faulkner and A. J. Bard, *Electroanal. Chem.* **10,** 1 (1977).
73. R. N. Noufi, P. A. Kohl, S. N. Frank, and A. J. Bard, *J. Electrochem. Soc.* **125,** 246 (1978).
74. M. M. Nicholson, *J. Electrochem. Soc.* **119,** 461 (1972).
75. J. D. Luttmer and A. J. Bard, *J. Electrochem. Soc.* **126,** 414 (1979).

76. W. Gissler and R. Memming, *J. Electrochem. Soc.* **124,** 1710 (1977).
77. S. K. Mohapatra, *J. Electrochem. Soc.* **125,** 285 (1978).
78. R. P. Buck and D. E. Hackleman, *Anal. Chem.* **49,** 2315 (1977).
79. J. N. Zemel, *Anal. Chem.* **47,** 255A (1975).
80. J. N. Zemel, *Surf. Sci.* **86,** 322 (1979).
81. J. Janata and S. D. Moss, *Biomed. Eng.* **11,** 241 (1976).
82. J. N. Zemel, *Res./Dev.* **28,** 38 (1977).
83. T. W. Healy, D. E. Yates, L. R. White, and D. Chan, *J. Electroanal. Chem.* **80,** 57 (1977).
84. J. F. Schenck, *J. Colloid Interface Sci.* **61,** 569 (1977).
85. R. G. Kelly, *Electrochim. Acta* **22,** 1 (1977).
86. M. T. Pham, *Phys. Stat. Solidi* **a49,** 261 (1978).
87. R. Memming and G. Schwandt, *Surf. Sci.* **4,** 109 (1966).
88. R. L. Meek, *J. Electrochem. Soc.* **118,** 437 (1971); *Surf. Sci.* **25,** 526 (1971).
89. L. Hollan, J. C. Tranchard, and R. Memming, *J. Electrochem. Soc.* **126,** 855 (1979).
90. M. M. Faktor, D. G. Fiddyment, and M. R. Taylor, *J. Electrochem. Soc.* **122,** 1566 (1975).
91. D. L. Rode, B. Schwartz, and J. V. DiLorenzo, *Solid State Electron.* **17,** 1119 (1974).
92. V. A. Tyagai, V. A. Sterligov, and G. Ya. Kolbasov, *Electrochim. Acta* **22,** 819 (1977).
93. B. J. Baliga and S. K. Ghandi, *J. Electrochem. Soc.* **124,** 1059 (1977).
94. A. L. Dalisa, W. K. Zwicker, D. J. DeBitetto, and P. Harnack, *Appl. Phys. Lett.* **17,** 208 (1970).
95. W. D. Johnston, Jr., H. J. Leamy, B. A. Parkinson, A. Heller, and B. Miller, *J. Electrochem. Soc.* **127,** 90(1980).

References To Review Articles and Books

R1. R. H. Tredgold, *Space Charge Conduction in Solids* (Elsevier, Amsterdam, 1966).
R2. W. C. Johnson, "Electronic Transport in Insulating Films," *IEEE Trans. Nucl. Sci.* **NS19,** 33 (1972).
R3. M. A. Lampert and P. Mark, *Current Injection in Solids* (Academic Press, New York, 1970).
R4. S. R. Morrison, *The Chemical Physics of Surfaces* (Plenum Press, New York, 1977).
R5. K. Tanabe, *Solid Acids and Bases* (Academic Press, New York, 1970).
R6. H. Gerischer, "Reactions at Semiconductor-Electrolyte Interfaces—Theory," NATO Summer School Notes, Ghent, 1968.
R7. L. Young, *Anodic Oxide Films* (Academic Press, New York, 1961).
R8. J. P.Hunt, *Metal Ions in Aqueous Solutions* (W. A. Benjamin, New York, 1963).
R9. T. Erdey-Grúz, *Kinetics of Electrode Processes* (Wiley Interscience, New York, 1972).
R10. S. R. Morrison, "Surface Phenomena Associated with the Semiconductor/Electrolyte Interface," *Progr. Surf. Sci.* **1,** 106 (1971).
R11. G. A. Parks, "The Isoelectronic Points in Solid Oxides, Solid Hydroxides, and Aqueous Hydroxo Complex Systems," *Chem. Rev.* **65,** 177 (1965).
R12. Yu. V. Pleskov, "The Electric Double Layer at the Semiconductor/Electrolyte Interface," *Prog. Surf. Membrane Sci.* **7,** 57 (1973).
R13. D. A. Vermilyea, "Corrosion Science 1970," *Ann. Rev. Corros. Sci.* **7,** 373 (1971).
R14. Yu. V. Pleskov, "The Electric Double Layer at the Semiconductor/Electrolyte Interface," *Croatica Chem. Acta* **44,** 179 (1972).
R15. A. Many, Y. Goldstein, and N. B. Grover, *Semiconductor Surfaces* (North-Holland, Amsterdam, 1965).

R16. D. R. Frankl, *Electrical Properties of Semiconductor Surfaces* (Pergamon Press, London, 1967).

R17. V. A. Myamlin and Yu. V. Pleskov, *Electrochemistry of Semiconductors* (Plenum Press, New York, 1967).

R18. D. T. Sawyer and J. L. Roberts, Jr., *Experimental Electrochemistry for Chemists* (Wiley, New York, 1974).

R19. B. E. Conway and H. Angerstein-Kozlowska, "The Study of Electrode Surfaces by Electrochemical Methods," NBS Spec. Publ. #455 (1976).

R20. Yu. V. Pleskov and V. Yu. Filinovskii, *The Rotating Disc Electrode* (Consultants Bureau, New York, 1976).

R21. H. Gerischer, "Electrolytic Decomposition and Photodecomposition of Compound Semiconductors in Contact with Electrolytes," *J. Vac. Sci. Technol.* **15,** 1422 (1978).

R22. R. R. Dogonadze and A. M. Kuznetsov, "Theory of Charge Transfer Kinetics at Solid-Polar Liquid Interfaces," *Progr. Surf. Sci.,* **VI,** (1976).

R23. S. R. Morrison, "Study of Semiconductors using Electrochemical Techniques," *J. Vac. Sci. Technol.* **15,** 1417 (1978).

R24. E. A. Efimov and I. G. Erusalimchik, *Electrochemistry of Germanium and Silicon* (Sigma Press, Washington, 1963).

R25. A. J. Nozik, Photoelectrochemistry: Applications to Solar Energy Conversion, *Ann. Rev. Phys. Chem.* **29,** 189 (1978).

R26. W. Shockley, *Electrons and Holes in Semiconductors* (Van Nostrand, New York, 1950).

R27. L. A. Harris and R. H. Wilson, "Semiconductors for Photoelectrolysis," *Ann. Rev. Mat. Sci.* **8,** 99 (1978).

R28. J. G. Mavroides, "Electrode Materials for the Electrolysis of Water," *Mater. Res. Bull.* **13,** 1379 (1978).

R29. R. Memming, "Reactions on Semiconductor Electrodes," in *Topics in Surface Chemistry* eds. E. Kaz and P.S. Bagus (Plenum, New York, 1978).

R30. H. Gerischer, "Charge Transfer Processes at the Semiconductor–Electrolyte Interface in Connection with Problems of Catalysis," *Surf. Sci.* **18,** 97 (1969).

R31. T. Freund and W. P. Gomes "Electrochemical Methods for Investigating Catalysis by Semiconductors," *Catal. Rev.* **3,** 1 (1969).

R32. H. A. Laitinen, "Electrochemistry of SnO_2 Electrodes," *Denki Kagaku* **44,** 626 (1976).

R33. K. Hauffe, "Some Current Problems on Semiconductor/electrolyte Interfaces," in *Electrochemistry: The Past Thirty and the Next Thirty Years,* eds. H. Bloom and F. Gutmann (Plenum Press, New York, 1977).

R34. A. K. Vijh, *Electrochemistry of Metals and Semiconductors* (Marcel Dekker, New York, 1973).

R35. H. Gerischer and H. Rösler, "Zum Mechanismus der Oxidation organischer Moleküle durch Defektelektronen an Halbleiter Oberflächen," *Chem. Ing. Tech.* **4,** 176 (1970).

R36. K. Hauffe, "Mechanism of Catalysis on Semiconductor Catalysts," *Rev. Pure Appl. Chem.* **18,** 79 (1968).

R37. A. J.Bard, "Photoelectrochemistry and Heterogeneous Photocatalysis at Semiconductors," *J. Photochem.* **10,** 59 (1979).

R38. H. Gerischer, D. M. Kolb, and J. K. Sass, "The Study of Solid Surfaces by Electrochemical Methods," *Adv. Phys.* **27,** 437(1978).

R39. Th. Volkenshtein, "The Electronic Theory of Photocatalytic Reactions on Semiconductors," *Adv. Catal.* **23,** 157 (1973).

R40. K. Hauffe, "On the Spectral Sensitization of ZnO," *Phot. Sci. Eng.* **20,** 124 (1976).

R41. K. Hauffe, H. Pusch, J. Range, and D. Rein, "Dye Sensitization of Zinc Oxide by

Means of the Electrochemical Cell Technique," in *Current Problems in Electrophotography*, eds. K. Hauffe and W. Berg (de Gruyter, Berlin, 1972).

R42. E. Inoue, "Dye Sensitization Problems in the Photoconduction of Zinc Oxide," in *Current Problems in Electrophotography*, eds. K. Hauffe and W. Berg (de Gruyter, Berlin, 1972).

R43. S. R. Morrison, "Surface and Related Phenomena on ZnO," in *Current Problems in Electrophotography*, eds. K. Hauffe and W. Berg (de Gruyter, Berlin, 1972).

R44. B. E. Conway, "Some Chemical Factors in the Kinetics of Processes at Electrodes," *Progr. Reaction Kinet.* **4**, 399 (1967).

R45. M. S. Wrighton, J. M. Bolts, A. B. Bocarsly, M. C. Palazzoto, and E. G. Walton, "Stabilization of *n*-type Semiconductors to Photoanodic Dissolution: II–VI and II–V Compound Semiconductors and Recent Results for *n*-type Silicon," *J. Vac. Sci. Technol.* **15**, 1429 (1978).

R46. R. Memming, "Electrochemical Surface Reactions on Non-Metals," in NBS Spec. Publ. #455 (1976).

R47. T. P. Hoar, in *Modern Aspects of Electrochemistry*, Vol. II, ed. J. O'M. Bockris (Butterworths, London, 1953).

R48. F. C. Frank, *Growth and Perfection of Crystals*, eds. R. H. Doremus, B. H. Roberts, and D. Turnbull (Wiley, New York, 1958).

R49. D. J. DiMaria, "Defects and Impurities in Thermal SiO_2," in *Physics of SiO_2 and its Interfaces*, ed. S.T. Pantelides (Pergamon, New York, 1978).

R50. D. A. Vermilyea, "Anodic Films," *Adv. Electrochem. Electrochem. Eng.* **3**, 211 (1963).

R51. J. W. Diggle, T. C. Downie, and C. W. Goulding, "Anodic Films on Aluminum," *Chem. Rev.* **69**, 365 (1969).

R52. C. J. Dell'Oca and P. J. Fleming, "Initial Stages of Oxide Growth and Pore Initiation in the Porous Anodization of Aluminum," *J. Electrochem. Soc.* **123**, 1487 (1976).

R53. G. Eisenman, "Glass Electrodes for Hydrogen and Other Cations," (Marcel Dekker, New York, 1967).

R54. D. S. Campbell and A. R. Morley, "Electrical Conduction in Thin Metallic, Dielectric, and Metallic–Dielectric Films," *Rep. Prog. Phys.* **34**, 283 (1971).

R55. J. W. Shultze and M. A. Habib, "Principles of Electrocatalysis and Inhibition by Electrosorbates and Protective Layers," *J. Appl. Electrochem.* **9**, 255 (1979).

R56. P. F. Schmidt, "On the Mechanisms of Electrolytic Rectification," *J. Electrochem. Soc.* **115**, 167 (1968).

R57. A. Heller, editor, "Semiconductor Liquid-Junction Solar Cells," Proceedings Airlie Conf., 1977 (Proceedings Volume 77–3, The Electrochemical Society, Princeton, New Jersey, 1977).

R58. M. D. Archer, "Electrochemical Aspects of Solar Energy Conversion," *J. Appl. Electrochem.* **5**, 17 (1975).

R59. H. Gerischer, "Electrochemical Systems for H_2O Photolysis with Solar Energy," *Ber. Bunsenges. Phys. Chem.* **80**, 1046 (1976).

R60. H. Gerischer, "Electrochemical Photo and Solar Cells Principles and some Experiments," *Electroanal. Chem. Interf. Chem.* **58**, 263 (1975).

R61. D. Haneman, "Surfaces for Photoelectrochemical Cells," Proc. Intl. Conf. on Solid Films and Surfaces, Tokyo, 1978; *Surf. Sci.* **86**, 462 (1979).

R62. W. A. Gerrard and L. M. Rouse, "Solar Energy Conversion using Semiconducting Photo Anodes," *J. Vac. Sci. Technol.* **15**, 1155 (1978).

R63. G. W. Murphy, "Model Systems in Photoelectrochemical Energy Conversion," *Solar Energy* **21**, 403 (1978).

R64. R. Memming, "Energy Production by Photoelectrochemical Processes," (Philips Tech. Rev. **38**, 160 (1978/79).

R65. M. S. Wrighton, A. B. Ellis, and S. W. Kaiser, *Solid State Chemistry of Energy Conversion and Storage*, eds. J. B. Goodenough and M. S. Whittingham, *Adv. Chem. Ser.* **163,** (1977).

R66. K. Rajeshwar, P. Singh, and J. Dubow, "Energy Conversion in Photoelectrochemical Systems—a Review," *Electrochim. Acta* **23,** 1117 (1978).

R67. P. Stonehart and P. N. Ross, "The Commonality of Surface Processes in Electrocatalysis and Gas Phase Heterogeneous Catalysis," *Catal. Rev. Sci. Eng.* **12,** 1 (1975); P. Stonehart, "Electrocatalysis at Transition Metal Oxide and Carbide Surfaces," in NBS Spec. Publ. #455 (1976).

R68. A. C. C. Tseung, "Semiconducting Oxide Oxygen Electrodes," *J. Electrochem. Soc.* **125,** 1660 (1978); "Oxygen Evolution on Semiconducting Oxides," in NBS Spec. Publ. #455 (1976).

R69. P. J. Holms, editor, *The Electrochemistry of Semiconductors* (Academic Press, London, 1962).

R70. G. L. Schnable and P. F. Schmidt, "Application of Electrochemistry to Fabrication of Semiconductor Devices," *J. Electrochem. Soc.* **123,** 310C (1976).

R71. P. Cheung, editor, *Proceedings of a Workshop on the Theory, Design, and Biomedical Applications of Solid State Chemical Sensors* (CRC Press, Cleveland, 1977).

R72. W. M. Latimer, *Oxidation Potentials* (Prentice Hall, New York, 1952).

R73. A. J. Nozik, editor, *Photoeffects at Semiconductor–Electrolyte Interfaces, A.C.S. Symposium Series* (in press, 1980).

R74. S. R. Morrison, M. J. Madou, and K. W. Frese, Jr., "Photocorrosion in Solar Cells: The Enhanced Effectiveness of Stabilizing Agents due to Oxide Films," *A.C.S. Symposium Series* (in press, 1980).

R75. M. A. Butler and D. S. Ginley, "Principles of Photoelectrochemical Solar Energy Conversion," *J. Mater. Sci.* **15,** 1 (1980).

R76. R. Memming, "Charge Transfer Processes at Semiconductor Electrodes," *Electroanal. Chem.* **11,** 1 (1979).

R77. H. P. Maruska and A. K. Ghosh, "Photocatalytic Decomposition of Water at Semiconductor Electrodes," *Sol. Energy* **20,** 443 (1978).

Author Index

The numbers in italics refer to text pages on which the reference appears, the boldface numbers refer to the chapter numbers, and the arabic numbers to the reference numbers. An R inside parentheses indicates the reference is found in the list of Reviews and Books.

* Reference R4, being most familiar to the author, is referred to repeatedly in the text. To save space, references to it are omitted from this listing.

Subject Index

Entries in italics indicate the subject is continued for more than one page.

Made in the USA
Monee, IL
13 September 2026

62976976R00234